TRANSMUTATION AND OPERATOR DIFFERENTIAL EQUATIONS

NORTH-HOLLAND
MATHEMATICS STUDIES **37**

Notas de Matemática (67)
Editor: Leopoldo Nachbin

Universidade Federal do Rio de Janeiro
and University of Rochester

Transmutation and Operator Differential Equations

R. W. Carroll
Mathematics Department
University of Illinois

1979

NORTH-HOLLAND PUBLISHING COMPANY – AMSTERDAM • NEW YORK • OXFORD

ISBN: 0 444 85328 6

Publishers:
NORTH-HOLLAND PUBLISHING COMPANY
AMSTERDAM • NEW YORK • OXFORD

Sole distributors for the U.S.A. and Canada:
ELSEVIER NORTH-HOLLAND, INC.
52 VANDERBILT AVENUE, NEW YORK, N.Y. 10017

Library of Congress Cataloging in Publication Data

Carroll, Robert Wayne, 1930-
Transmutation and operator differential equations.

(Notas de matemáticas ; 67) (North-Holland mathematics studies ; 37)
1. Differential equations, Partial.
2. Operator equations. I. Title. II. Series.
QA1.N86 no. 67 [QA374] 510'.8s [515'.353] 79-12341
ISBN 0-444-85328-6

PRINTED IN THE NETHERLANDS

PREFACE

With the advent of modern functional analysis and the theory of distributions it became possible in the early 1950's to systematically explore and develop the theory of partial differential equations (PDE) to a degree previously unimaginable. One important feature of this work was the determination of "natural" spaces of functions of distributions in which to pose various differential problems. Then linear differential operators were treated as abstract linear operators A mapping a domain of definition $D(A) \subset F$ into G for suitable locally convex topological vector spaces F and G. The abstract operational properties of A were studied and existence-uniqueness theorems for example were then deduced as results in operator theory. This kind of approach was extensively developed also for certain types of nonlinear problems beginning in the early 1960's. In theoretical and applied work in other aspects of PDE one frequently has recourse to these now well established operational methods and distribution techniques. It is no accident that this material interacts naturally with various geometrical and variational points of view for example and enriches studies in these contexts. Much of this material already appears in book form but research continues and we will report on some recent work as indicated below.

One fascinating area of study which took form in these years involves abstract evolution equations of the form $du/dt = A(t)u$ with $u(0) = u_o$. Here $t \to u(t)$ is an F valued function of t with $u(t) \in D(A(t))$ where we shall confine our attention to families of densely defined <u>linear</u> operators $A(t)$ in F. Investigations of such problems played off operator properties of the $A(t)$ against properties of d/dt to produce a variety of powerful results and in this spirit one is led to study ordinary differential equations (ODE) with operator coefficients of the form $(D_t = d/dt;\ u^{(j)} = D_t^j u)$

$$(1.1) \qquad \sum_{j=0}^{m} A_j(t) D_t^j u = f$$

where the $A_j(t)$ could be thought of as arising from differential expressions in n "space" variables $(x_1,\dots,x_n) = x$ but could of course also be more general operators (e.g. pseudodifferential operators). These could represent evolution type problems with Cauchy data $u^{(j)}(0) = u_j$ $(0 \le j \le m-1)$ prescribed (or incomplete Cauchy data $u^{(j)}(0) = u_j$ for some $j \in \{0, m-1\}$); although we will usually take $A_m(t) = I$ this is not necessary. Similarly systems of the type

$$(1.2) \qquad \frac{d\vec{u}}{dt} = A(t)\vec{u} + \vec{f}; \quad \vec{u}(0) = \vec{u}_0$$

can be studied where $\vec{u}$ is an F^m valued m vector and $A(t) = ((A_{ij}(t)))$ with linear operator entries $A_{ij}(t)$. To formulate guidelines under which problems of the form (1.1) - (1.2) will be tractable let us think of the $A_j(t)$ as arising from differential expressions of the form $\Sigma a_{j\alpha}(x,t)D_x^\alpha$ where $D_x^\alpha = D_1^{\alpha_1}\dots D_n^{\alpha_n}$ with $D_k = \partial/\partial x_k$. Now the D_k commute in an obvious sense and generate groups $T_k(\xi) = \exp \xi D_k$ in say $C^0(\mathbb{R}^n)$ where $T_k(\xi)f(x) = f(x_1,\dots,x_k + \xi,\dots,x_n)$. Not surprisingly these features interact naturally with d/dt and corresponding to the case where the $a_{j\alpha}(x,t) = a_{j\alpha}(t)$ are independent of x one considers in general spaces F a family A_k of densely defined commuting operators generating one parameter groups $T_k(\xi) = \exp \xi A_k$ in F with $A_j(t) = \Sigma a_{j\alpha}(t)A^\alpha = \Sigma a_{j\alpha}(t)A_1^{\alpha_1}\dots A_n^{\alpha_n}$. Such situations as well as systems with $A_{ij}(t) = \Sigma a_{ij\alpha}A^\alpha$ lead to a variety of interesting solvable problems which will be discussed below. Some siutations involving noncommutative families A_k are also discussed. Another related tractable situation occurs when F is a Hilbert space and spectral methods may be employed. For example let $F = L^2(\mathbb{R}^n)$ and recall that $i\,\partial/\partial x_k = (-\Delta)^{\frac{1}{2}}R_k$ where the Riesz operators R_k are continuous in F and $A = 1 + (-\Delta)^{\frac{1}{2}}$ is a densely defined positive selfadjoint operator. Assuming that $a_{j\alpha}(x,t) = a_{j\alpha}(t)$ again we have $\Sigma a_{j\alpha}(t)D_x^\alpha = \Sigma a_{j\alpha}(t)(-i)^{|\alpha|}R^\alpha(A-1)^{|\alpha|} = \Sigma\tilde{a}_{jk\alpha}(t)R^\alpha A^k$ where $R^\alpha = R_1^{\alpha_1}\dots R_n^{\alpha_n}$. Thus in terms of one nice unbounded operator A and a commuting family R_k of bounded operators we can write $A_j(t) = \Sigma\hat{a}_{jk}(t,R)A^k$ with continuous polynomial valued operator functions $t \to \hat{a}_{jk}(t,R) = \sum_\alpha \tilde{a}_{jk\alpha}(t)R^\alpha$. Similar remarks apply to systems. We will present some recent results for ODE with operator coefficients of the types indicated and relate some of the results to formulas in operational calculus. Some variable domain operator problems with $A_k = A_k(t)$ are also treated as well as various other techniques for higher order equations.

We wish to emphasize here that the use of operator methods to solve differential problems represents a great achievement of mathematics in the last 25 years or so. It is not an artifical construction of effete snobs to avoid messy analysis in L^p (and generate publications) but rather a synthesis of such contexts which isolates the important natural features and produces elegant general theorems. Anyone who struggled with differential problems in the early 1950's surely must be impressed and gratified by the power and economy of these methods. Some of that messy analysis was misguided, repetitive, or unnecessary and came about because the wrong problem was being studied or the important features were obscure. One difficulty was to find the right spaces in which the differential expressions would have good properties when realized as operators in these spaces. Today some messy analysis is still usually necessary (and healthy) but it can be better focused.

Another aspect of the material discussed in this book can be illustrated by the following examples of "related" differential equations. Consider the problems

(1.3) $u_t - A^2u = 0; \quad u(0) = u_0$

(1.4) $w_{tt} - A^2w = 0; \quad w(0) = u_0; \quad w_t(0) = 0$

(1.5) $z_{tt} + A^2z = 0; \quad z(0) = u_0$

where A is an appropriate linear operator in a separated, complete, locally convex topological vector space F (wherein $u(t)$, $w(t)$, and $z(t)$ take their values) and $u_0 \in D(A^2)$. For example F could be $\mathcal{E} = \mathcal{E}(\mathbb{R})$ (= $C^\infty(\mathbb{R})$ with the Schwartz topology), or $\mathcal{D}' = \mathcal{D}'(\mathbb{R})$, with $A = d/dx$, in which case we would be dealing with certain Cauchy problems for the heat and wave equations and a half plane Dirichlet problem for the Laplace equation. As will be proved later, under suitable hypotheses it is possible to connect the solutions of (1.3) - (1.5) by integral formulas of the form

(1.6) $$u(t) = \int_0^\infty K(t,\tau)w(\tau)d\tau$$

(1.7) $$z(t) = \int_0^\infty I(t,\tau)u(\tau)d\tau$$

(1.8) $$z(t) = \int_0^\infty H(t,\tau)w(\tau)d\tau$$

where the kernels K, I, and H are given by

(1.9) $$K(t,\tau) = (\pi t)^{-\frac{1}{2}}\exp(-\tau^2/4t)$$

(1.10) $$I(t,\tau) = \frac{1}{2}\pi^{-\frac{1}{2}}t\tau^{-3/2}\exp(-t^2/4\tau)$$

(1.11) $$H(t,\tau) = 2t\pi^{-1}(t^2 + \tau^2)^{-1}$$

(and one checks easily that (1.8) follows from (1.6) and (1.7)). On the other hand in terms of operational calculus the solutions can be expressed formally as

(1.12) $$u(t) = \exp(A^2t)u_0; \quad w(t) = \cosh(At)u_0; \quad z(t) = \exp(iAt)u_0$$

such that the formulas (1.6) - (1.8) are known integral transforms when A is replaced by a parameter $\pm\lambda$. One objective in this monograph will be to report on some relationships of this sort and show how they arise via transmutation. In fact we develop in Chapter 2 a formalism based on the transmutation concept which connects eigenfunction transform kernels of ODE and displays the transmutation operator accordingly; numerous examples are given. As a byproduct one obtains a new derivation of inversion formulas such as the Hankel transform. Using the idea of a generalized convolution this also yields elegant expressions for the solutions of certain PDE. An abstract treatment of generalized translation operators is also given and the application of transmutation methods to problems in inverse scattering theory is indicated. An abstract transform theory is also developed.

In order to make the abstract presentation of differential problems more accessible we have included an appendix wherein certain basic facts from functional analysis, distribution theory, etc. are outlined. Such material is of course standard knowledge for specialists but it should be helpful for graduate students or others to have it readily available here. We suggest that the reader glance at the appendix before reading the text since specific references to the appendix will not be made. Standard notations will be used throughout and in this regard if a reference to formula (x.y) appears in a given chapter it means (x.y) of that chapter. If a reference (x·y·z) appears it means formula (y·z) of Chapter x while a reference (I·y·z) (resp. (A·y·z)) means formula (y·z) of this Preface = Introduction (resp. Appendix 1). Theorems, Lemmas, Remarks, etc. will be labeled consecutively. We have also found it useful to state various "formal" theorems wherein nothing is really proved or assumed beyond the formalism but precise versions of the result appear later as theorems or examples; such theorems are denoted by the symbol Theorem x.y(F).

The author would like to express his gratitude to Leopoldo Nachbin for his encouragement to write a monograph for the Notas de Matematica series. This resulting contribution is dedicated to my children David and Malcolm.

TABLE OF CONTENTS

CHAPTER 1

GENERAL TECHNIQUES

1.1 Preliminary results. In order to deal with ODE having operator coefficients there are many techniques available when F is a Banach or Hilbert space (see Remark 1.9 for some references). However for various reasons, some of which are connected with properties of solutions, it is essential to work in "larger" spaces F, which for convenience will always be assumed to be complete, separated, locally convex topological vector spaces. It will be instructive to begin our discussion with some constructions having immediate application to problems of obvious interest and to proceed from such particulars to more general theorems. For background material on functional analysis let us refer to Bourbaki [1; 2], Carroll [4], Horváth [1], Köthe [1], Reed-Simon [1], Schaeffer [1], and Treves [1]. First let $h \in C^N(\mathbb{R}; F)$ (i.e. $h(x) \in F$ for $x \in \mathbb{R}$, and as an F valued function of $x \in \mathbb{R}$, h is N times continuously differentiable in the topology of F). As a topological vector space $C^N(\mathbb{R}; F)$ is assumed to have the topology of uniform convergence of the functions and their derivatives of order $\leq N$ on compact subsets of $\mathbb{R}$. It is known that $C^N(\mathbb{R}; F) = C^N(\mathbb{R}) \hat{\otimes}_\varepsilon F$ where $C^N(\mathbb{R}) = C^N(\mathbb{R}; \mathbb{C})$ and $\hat{\otimes}_\varepsilon$ denotes the completion of $C^N(\mathbb{R}) \otimes F$ in the ε topology of uniform convergence on products of equicontinuous sets in $C^N(\mathbb{R})' \times F'$ (cf. Schwartz [2; 3] and Treves [1]). If $K \subset \mathbb{R}$ is compact then $C^N(K) = C^N(K; \mathbb{C})$ has the natural Banach topology with $C^N(K; F) = C^N(K) \hat{\otimes}_\varepsilon F$ as above. Let now $h_\alpha = \Sigma h_\alpha^\nu \otimes f_\alpha^\nu \to h$ in $C^N(\mathbb{R}) \hat{\otimes}_\varepsilon F$ where $h_\alpha^\nu \in C^N(\mathbb{R})$ and $f_\alpha^\nu \in F$ (finite sum in ν). Then we are able to define $\langle S,h\rangle = \langle h,S\rangle = \lim\langle h_\alpha,S\rangle = \lim \Sigma\langle h_\alpha^\nu,S\rangle f_\alpha^\nu$ for $S \in C^N(\mathbb{R})'$ since taking $B' \subset F'$ equicontinuous, one knows that $\langle h_\alpha, Sxf'\rangle = \langle\langle h_\alpha,S\rangle,f'\rangle = \Sigma\langle h_\alpha^\nu,S\rangle \langle f_\alpha^\nu,f'\rangle \to \langle h,Sxf'\rangle = \langle Sxf',h\rangle$ uniformly for $f' \in B'$. But this asserts that $\langle h_\alpha,S\rangle$ converges in F, to what we define as $\langle h,S\rangle$, since if V is a closed convex disced neighborhood (nbh) of 0 in F then $V = V^{oo}$ with

$V^o = B'$ equicontinuous (V^o denotes $F - F'$ polarity) and the topology of F coincides with the topology of uniform convergence on equicontinuous subsets of F'. This leads to

<u>Definition 1.1</u> If $h \in C^N(\mathbb{R}; F)$ and $S \in C^N(\mathbb{R})'$ then $\langle S,h\rangle = \langle h,S\rangle \in F$ is well defined.

<u>Remark 1.2</u> It is of particular interest to apply this construction to the case where $S \in \mathcal{E}' = \mathcal{E}'(\mathbb{R})$ is of order $\leq N$ with supp $S \subset L = \{x : |x| \leq \alpha < \infty\}$. Set $K = \{x : |x| \leq \alpha + \varepsilon\}$ and let $h \in C^N(K; F)$. Then by Schwartz [1] $S \in C^N(K)'$ and we can define $\langle h,S\rangle = \langle S,h\rangle \in F$ as above. This will be of use when working with hyperbolic problems (cf. Carroll [5; 6; 12; 13; 14] and Carroll-Showalter [8]).

<u>Lemma 1.3</u> For $L \subset \mathbb{R}$ compact the composition $S \to \langle S,h\rangle : C^N(L)' \to F$ is continuous for any $h \in C^N(L; F)$ fixed.

<u>Proof</u>: $C^N(L)$ is a Banach space, hence barreled, and bounded sets in $C^N(L)'$ are consequently equicontinuous. Let $S_n \to 0$ in $C^N(L)'$ and set $A' = \{S \in C^N(L)' : \|S\| \leq 1\}$ with $B' \subset F'$ also equicontinuous. Then if $h_\alpha = \sum_\nu h_\alpha^\nu \otimes f_\alpha^\nu \to h$ in $C^N(L; F)$ (finite sum in ν) one has the quantity $|\langle h_\alpha, S_n \times f'\rangle - \langle h, S_n \times f'\rangle|$ uniformly small for $\alpha \geq \alpha_o$ and $n \geq n_o$ (so that $S_n \in A'$) while $f' \in B'$. This means that $\langle h,S_n\rangle - \langle h_\alpha,S_n\rangle$ can be put in any closed convex disced nbh V of 0 in F by taking $B' = V^o$, $n \geq n_o$, and $\alpha \geq \alpha_o(V)$. But for α fixed one can make each $|\langle h_\alpha^\nu, S_n\rangle|$ arbitrarily small for n sufficiently large so that $\langle h_\alpha, S_n\rangle = \sum_\nu \langle h_\alpha^\nu, S_n\rangle f_\alpha^\nu \in V$ for $n \geq n_1(\alpha,V)$. Consequently $\langle S_n,h\rangle \in 2V$ for n sufficiently large. QED

Since our construction of $\langle S,h\rangle$ for $S \in C^N(L)'$ fixed shows that $h \to \langle S,h\rangle = \langle h,S\rangle : C^N(L; F) \to F$ is continuous one has established now that the bilinear map $(S,h) \to \langle S,h\rangle : C^N(L)' \times C^N(L; F) \to F$ is separately continuous. But the Banach space $C^N(L)'$ is barreled so we know for example (cf. Bourbaki [2] and Horváth [1])

<u>Lemma 1.4</u> The bilinear map $(S,h) \to \langle S,h \rangle : C^N(L)' \times C^N(L;\ F) \to F$ is hypocontinuous relative to bounded subsets of $C^N(L;\ F)$.

In particular this means that if $S_n \to S$ in $C^N(L)'$ and h_n is a sequence with $h_n \to h$ in $C^N(L;\ F)$ then $\langle S_n, h_n \rangle \to \langle S,h \rangle$ in F. We note in this respect that h_n will be a Cauchy sequence automatically and hence bounded (cf. Bourbaki [1; 2], Carroll [4], Horváth [1], etc.). One remarks also that the continuous map $d/dx : C^N(L) \otimes_\varepsilon F \to C^{N-1}(L) \otimes_\varepsilon F$ defined by $d/dx\ \sum_\nu h_\alpha^\nu \otimes f_\alpha^\nu = (d/dx \otimes 1)\sum_\nu h_\alpha^\nu \otimes f_\alpha^\nu = \sum_\nu d/dx\ h_\alpha^\nu \otimes f_\alpha^\nu$ extends by continuity to be a continuous linear map $C^N(L;\ F) \to C^{N-1}(L;\ F)$ for $N \geq 1$.

When working with abstract differential problems in general spaces F as indicated it is extremely useful to utilize the concept of locally equicontinuous group or semigroup as developed by T. Komura [1] and Dembart [1] (cf. also Bablola [1], Komatsu [1], Lions [1], Löfström [2], Miyadera [1], Oucii [1], Schwartz [4], Waelbroeck [1], Yosida [1], etc. for various other general semigroups - for the standard strongly continuous semigroups in Banach spaces see Hille-Phillips [1]). We recall that a closed densely defined linear operator A in F is said to generate a group $T(x) \in L(F)$ provided $T(x)T(y) = T(x+y)$ for $x,y \in \mathbb{R}$, $T(0) = I$, and $\lim(T(x) - I)f/x = Af$ as $x \to 0$ for $f \in D(A)$. We will be concerned primarily with groups $T(x)$ and simply mention that semigroups arise in the same way except that $T(x)$ is only defined for $x \geq 0$. The group $T(x)$ is said to be locally equicontinuous if $x \to T(x)f \in C^0(F)$ for any $f \in F$ while given any (continuous) seminorm p on F there exists a (continuous) seminorm q on F such that $p(T(x)f) \leq q(f)$ for $|x| \leq x_0 < \infty$ (any finite x_0 will do). The group is equicontinuous if this last condition holds for all $x \in \mathbb{R}$ and will be called quasi-equicontinuous if $p(T(x)f) \leq q(f)\exp \alpha|x|$ for some α $(x \in \mathbb{R})$. In general with any of these groups or semigroups it is well known that $d/dx\ T(x)f = AT(x)f = T(x)Af$ when $f \in D(A)$.

<u>Example 1.5</u> Let $F = C^0(\mathbb{R})$ with the topology of uniform convergence on compact subsets of $\mathbb{R}$. The operator $A = d/dx$ generates a locally equicontinuous group

$T(x)f(y) = f(x+y)$ which is not equicontinuous.

We will now sketch formally an important technique due to Hersh [1; 2] for treating differential equations with operator coefficients (precise theorems are spelled out when needed). The method was first developed by Hersh for Banach spaces F and later extended to general F by Carroll [5; 6] using the construction $\langle S,h\rangle$ of Definition 1.1; it has proved very useful in the study of various abstract differential problems (cf. Bobisud-Hersh [1], Bragg [1], Carroll [5; 6; 12; 13; 14; 15; 16; 17], Carroll-Showalter [8], Donaldson [1; 2; 3; 4; 5; 7; 8], Donaldson-Hersh [6], Hersh [1; 2; 3]). The idea goes as follows, assuming everything makes sense. In (I.1.1) let $A_j(t) = \sum_0^{\ell} a_{jk}(t)A^k$ where A is a densely defined linear operator in F generating a suitable one parameter group $T(x) = \exp Ax$. We set

$$(1.1) \qquad P(t,D_t,A) = \sum_{j=0}^{m} \sum_{k=0}^{\ell} a_{jk}(t)D_t^j A^k$$

so $P(t,D_t,A) = \sum_0^m A_j(t)D_t^j$ and consider the abstract Cauchy problem for $t \geq 0$, $u \in C^m(F)$,

$$(1.2) \qquad P(t,D_t,A)u = 0; \quad u^{(j)}(0) = u_j \quad (0 \leq j \leq m-1)$$

where $u_j \in D(A^{M_j})$ for suitable M_j. Now replace A by $-D_x = -d/dx$ and look for distributions $G_n(t,x)$, $0 \leq n \leq m-1$, satisfying

$$(1.3) \qquad P(t,D_t,-D_x)G_n = 0; \quad D_t^k G_n\big|_{t=0} = \delta_{nk}\delta(x)$$

where δ_{nk} is the Kronecker symbol and $0 \leq k \leq m-1$ (cf. Carroll [4], A. Friedman [1], Gelfand-Šilov [1; 2; 3]). There is an obvious abuse of notation in writing $G_n(t,x)$ for a distribution $G_n \in \mathcal{D}'_{t,x}$ but we will employ this notation anyway since no confusion is likely to arise. Assume now such G_n exist and that a suitable distribution pairing or bracket $\langle G_n(t,\cdot), h(\cdot)\rangle \in F$ can be defined for the corresponding $G_n(t,\cdot) \in \mathcal{D}'(\mathbb{R})$ and appropriate $h(\cdot) \in C^Q(\mathbb{R}; F)$ of the form $h(x) = T(x)u_n$ for large enough Q (this may involve growth at ∞, etc., and will be illustrated by examples below). Then formally a solution of (1.2) can be written as

(1.4) $$u(t) = \sum_{n=0}^{m-1} \langle G_n(t,\cdot),\ T(\cdot)u_n\rangle$$

since we have then formally

(1.5) $$D_t^k u|_{t=0} = \sum_{n=0}^{m-1} \langle D_t^k G_n(t,\cdot),\ T(\cdot)u_n\rangle|_{t=0}$$

$$= \sum_{n=0}^{m-1} \langle \delta_{nk}\delta(\cdot),\ T(\cdot)u_n\rangle = u_k$$

for $0 \leq k \leq m-1$ while, since $A^k T(\cdot)u_n = D_x^k T(\cdot)u_n$, the differential equation in (1.2) takes the form

(1.6) $$P(t,D_t,A)u = \sum_{j=0}^{m} A_j(t)D_t^j u =$$

$$\sum_{j=0}^{m} \sum_{k=0}^{\ell} a_{jk}(t)D_t^j A^k \sum_{n=0}^{m-1} \langle G_n(t,\cdot),\ T(\cdot)u_n\rangle =$$

$$\sum_{n=0}^{m-1} \langle \sum_{j,k} a_{jk}(t)D_t^j G_n(t,\cdot),\ A^k T(\cdot)u_n\rangle =$$

$$\sum_{n=0}^{m-1} \langle \sum_{j,k} a_{jk}(t)D_t^j(-D_x)^k G_n(t,\cdot),\ T(\cdot)u_n\rangle =$$

$$\sum_{n=0}^{m-1} \langle P(t,D_t,-D_x)G_n(t,\cdot),\ T(\cdot)u_n\rangle = 0$$

The key fact of course is that $AT(\cdot)u_n = D_x T(\cdot)u_n$ while $\langle G_n(t,\cdot),\ D_x h(\cdot)\rangle = \langle(-D_x)G_n(t,\cdot),\ h(\cdot)\rangle$ for a reasonable distribution pairing. We summarize this in

Theorem 1.6(F). Under the types of hypotheses indicated let the G_n satisfy (1.3) and let $u(t)$, given by (1.4), be well defined. Then u satisfies (1.2) formally.

Example 1.7 As a simple but important application of this technique let us apply it to the problem (I.1.4) when A generates a locally equicontinuous group $T(x)$ in F. We are led to consider

(1.7) $$G_{tt} - G_{xx} = 0;\quad G(0,x) = \delta(x);\quad G_t(0,x) = 0$$

The (unique) solution of (1.7) is evidently $G(t,x) = \frac{1}{2}[\delta(x+t) + \delta(x-t)]$ and for $u_o \in D(A^2)$ we write the solution of (I.1.4) as $w(t) = \langle G(t,\cdot),\ T(\cdot)u_o\rangle$ where upon restricting t to be suitably finite (e.g. $t \le T < \infty$) the bracket can be defined rigorously as in Remark 1.2. Indeed in this event the δ-functions in G have support at $x + t = 0$ and $x - t = 0$ so $|x| \le T$ where we will assume $T < x_0$ and take say $L = \{x : |x| \le T\}$ with $K = \{x : |x| \le x_0\}$ where x_0 is the number associated with the locally equicontinuous group $T(x)$. Upon evaluation of $\langle G(t,\cdot), T(\cdot)u_o\rangle$ we obtain $w(t) = \frac{1}{2}[T(t) + T(-t)]u_0 = \cosh(At)u_0$ as indicated in (I.1.12). Note here that from the specific nature of $\langle G(t,\cdot), T(\cdot)u_o\rangle$ in this case it is clear that $A^2\langle G(t,\cdot), T(\cdot)u_o\rangle = \langle G(t,\cdot), A^2T(\cdot)u_o\rangle$ for $u_0 \in D(A^2)$. Finally in order to obtain $D_t\langle G(t,\cdot), T(\cdot)u_o\rangle$ for example we write $\Delta G = G(t,\cdot) - G(t_0,\cdot)$ and look at $\lim\langle \Delta G/\Delta t - G_t(t_0,\cdot), T(\cdot)u_0\rangle$. Since $\Delta G/\Delta t - G_t(t_0,\cdot) \to 0$ in $C^1(K)'$ it follows by Lemma 1.3 that $D_t \langle G(t,\cdot), T(\cdot)u_o\rangle = \langle G_t(t,\cdot), T(\cdot)u_o\rangle$ in F.

To show that the $w(t)$ constructed above is the unique solution of (I.1.4) suppose w is <u>any</u> solution of $w_{tt} - A^2w = 0$ in $C^2(F)$ with $w(0) = w_t(0) = 0$. Let us write $G(t,s,x)$ for the solution of $G_{tt} - G_{xx} = 0$, $G(s,s,x) = \delta(x)$, and $G_t(s,s,x) = 0$ $(0 \le s \le t)$ so that evidently $G(t,s,x) = \frac{1}{2}[\delta(x + t - s) + \delta(x - t + s)]$. Similarly let $H(t,s,x) = \frac{1}{2}\int_{x-t+s}^{x+t-s} \delta(\xi)d\xi$ (integral in $\mathcal{E}'(\mathbb{R})$ - see e.g. Bourbaki [3; 4; 5] or Carroll [4] for vector valued integration) be the solution of $H_{tt} - H_{xx} = 0$, $H(s,s,x) = 0$, and $H_t(s,s,x) = \delta(x)$ $(0 \le s \le t)$. It is easily checked that $H_s = -G$ and $G_s = -H_{tt} = -H_{xx}$. We use now a technique for uniqueness in this simple example which is a substantial variation of a procedure of Fattorini [1] and Hille [2] and was developed by Carroll [5; 6; 12; 13; 14] and Carroll-Showalter [8] to apply to much more general second order singular and degenerate hyperbolic problems (cf. also Carroll [16; 17], Donaldson [4; 5; 7], and later sections of this chapter). We consider the function $\phi(t,s) = \phi_1(t,s) + \phi_2(t,s)$ where

(1.8) $$\phi_1(t,s) = \langle H(t,s,\cdot), T(\cdot)w_s(s)\rangle;$$

$$\phi_2(t,s) = \langle G(t,s,\cdot), T(\cdot)w(s)\rangle.$$

The brackets make sense as before and for completeness we will check the continuity and differentiability of ϕ_1 and ϕ_2 as functions of s $(0 \le s \le t)$. Here we recall that $w_{ss} - A^2w = 0$ with $w \in C^2(F)$ and $w(0) = w_s(0) = 0$. First, recall that one need only consider sequences $s_n \to s_o$ to determine continuity, etc. of functions defined on $\mathbb{R}$ (cf. Carroll [4]), so let $s_n \to s_o$ and consider for example $\Delta\phi_2 = \langle G(t,s_n,\cdot),\ T(\cdot)w(s_n)\rangle - \langle G(t,s_o,\cdot),\ T(\cdot)w(s_o)\rangle =$ $\langle G(t,s_n,\cdot),\ T(\cdot)w(s_n)\rangle - \langle G(t,s_o,\cdot),\ T(\cdot)w(s_n)\rangle + \langle G(t,s_o,\cdot),\ T(\cdot)w(s_n)\rangle -$ $\langle G(t,s_o,\cdot),\ T(\cdot)w(s_o)\rangle$. Now is p is a (continuous) seminorm on F we have $p(T(x)w(s_n)) \le q(w(s_n))$ for some (continuous) seminorm q on F (here $|x| \le x_o < \infty$ for suitable x_o when $t \le T < x_o$ as above). But $w(s_n)$ is a Cauchy sequence in F, hence bounded, so $q(w(s_n))$ will be bounded. Thus $T(x)w(s_n)$ is bounded in $C^o(K;\ F)$ for suitable compact K as above and $\Delta G = G(t,s_n,\cdot) - G(t,s_o,\cdot) \to 0$ in $C^o(K)'$; hence by Lemma 1.4 $\langle \Delta G,\ T(\cdot)w(s_n)\rangle \to 0$. Similarly for $\Delta w = w(s_n) - w(s_o), p(T(x)\Delta w) \le q(\Delta w) \to 0$ for $x \in K$ (i.e. $T(x)\Delta w \to 0$ in $C^o(K;\ F)$) and hence $\langle G(t,s_o,\cdot),\ T(\cdot)\Delta w\rangle \to 0$. This proves the continuity of $\phi_2(t,\cdot)$ and that of $\phi_1(t,\cdot)$ follows by a similar argument. For differentiability we consider first $(\partial/\partial s)\phi_2(t,s)$ which formally can be written $(\partial/\partial s)\phi_2(t,s) =$ $\langle G_s(t,s,\cdot),\ T(\cdot)w(s)\rangle + \langle G(t,s,\cdot),\ T(\cdot)w_s(s)\rangle$. Then with $\Delta s = s_n - s_o$ we look at

$$(1.9)\qquad \frac{\Delta\phi_2}{\Delta s} = \left\langle \frac{\Delta G}{\Delta s},\ T(\cdot)w(s_n)\right\rangle + \left\langle G(t,s_o,\cdot),\ T(\cdot)\,\frac{\Delta w}{\Delta s}\right\rangle$$

Now, with a view to applying Lemma 1.4 again, $\Delta G/\Delta s \to G_s(t,s_o,\cdot)$ in $C^1(K)'$ and $T(x)w(s_n)$ will be bounded in $C^1(K;\ F)$ since $(d/dx)T(x)w(s_n) = AT(x)w(s_n) =$ $T(x)Aw(s_n)$ will be bounded in $C^o(K;\ F)$ as well as $T(x)w(s_n)$. To see this it is clearly sufficient to verify that $Aw(\cdot)$ is continuous when $w(\cdot)$ and $A^2w(\cdot)$ are continuous. This "obvious" fact actually needs a little proof which we supply for completeness. Thus (cf. T. Komura [1]) when $x \le x_o$ for example one can write $(T(x) - I)w(s_n) = \int_o^x T(\eta)Aw(s_n)d\eta$ and $(T(\eta) - I)Aw(s_n) = \int_o^\eta T(\xi)A^2w(s_n)d\xi$ so that

$$(1.10)\qquad (T(x) - I)w(s_n) = \int_0^x [Aw(s_n) + \int_0^\eta T(\xi)A^2w(s_n)d\xi]d\eta$$

$$= xAw(s_n) + \int_0^x \int_0^\eta T(\xi)A^2w(s_n)d\xi d\eta.$$

But $(T(x) - I)w(s_n) \to (T(x) - I)w(s_o)$ and evidently $\int_0^x \int_0^\eta T(\xi)A^2w(s_n)d\xi d\eta \to \int_0^x \int_0^\eta T(\xi)A^2w(s_o)d\xi d\eta$. Thus $xAw(s_n)$ and hence $Aw(s_n)$ converges with $w(s_n)$ so that $Aw(s_n) \to Aw(s_o)$ since A is closed and consequently $Aw(\cdot) \in C^o(K; F)$. Therefore Lemma 1.4 applies to the first bracket. For the second bracket in (1.9) we know $\Delta w/\Delta s \to w_s(s_o)$ in $C^o(K; F)$ since $w \in C^2(F)$ and hence the bracket tends to the desired expression. Next we look at $(\partial/\partial s)\phi_1(t,s) = \langle H_s(t,s,\cdot), T(\cdot)w_s(s)\rangle + \langle H(t,s,\cdot), T(\cdot)w_{ss}(s)\rangle$. In view of previous arguments there is obviously no problem in justifying this formula. Since now $H_s = -G$ and $G_s = -H_{xx}$ we obtain

$$(1.11)\qquad \frac{\partial\phi}{\partial s}(t,s) = \langle H_s(t,s,\cdot), T(\cdot)w_s(s)\rangle + \langle H(t,s,\cdot), T(\cdot)w_{ss}(s)\rangle + \langle G_s(t,s,\cdot), T(\cdot)w(s)\rangle$$

$$+ \langle G(t,s,\cdot), T(\cdot)w_s(s)\rangle = \langle H(t,s,\cdot), T(\cdot)w_{ss}(s)\rangle$$

$$- \langle H_{xx}(t,s,\cdot), T(\cdot)w(s)\rangle = \langle H(t,s,\cdot), T(\cdot)[w_{ss} - A^2w]\rangle = 0.$$

Here we observe that for suitable K as above with $\text{supp}\, H(t,s,\cdot) \subset L \subset K$ we can transport D_x around under the bracket $\langle\ ,\ \rangle$ in a distribution sense as was done formally in (1.6). Therefore $\phi(t,t) = w(t) = \phi(t,0) = 0$ and $w(t) \equiv 0$. This concludes our discussion of Example 1.7 and we state the result as

<u>Theorem 1.8</u> Let F be a complete separated locally convex space and let A generate a locally equicontinuous group in F. Then the unique solution of (I.1.4) in $C^2(F)$ is given by $w(t) = \cosh(At)u_o$.

This theorem is a special case of more general results to follow but it has been proved as an illustrative example to introduce various techniques. We will give another such result for (I.1.3) before proceeding to general equations and systems in Section 2 (see Example 1.10).

Remark 1.9 Differential equations with operator coefficients and operational differential equations in various spaces have been studied extensively in the last 25 years or so and a complete bibliography would be virtually impossible to assemble. In addition to the references already cited in this respect, let us mention however a few others which contain valuable background material, provide other points of view, or have in some way influenced our thinking. For material in book or lecture note form, information involving both ordinary and partial differential operators in Hilbert space (linear and nonlinear) can be found in Akhiezer-Glazman [1], Berezanskij [1], Brézis [1], Calderón [1], Coddington-Levinson [1], Ekeland-Teman [1], Foias-St. Nagy [2], Garding [1], Garnir [1], Gohberg-Krein [1], Lax-Phillips [1], Naimark [1], Showalter [1], and Stone [1]; developments of the theory in Banach and other spaces appear in Aubin [1], Barbu [1], Berger [1], Browder [1], Colojoara-Foias [1], Courant-Hilbert [1], Daleckij-Krein [1], Dunford-Schwartz [1], Duvaut-Lions [1], Ehrenpreis [1], A. Friedman [2; 3], B. Friedman [1], Goldstein [4], Hille [3], Hörmander [1], Kato [4], Krein [1], Ladas-Lakshmikantham [1], Ladyženskaya-Uraltzeva [1], Ladyženskaya-Solonnikov-Uraltzeva [2], Lakshmikanthem-Leela [1], Lions [4; 5; 6; 7], Lions-Magenes [8], Martin [1], Massera-Schaeffer [1], Maurin [1; 2], Palamodov [1], Pascali [1], Schechter [1], Sobolev [1], Strauss [1], Treves [2; 3; 4; 5], and Zaidman [1]. Some further papers dealing with differential problems in general spaces or in distribution spaces are Carroll [1; 2; 3], Dettman [2; 4], DuChateau [1], Fattorini [2], Foiaş-Gussi-Poenaru [1], Millionščikov [1], Ovsyannikov [1], Schwartz [5], Sebastiao-Silva [1; 2], Steinberg-Treves [1], and Treves [5; 6; 7]. We will omit any discussion of hyperfunctions, ultradistributions, microfunctions, etc. beyond mention of Komatsu [2] and Shapira [1]. For problems in abstract operator equations and abstract boundary value problems see Aubin [2], Browder [2; 5; 6; 7], Calkin [1], Carroll [7; 27; 28; 30], Cordes [1], DaPrato [1; 2; 4], Dezin [1], Milman [1], Višik [2], and Wyler [1]. For linear abstract evolution problems and semi-groups see for example Balakrishnan [1; 2], Beals [1; 2], Emami-Rad [1], Feller [1], Goldstein [1; 2; 3; 5], Goldstein-Sandefur

[7; 8], Grisvard [1], Hackman [1], Kato-Tanabe [1], Kato [2; 3], Kisynski [1], Ladyženskaya [3; 4], Lyubič [1], Phillips [1], Poulsen [1], Raskin-Sobolevskij [1], Sobolevskij [1], Tanabe [1], Višik [1], and Višik-Ladyženskaya [3]. Some abstract Cauchy problems are treated by spectral methods in Carroll [18; 19; 20; 21; 22], Carroll-Neuwirth [23], Carroll-Wang [29], Sandefur [1], and Walker [1]. Variable domain abstract evolution problems are studied in Baiocchi [1; 2], Bernardi [1], Bernardi-Brezzi [2], Carroll [31], Carroll-Cooper [24], Carroll-State [25], Carroll-Mazumdar [32], Cooper [1], Kato [6; 7], and Mazumdar [1; 2; 3; 4]. Some fundamental papers on nonlinear evolution problems, nonlinear semigroups, variational inequalities, etc. are Bardos-Brézis [1], Brézis [3], Browder [3; 4], Crandall [1], Crandall-Pazy [2], Crandall-Liggett [3], Dorroh [1], Kato [5], Y. Komura [1], Lagnese [1; 2], Leray-Lions [1], and Segal [1]. For Sobolev equations see the book of Carroll-Showalter [8] and for example Galpern [1], Lagnese [3; 4; 5], Levine [4], Rao [1], Showalter [6; 7; 8; 9; 10], Sobolev [2], Ting [1]. For operational calculus let us mention Berg [1], Božinov [1], Božinov-Dimovski [2], Churchill [1], Ditkin-Prudnikov [1], Doetsch [1], T. Donaldson [1], Erdelyi [2], B. Friedman [2], Groetsch [1], Hirschman-Widder [1], Jury [1], Krabbe [1], Laptev [1], Lighthill [1], Maslov [2], Mikusinski [1], van der Pol-Bremmer [1], Rubel [1], Sneddon [1], Struble [1; 2; 3], Ungar [1; 2; 3], Zemanian [1].

Example 1.10 We will now solve (I.1.3) and produce the connection formula (I.1.6) relating solutions of (I.1.3) and (I.1.4). Such formulas have led to the concept of related differential equations studied mainly in a concrete analytical setting by Bragg [1; 2; 3; 4; 5], Bragg-Dettman [6; 7], and Dettman [1; 2; 3; 4]. Such problems are also connected with the idea of transmutation of operators developed by Delsarte [1], Delsarte-Lions [2; 3], Hutson-Pym [1; 2], Levitan [1; 3], Lions [2; 3; 4], Marčenko [1; 3], and Povzner [1]; this will be discussed in detail later and further references given. First let us look at (I.1.3) and apply the Hersh technique as in Example 1.7. Thus one considers

$$(1.12)\qquad G_t - G_{xx} = 0; \quad G(0,x) = \delta(x).$$

A solution to this is easily seen to be $G(t,x) = \frac{1}{2} K(t,x)$ where K is given by (I.1.9). This can be derived for example by taking Fourier transforms in (1.12) in x to obtain $\hat{G}_t + s^2\hat{G} = 0$ with $\hat{G}(0,s) = 1$ where we use the Fourier transform $\hat{\phi}(s) = F\phi(s) = \int_{-\infty}^{\infty} \phi(x)\exp isxdx$ so that $(F\phi')(s) = -is\hat{\phi}(s)$. Thus $\hat{G}(t,s) = \exp(-s^2t)$ and it is known that $F_x \frac{1}{2} K(t,x) = \exp(-s^2t)$ (cf. Titchmarsh [1]). Let now A generate a quasiequicontinuous or equicontinuous group $T(x)$ in a complete separated space F (recall $p(T(x)f) \le q(f)\exp \alpha|x|$ in this event). We consider formally then as a solution of (I.1.3)

$$(1.13) \qquad u(t) = \langle G(t,\cdot),\ T(\cdot)u_o\rangle = \frac{1}{2}\int_{-\infty}^{\infty} K(t,x)T(x)u_o dx$$

$$= \int_0^{\infty} K(t,x)\cosh(Ax)u_o dx = \int_0^{\infty} K(t,x)w(x)dx$$

where $w(x)$ is the unique solution of (I.1.4) obtained in Theorem 1.8. Because of the nature of $K(t,x) = (\pi t)^{-\frac{1}{2}}\exp(-x^2/4t)$ as $x \to \infty$ and the growth estimates for $T(x)u_o$ it is easily seen that the integral in (1.13) makes sense and one can differentiate under the integral sign for $t > 0$. This leads us to state

Theorem 1.11 Let F be a complete separated locally convex space and let A generate a quasiequicontinuous group in F. Then the unique solution of (I.1.3) in $C^1(F)$ is given by (1.13).

Proof: It remains to prove uniqueness so we look at the solution of $G_t = G_{xx}$ for $t \ge s$ with $G(s,s,x) = \delta(x)$ given by $G(t,s,x) = \frac{1}{2}[\pi(t-s)]^{-\frac{1}{2}}\exp(-x^2/4(t-s))$. Then let $u \in C^1(F)$ be any solution of (I.1.3) with $u(0) = 0$ and consider for $t \ge s \ge 0$

$$(1.14) \qquad \phi(t,s) = \langle G(t,s,\cdot),\ T(\cdot)u(s)\rangle =$$

$$\int_{-\infty}^{\infty} G(t,s,x)T(x)u(s)dx.$$

It is clear that $G_s = -G_t$ so formally

(1.15) $$\phi_s = \langle G_s(t,s,\cdot),\ T(\cdot)u(s)\rangle +$$

$$\langle G(t,s,\cdot),\ T(\cdot)u_s(s)\rangle = \langle G(t,s,\cdot),\ A^2T(\cdot)u(s)\rangle$$

$$- \langle G_{xx}(t,s,\cdot),\ T(\cdot)u(s)\rangle = 0$$

since $\langle G_{xx}(t,s,\cdot),\ T(\cdot)u(s)\rangle = \langle G(t,s,\cdot),\ D_x^2T(\cdot)u(s)\rangle$. The calculations evidently make sense for quasiequicontinuous $T(x)$ and now from $\phi(t,s)$ = constant we have $\phi(t,t) = u(t) = \phi(t,0) = 0$. QED

There is no need to discuss a fancy bracket $\langle\ ,\ \rangle$ here in the manner of Example 1.7 since we put everything immediately in the form of integrals. This could also have been done in Example 1.7 of course because of the nature of the distributions G and H. However in general various brackets clearly must be defined and we continue in this spirit in Chapter 2.

1.2. General systems. For nonhyperbolic problems the distributions G_n which arise in (1.3) will not have compact support in x and one is obliged to define pairings $\langle G_n(t,\cdot),\ T(\cdot)u_n\rangle$ relative to suitable spaces of F valued functions in x and distributions in x delimited by growth conditions at ∞ (cf. for example the construction in Example 1.10). In this section we utilize pairings of the type employed by Donaldson [4; 5; 7] when F was a Banach space. We will continue to work in general complete separated locally convex F, thus generalzing Donaldson's constructions, and will spell out the details more completely as in Carroll [16; 17]. For heuristic purposes it seems more appropriate to deal with single equations separately and then extend the treatment to systems. Thus let $L_s(F)$ be the space of continuous linear maps $F \to F$ provided with the strong operator topology (i.e. $Q_\alpha \to Q$ in $L_s(F)$ means $Q_\alpha f \to Qf$ for each fixed $f \in F$). We consider (1.2) (with $u \in C^m(F)$) and for convenience take $a_{jk}(t) = a_{jk}$ independent of t with $A_m(t) \equiv 1$; extensions of the theory for continuous $a_{jk}(\cdot)$ are immediate (cf. Theorems 2.6 and 2.8). Let A generate a quasiequicontinuous group $T(x)$ in F for example so that in particular $T(\cdot) \in C^0(L_s(F))$.

Set $U(t,x) = T(x)u(t)$ so $U(t,x)$ is an F valued function of (t,x) and let us write $U(t,x) - U(t_o,x_o) = T(x)u(t) - T(x)u(t_o) + T(x)u(t_o) - T(x_o)u(t_o)$. Let p be any (continuous) seminorm in F and observe that $p(T(x)(u(t) - u(t_o)) \leq q(u(t) - u(t_o)) \exp \alpha|x|$ for some (continuous) seminorm q in F. As $t \to t_o$, $u(t) \to u(t_o)$ for $u \in C^o(F)$ and hence $q(u(t) - u(t_o)) \exp \alpha|x| \to 0$ uniformly for x near x_o (x_o finite). On the other hand since $x \to T(x) \in C^o(L_s(F))$ the term $(T(x) - T(x_o))u(t_o) \to 0$. This shows that $(t,x) \to U(t,x)$ is continuous with values in F for $t \geq 0$ and x finite. Next it is evident that $T(x)D_t u(t) = D_t T(x)u(t)$ whereas for t fixed $T(x)A^j u(t) = D_x^j T(x)u(t)$ and we want $u(t) \in D(A^\ell)$ for all terms in (1.2) to make sense. Now operate on (1.2) with $T(x)$ to obtain

$$(2.1) \qquad P(D_t,D_x)U(t,x) = 0; \quad D_t^k U(0,x) = T(x)u_k$$

for $0 \leq k \leq m-1$ where we have written $P(D_t,A) = P(t,D_t,A)$. Note here we will be working with $P(D_t,D_x)$ instead of $P(D_t,-D_x)$ as in the Hersh technique in Section 1.1.

Now if (2.1) can be solved then $u(t) = T(-x)U(t,x)$ will be formally a solution of (1.2) so let us concentrate first on (2.1). For fixed t it is useful here to regard $U(t,\cdot) \in L_s(\Phi; F)$ for some suitable space $\Phi \supset D(\mathbb{R})$ of C^∞ test functions. Here $L_s(\Phi; F)$ denotes $L(\Phi; F)$ with the strong operator topology (i.e. $Q_\alpha \to Q$ in $L_s(\Phi; F)$ means $Q_\alpha(\phi) \to Q(\phi)$ in F for each $\phi \in \Phi$). The pairing will be denoted by $\langle U(t,\cdot), \phi(\cdot)\rangle \in F$ and since $U(t,\cdot)$ is a continuous F valued function we can write $\langle U(t,\cdot), \phi(\cdot)\rangle = \int_{-\infty}^{\infty} U(t,x)\phi(x)dx$ (Riemann type integral in F) and this can also be written for fixed t as $(\int_{-\infty}^{\infty} T(x)\phi(x)dx)u(t)$ with the integral in $L_s(F)$. The space Φ must be chosen in accordance with the nature of the group $T(x)$ and with the growth properties of certain multipliers $\hat{G}_n(t,s)$ and $\hat{G}_n(t,\tau,s)$ defined below - examples will be given. We will want to take Fourier transforms so let us assume $\Phi \subset S'$ with $F : \Phi \to \hat{\Phi}$ a topological isomorphism and it will be useful to make a blanket hypothesis that Φ and $\hat{\Phi}$ are complete and barreled. Let us recall next the Parseval formula (cf. Schwartz [1])

$\langle FQ,\phi\rangle = \langle Q,F\phi\rangle$ for $\phi \in S$ and $Q \in S'$ (S and S' are the standard Schwartz spaces. If $\phi = F(\psi)$ now one obtains $\langle FQ,F\psi\rangle = \langle Q,FF\psi\rangle$ and we observe that $(F\psi)(s) = \int_{-\infty}^{\infty} \psi(x)\exp isx\, dx = \int_{-\infty}^{\infty} \psi(-\xi)\exp(-is\xi)d\xi = 2\pi(F^{-1}\psi(-x))(s)$. Hence $(FF\psi)(x) = 2\pi\psi(-x)$ and one has $\langle FQ,F\psi\rangle = 2\pi\langle Q,\psi(-x)\rangle$ (cf. Gelfand-Šilov [1]). It is in this way that one defines $\hat{Q} = FQ$ formally as an F-valued linear operator on $F\Phi = \hat{\Phi}$ given that $Q \in L_s(\Phi; F)$.

<u>Definition 2.1</u> Let $Q \in L_s(\Phi; F)$ and define $FQ = \hat{Q} \in L_s(\hat{\Phi}; F)$ by the rule $\langle\hat{Q},\hat{\phi}\rangle = 2\pi\langle Q,\phi(-x)\rangle \in F$ for $\phi \in \Phi$ where $\hat{\phi} = F\phi$ and F is a topological isomorphism $\Phi \to \hat{\Phi}$.

This definition was used by Donaldson [4; 5; 7] for F a Banach space and we will show now that it makes sense for general F. First let $\hat{\phi}_\alpha \to \hat{\phi}$ in $\hat{\Phi}$ so that $\phi_\alpha = F^{-1}\hat{\phi}_\alpha \to F^{-1}\hat{\phi} = \phi$ in Φ. Then $\langle Q,\phi_\alpha(-x)\rangle \to \langle Q,\phi(-x)\rangle$ in F since $Q \in L_s(\Phi; F)$ and hence $\phi \to 2\pi\langle Q,\phi(-x)\rangle$ is a (uniquely defined) continuous linear map $\hat{Q} : \hat{\Phi} \to F$ the action of which we denote by $\langle\hat{Q},\hat{\phi}\rangle$. Consequently $\hat{Q}$ is well defined and $Q \to \hat{Q} : L_s(\Phi; F) \to L_s(\hat{\Phi}; F)$ is clearly onto. Now let $Q_\alpha \to Q$ in $L_s(\Phi; F)$ so that for fixed $\phi \in \Phi$ one has $\langle Q_\alpha,\phi(-x)\rangle \to \langle Q,\phi(-x)\rangle$ in F; it follows that $\langle\hat{Q}_\alpha,\hat{\phi}\rangle \to \langle\hat{Q},\hat{\phi}\rangle$ in F and the map $Q \to \hat{Q} : L_s(\Phi; F) \to L_s(\hat{\Phi}; F)$ is continuous. Similarly if $\hat{Q}_\alpha \to \hat{Q}$ in $L_s(\hat{\Phi}; F)$ then $Q_\alpha \to Q$ in $L_s(\Phi; F)$ so we can assert

<u>Proposition 2.2</u> The map $Q \to \hat{Q} : L_s(\Phi; F) \to L_s(\hat{\Phi}; F)$ of Definition 2.1 is a topological isomorphism for any complete separated locally convex F.

We observe in passing that the standard formulas $(D_xQ)^\wedge = -is\hat{Q}$, etc., are preserved since one has for example $\langle(D_xQ)^\wedge,\hat{\phi}\rangle = 2\pi\langle D_xQ,\phi(-x)\rangle = -2\pi\langle Q,D_x\phi(-x)\rangle = 2\pi\langle Q,\phi'(-x)\rangle = \langle\hat{Q},(D_x\phi)^\wedge\rangle = -is\langle\hat{Q},\hat{\phi}\rangle$. Now apply this Fourier transform $Q \to \hat{Q}$ to $U(t,\cdot)$ for t fixed in (2.1) so that writing $\hat{U}(t,s)$ for $F_xU(t,\cdot)$ we have

$$P(D_t,-is)\hat{U}(t,s) = 0; \quad D_t^k\hat{U}(0,s) = \widehat{T(\cdot)u_k}(s) \tag{2.2}$$

for $0 \le k \le m-1$. In conjunction with this problem we consider the ordinary differential equations (cf. (1.3))

(2.3) $P(D_t,-is)\hat{G}_n(t,s) = 0;\quad D_t^k\hat{G}_n(0,s) = \delta_{nk}$

$(0 \le n,k \le m-1)$. Assume that the equations (2.3) can be solved for $\hat{G}_n(t,s)$ which are multipliers in $\hat{\Phi}$ for each t and suitably differentiable in t with values in $L_s(\hat{\Phi})$ as multipliers (examples will be given later). Then the solution of (2.2) can be written formally as

$$\hat{U}(t,s) = \sum_{n=0}^{m-1} \hat{G}_n(t,s)\widehat{T(\cdot)u_n}(s) \tag{2.4}$$

and taking inverse Fourier transforms we obtain the formula $(G_n(t,\cdot) = F^{-1}\hat{G}_n(t,s))$

$$U(t,x) = \sum_{n=0}^{m-1} G_n(t,\cdot)*T(\cdot)u_n = \sum_{n=0}^{m-1}\int_{-\infty}^{\infty} G_n(t,\xi)T(x-\xi)u_n d\xi \tag{2.5}$$

There is some conscious abuse of notation in writing integrals for distribution action but this should cause no confusion. We will continue in a formal manner to prove the following theorem; a more precise statement and a rigorous proof are contained in Theorem 2.6 to follow but this result seems worth stating separately.

<u>Theorem 2.3(F)</u>. Let Φ be a space of C^∞ test functions where $\mathcal{D} \subset \Phi \subset \mathcal{S}'$ with $F : \Phi \to \hat{\Phi}$ a topological isomorphism; Φ and $\hat{\Phi}$ are also assumed to be barreled and complete. Let $T(x) \in L(F)$ be a suitable group generated by a closed densely defined operator A in the complete separated locally convex space F with $u_n \in D(A^\ell)$ and $T(\cdot)u_n \in L_s(\Phi; F)$. Suppose $t \to \hat{G}_n(t,s) \in C^m(L_s(\hat{\Phi}))$ as multiplier functions (i.e. the $D_t^k\hat{G}_n(t,s)$ are multipliers in $\hat{\Phi}$ for $0 \le k \le m$ depending continuously on t in $L_s(\hat{\Phi})$). Then $\hat{U}(t,s)$ is a solution of (2.2) in $L_s(\hat{\Phi}; F)$, $U(t,x)$ satisfies (2.1) in $L_s(\Phi; F)$, and a solution of (1.2) is given formally by the formula

$$u(t) = T(-x)U(t,x) = \sum_{n=0}^{m-1} \langle G_n(t,\cdot), T(-\cdot)u_n\rangle = \sum_{n=0}^{m-1}\int_{-\infty}^{\infty} G_n(t,\xi)T(-\xi)u_n d\xi \tag{2.6}$$

provided $T(-x)[G_n(t,\cdot)*T(\cdot)u_n] = [G_n(t,\cdot)*T(\cdot)u_n]_{x=0}$ makes suitable sense.

Proof: Calculating formally one has from (2.6) $(A_j = A_j(t) = \sum_o^{\ell} a_{jk}A^k = P_j(A))$

(2.7) $$P(D_t,A)u(t) = \sum_{i=0}^{m} A_i D_t^i =$$

$$\sum_{n=0}^{m-1} < \sum_{i=0}^{m} D_t^i G_n(t,\cdot),\ P_i(A)T(-\cdot)u_n > =$$

$$= \sum_{n=0}^{m-1} < \sum_{i=0}^{m} D_t^i G_n(t,\cdot),\ P_i(-D_x)T(-\cdot)u_n > =$$

$$\sum_{n=0}^{m-1} < \sum_{i=0}^{m} D_t^i P_i(D_x)G_n(t,\cdot),\ T(-\cdot)u_n > =$$

$$\sum_{n=0}^{m-1} <P(D_t,D_x)G_n(t,\cdot),\ T(-\cdot)u_n> = 0$$

since from (2.3) $P(D_t,D_x)G_n(t,x) = 0$. Further because $D_t^k G_n(0,x) = \delta_{nk}\delta(x)$ by (2.3) we obtain $D_t^k u(0) = u_k$ for $0 \leq k \leq m-1$. QED

To obtain uniqueness theorems for the solution $u(t)$ of (2.6) for (1.2) we look at uniqueness for (2.2) and employ a variation of a technique developed in Carroll [12; 13; 14] and Carroll-Showalter [8] which was indicated here already for the wave equation in Example 1.7. The uniqueness argument in Donaldson [4; 5; 7] for Banach spaces is incomplete (although correctly displayed in Donaldson [8]) and the present exposition following Carroll [16; 17], in the rigorous form given below for systems, serves to generalize and complete this argument, since in the present context the two approaches are shown to be equivalent. The uniqueness question had been open for a number of years and earlier proofs by Hersh [1; 2] and Carroll [5: 6] required (unnecessary) hypotheses on F' and A*. We proceed formally at first and will make everything rigorous later. Thus let $\hat{G}_n(t,\tau,s)$ satisfy

(2.8) $$P(D_t,-is)\hat{G}_n(t,\tau,s) = 0;\quad D_t^k\hat{G}_n(\tau,\tau,s) = \delta_{nk}$$

$(0 \leq n \leq m-1;\ 0 \leq \tau \leq t \leq T < \infty;\ 0 \leq k \leq m-1)$. One can think of these equations as arising from a system $(0 \leq k,n \leq m-1)$

(2.9) $$\frac{d}{dt}\hat{\vec{G}}_n(t,\tau,s) = \mathbb{P}(-is)\hat{\vec{G}}_n(t,\tau,s);$$

$$\hat{\vec{G}}_n(\tau,\tau,s) = (\overrightarrow{\delta_{bk}})$$

where $(\vec{\delta}_{nk})$ is the column vector with entries δ_{nk} and $\hat{\vec{G}}$ is the column vector $(D_t^k\hat{G}_n(t,\tau,s))$. We are assuming $A_m = A_m(t) \equiv 1$ so the matrix $\mathbb{P}(-is)$ has the form

(2.10) $$\mathbb{P}(-is) = \begin{pmatrix} 0 & 1 & 0 & \cdots & 0 \\ 0 & 0 & 1 & \cdots & 0 \\ & & \cdots & & \\ 0 & 0 & 0 & & 1 \\ -\hat{A}_o & -\hat{A}_1 & -\hat{A}_2 & \cdots & -\hat{A}_{m-1} \end{pmatrix}$$

where $\hat{A}_j = P_j(-is) = \sum_o^{\ell} a_{jk}(-is)^k$ (recall here that $P(D_t,A)$ was written in the form $\sum_o^m P_j(A)D_t^j$ - cf (2.7))

Now let $\hat{\mathbb{G}}(t,\tau,s) = (\hat{\vec{G}}_n(t,\tau,s))$ be the fundamental matrix with entries $D_t^k\hat{G}_n(t,\tau,s)$ $(0 \le k,n \le m-1)$ with k denoting the row index and n the column index. Then one has

(2.11) $$\frac{d}{dt}\hat{\mathbb{G}}(t,\tau,s) = \mathbb{P}(-is)\hat{\mathbb{G}}(t,\tau,s)$$

with $\mathbb{G}(\tau,\tau,s) = I$ and it is well known that

(2.12) $$\frac{d}{d\tau}\hat{\mathbb{G}}(t,\tau,s) = -\hat{\mathbb{G}}(t,\tau,s)\mathbb{P}(-is)$$

(cf. Carroll [4], Schwartz [5]). Such a formula generalizes easily to the case when $\mathbb{P}$ also depends on t (cf below). In particular, applied to the columns $\hat{\vec{G}}_n(t,\tau,s)$ of $\hat{\mathbb{G}}(t,\tau,s)$, we will have various formulas for the terms $D_\tau D_t^k\hat{G}_n(t,\tau,s)$ of which the most relevant are $(0 \le n \le m-1)$

(2.13) $$D_\tau\hat{G}_n(t,\tau,s) = -\hat{G}_{n-1}(t,\tau,s) + P_n(-is)\hat{G}_{m-1}(t,\tau,s)$$

where $\hat{G}_{-1} \equiv 0$. Suppose now that $\tau \to \hat{G}_n(t,\tau,s) \in C^1(L_s(\hat{\Phi}))$ as multiplier functions of τ. Let $\hat{U}(\tau,s)$ be <u>any</u> solution of (2.2) with t replaced by τ and

$D_\tau^k\hat{U}(0,s) = 0$ for $0 \le k \le m-1$. For t fixed consider the expression

$$(2.13') \qquad \phi(t,\tau) = \sum_{n=0}^{m-1} \hat{G}_n(t,\tau,s)D_t^n\hat{U}(\tau,s)$$

Clearly $\phi(t,t) = \hat{G}_o(t,t,s)\hat{U}(t,s)$ while $\phi(t,0) = 0$. However, formally, using (2.13)

$$(2.14) \qquad \phi_\tau(t,\tau) = \sum_{n=0}^{m-1} D_\tau\hat{G}_n(t,\tau,s)D_\tau^n\hat{U}(\tau,s) +$$

$$+ \sum_{n=0}^{m-1} \hat{G}_n(t,\tau,s)D_\tau^{n+1}\hat{U}(\tau,s) =$$

$$- \sum_{n=1}^{m-1} \hat{G}_{n-1}(t,\tau,s)D_\tau^n\hat{U}(\tau,s) + \sum_{n=0}^{m-1} P_n(-is)\hat{G}_{m-1}(t,\tau,s)D_\tau^n\hat{U}(\tau,s)$$

$$+ \sum_{n=0}^{m-1} \hat{G}_n(t,\tau,s)D_\tau^{n+1}\hat{U}(\tau,s)$$

But $\sum_{n=0}^{m-1}\hat{G}_n(t,\tau,s)D_\tau^{n+1}\hat{U}(\tau,s)$ can be written in the form $\hat{G}_{m-1}(t,\tau,s)D_\tau^m\hat{U}(\tau,s) + \sum_{k=1}^{m-1}\hat{G}_{k-1}(t,\tau,s)D_\tau^k\hat{U}(\tau,s)$ so that (2.14) becomes

$$(2.15) \qquad \phi_\tau(t,\tau) = \hat{G}_{m-1}(t,\tau,s)[D_\tau^m + \sum_{n=0}^{m-1} P_n(-is)D_\tau^n]\hat{U}(\tau,s)$$

which vanishes identically. Consequently $\hat{U}(t,s) = \hat{U}(t,0) = 0$ and $\hat{U}(t,s) \equiv 0$. Thus formally we can state (cf. also Theorem 2.8 below)

Theorem 2.4(F). Assume the hypotheses of Theorem 2.3(F) and that $\tau \to \hat{G}_n(t,\tau,s) \in C^1(L_s(\hat{\Phi}))$ as multipliers. Then the $u(t)$ of (2.6) is the unique solution of (1.2) in $C^m(F)$ with $T(\cdot)u_n \in L_s(\Phi;\ F)$.

We consider now the case of general first order systems and will be more rigorous in our treatment. Thus consider

$$(2.16) \qquad \frac{d\vec{u}}{dt} = \mathbb{P}(t,A)\vec{u}\ ; \quad \vec{u}(0) = \vec{u}_o$$

where $\vec{u}(t)$ is a column vector with entries $u_j(t)$ $(0 \le j \le m-1)$ and $\mathbb{P}(t,A)$ is an $m \times m$ matrix with entries

$$(2.17) \qquad p_{ij}(t,A) = \sum_k c_{ijk}(t)A^k$$

where $k \leq M_{ij} \leq M$ and the $c_{ijk}(\cdot)$ are continuous. Let the closed densely defined A generate a suitable group $T(x)$ as before and write $\vec{U}(t,x) = T(x)\vec{u}(t)$ for the vector with components $T(x)u_j(t)$. We assume $u_i(t) \in D(A^M)$, which can be assured by appropriate hypotheses on $\vec{u}_o$ (cf. (2.22)) and operate on (2.16) with $T(x)$ to obtain

$$\text{(2.18)} \qquad \frac{d}{dt}\vec{U}(t,x) = T(x)\mathbb{P}(t,A)\vec{u} = \mathbb{P}(t,D_x)\vec{U}(t,x) \ ;$$

$$\vec{U}(0,x) = T(x)\vec{u}_o$$

Denote by $L_s^m(\Phi;\ F)$ the space of m-vectors with components in $L_s(\Phi;\ F)$ and Fourier transform (2.18) in x to get

$$\text{(2.19)} \qquad \frac{d}{dt}\hat{\vec{U}}(t,s) = \mathbb{P}(t,-is)\hat{\vec{U}}(t,s);$$

$$\hat{\vec{U}}(0,s) = \widehat{T(\cdot)\vec{u}_o}$$

for $\hat{\vec{U}}(t,\cdot) \in L_s^m(\hat{\Phi};\ F)$. In conjunction with (2.19) we consider

$$\text{(2.20)} \qquad \frac{d}{dt}\hat{\mathbb{G}}(t,s) = \mathbb{P}(t,-is)\hat{\mathbb{G}}(t,s); \quad \hat{\mathbb{G}}(0,s) = I$$

where $\hat{\mathbb{G}}(t,s)$ is an $m \times m$ matrix. If the entries $\hat{g}_{ij}(t,s)$ of $\hat{\mathbb{G}}(t,s)$ are multipliers in $\hat{\Phi}$ for each t and belong to $C^1(L_s(\hat{\Phi}))$ as multiplier functions of t (which will be assured if $t \to \hat{g}_{ij}(t,\cdot) \in C^0(L_s(\hat{\Phi}))$ and polynomials are multipliers in $\hat{\Phi}$) then a solution of (2.19) in $L_s^m(\hat{\Phi};\ F)$ can be written as

$$\text{(2.21)} \qquad \hat{\vec{U}}(t,s) = \hat{\mathbb{G}}(t,s)\hat{\vec{U}}(0,s)$$

where $\hat{\vec{U}}(0,s) = \widehat{T(\cdot)\vec{u}_o}(s) \in L_s^m(\hat{\Phi};\ F)$. A solution of (2.18) in $L_s^m(\Phi;\ F)$ is then given by $\vec{U}(t,x) = F^{-1}\hat{\vec{U}}(t,s) = \mathbb{G}(t,\cdot)*\vec{u}(0,\cdot)$ and <u>formally</u>

$$\text{(2.22)} \qquad \vec{u}(t) = T(-x)\vec{U}(t,x) = T(-x)[\mathbb{G}(t,\cdot)*T(\cdot)u_o]$$

$$= [\mathbb{G}(t,\cdot)*T(\cdot)\vec{u}_o]_{x=0} = \int_{-\infty}^{\infty} \mathbb{G}(t,\xi)T(-\xi)\vec{u}_o d\xi$$

is a solution of (2.16).

Let us examine the constructions and see what can be asserted precisely. First note that (2.22) formally displays $\vec{u}(f) \in F$ and shows that $u_i(t) \in D(A^M)$ whenever the components of $\vec{u}_o$ belong to $D(A^M)$. Now under our assumptions $\hat{\mathbb{G}}(t,s) \in L_s(\hat{\Phi}^m)$ for fixed t and we have defined a map $(\hat{\mathbb{G}}(t,s), \hat{\vec{Q}}(s)) \to \hat{\mathbb{G}}(t,s)\hat{\vec{Q}}(s) : L_s(\hat{\Phi}^m) \times L_s^m(\hat{\Phi}; F) \to L_s^m(\hat{\Phi}; F)$ by the rule $<\hat{\mathbb{G}}(t,\cdot)\hat{\vec{Q}}(\cdot),\hat{\vec{\phi}}(\cdot)> = <\hat{\vec{Q}}(\cdot),\hat{\mathbb{G}}^T(t,\cdot)\hat{\vec{\phi}}(\cdot)>$ for $\hat{\vec{\phi}} \in \hat{\Phi}^m$ ($\hat{\mathbb{G}}^T$ denotes $\hat{\mathbb{G}}$ transpose). This corresponds to defining a map from $L_s(\Phi^m) \times L_s^m(\Phi; F)$ to $L_s^m(\Phi; F)$ by the action $(\mathbb{G}(t,x),\vec{Q}(x)) \to \mathbb{G}(t,\cdot) * \vec{Q}(\cdot)$ since $F(\mathbb{G}(t,\cdot) * \vec{Q}(\cdot)) = \hat{\mathbb{G}}(t,s)\hat{\vec{Q}}(s)$. We know $\hat{\mathbb{G}}(t,s)$ satisfies (2.20) in $L_s(\hat{\Phi}^m)$ so that $\mathbb{G}(t,x)$ satisfies

(2.23) $$\frac{d}{dt}\mathbb{G}(t,x) = \mathbb{P}(t, D_x)\mathbb{G}(t,x); \quad \mathbb{G}(0,x) = \delta(x)I$$

in $L_s(\Phi^m)$. But under our definitions it is then clear that $\hat{\mathbb{G}}$, acting by multiplication (resp. $\mathbb{G}$, acting by convolution) satisfies (2.20) (resp. (2.23)) in $L_s(L_s^m(\hat{\Phi}; F))$ (resp. $L_s(L_s^m(\Phi; F))$). Note here that $\hat{\mathbb{H}}_\alpha \to \hat{\mathbb{H}}$ in $L_s(L_s^m(\hat{\Phi}; F))$ means $\hat{\mathbb{H}}_\alpha(\hat{\vec{Q}}) \to \hat{\mathbb{H}}(\hat{\vec{Q}})$ in $L_s^m(\hat{\Phi}; F)$ for $\hat{\vec{Q}} \in L_s^m(\hat{\Phi}; F)$; but this is equivalent to $<\hat{\mathbb{H}}_\alpha(\hat{\vec{Q}}),\hat{\vec{\phi}}> \to <\hat{\mathbb{H}}(\hat{\vec{Q}}), \hat{\vec{\phi}}>$ in F for $\hat{\vec{\phi}} \in \hat{\Phi}^m$. Hence we can say that $\mathbb{G}_t(t,\cdot) * T(\cdot)\vec{u}_o = \mathbb{P}(t,D_x)\mathbb{G}(t,\cdot) * T(\cdot)\vec{u}_o = \mathbb{P}(t,A)\mathbb{G}(t,\cdot) * T(\cdot)\vec{u}_o$ in $L_s^m(\Phi; F)$. Let us explicitly observe here that one can pass the verification of continuity and differentiability for $\hat{\mathbb{G}}(t,s)$ in the above spaces to transpose action. Thus if $\hat{v}(t,s) = \hat{g}_{ij}(t,s)$ is a typical multiplier element and $\hat{w}(s) = T(\cdot)u_{ok}(s) \in L_s(\hat{\Phi}; F)$ then $\hat{v}(t,s)\hat{w}(s) \in L_s(\hat{\Phi}; F)$ since $<\hat{v}(t,s)\hat{w}(s),\hat{\phi}(s)> = <\hat{w}(s),\hat{v}(t,s)\hat{\phi}(s)>$. Setting $\Delta\hat{v} = \hat{v}(t,s) - \hat{v}(t_o,s)$ we have $<\Delta\hat{v}\hat{w},\hat{\phi}> = <\hat{w},\Delta\hat{v}\hat{\phi}>$ so that $t \to \hat{v}(t,\cdot) \in C^o(L_s(\hat{\Phi}))$ as a multiplier implies $\Delta\hat{v}\hat{w} \to 0$ in $L_s(\hat{\Phi}; F)$ as $t \to t_o$ or $t \to \hat{v}(t,\cdot) \in C^o(L_s(L_s(\hat{\Phi}; F)))$. Similarly one looks at $<(\Delta\hat{v}/\Delta t)\hat{w},\hat{\phi}> = <\hat{w},(\Delta\hat{v}/\Delta t)\hat{\phi}> \to <\hat{w},\hat{v}_t\hat{\phi}> = <\hat{v}_t\hat{w},\hat{\phi}>$ for $t \to \hat{v}(t,\cdot) \in C^1(L_s(\hat{\Phi}))$ as a multiplier.

<u>Remark 2.5</u> Let us note here that if $\vec{T} \in L_s^m(\Phi; F)$ and $\vec{\phi} \in \Phi^m$ then natural distribution pairings yield

(2.24) $$< \mathbb{P}(D_x)\vec{T},\vec{\phi}(-x)> = <\vec{T}, \mathbb{P}^T(-D_x)\vec{\phi}(-x)> = <\vec{T},[\mathbb{P}^T(D_x)\vec{\phi}](-x)>$$

and applying the Parseval formula one obtains

(2.25) $$< \mathbb{P}(-is)\hat{\vec{T}},\hat{\vec{\phi}}> = 2\pi< \mathbb{P}(D_x)\vec{T},\vec{\phi}(-x)> = <\hat{\vec{T}}, \mathbb{P}^T(-is)\hat{\vec{\phi}}>$$

(here $\mathbb{P}(D_x)$ is any matrix of the form $\mathbb{P}(t,D_x)$ above). Similarly using the notation $\hat{\mathbb{G}}(t,s) = \mathbb{G}(t,-is) = F\mathbb{G}(t,D_x) = F\mathbb{G}(t,x)$ we have

$$(2.26) \qquad <\mathbb{G}(t,-is)\hat{\vec{T}},\hat{\vec{\phi}}> = 2\pi<\mathbb{G}(t,D_x)\vec{T},\vec{\phi}(-x)>$$

$$= 2\pi<\vec{T},[\mathbb{G}^T(t,D_x)\vec{\phi}](-x)> = <\hat{\vec{T}},\mathbb{G}^T(t,-is)\hat{\vec{\phi}}> .$$

These formulas show for example that using the multiplier "adjoints" $\hat{\mathbb{G}}*(t,s) = G*(t,-is) = \mathbb{G}^T(t,-is)$ in $L_s^m(\hat{\Phi};\ F)$ is consistent with using the operator "adjoints" $\mathbb{G}*(t,x) = \mathbb{G}*(t,D_x) = \mathbb{G}^T(t,-D_x)$ in $L_s^m(\Phi;\ F)$.

Now we have shown that $\mathbb{G}_t(t,\cdot)*T(\cdot)\vec{u}_o = \mathbb{P}(t,A)\mathbb{G}(t,\cdot)*T(\cdot)\vec{u}_o$ in $L_s^m(\Phi;\ F)$ with $\mathbb{G}(0,\cdot)*T(\cdot)\vec{u}_o = \vec{u}_o$ and it remains to particularize to $x = 0$ in order to obtain (2.22). It is to be noted that in this process the space Φ virtually disappears from consideration in (2.22) but it really occurs implicitly in two ways in that $T(\cdot)\vec{u}(t) \in L_s^m(\Phi;\ F)$ and $\mathbb{G}(t,\cdot) \in L_s(\Phi^m)$ under convolution (satisfying (2.23)). In general there may be many suitable Φ, or perhaps none, for which this is the case, and this question will be examined later. The requirement $T(\cdot)\vec{u}(t) \in L_s^m(\Phi;\ F)$ is obviously not too much of a restriction in any event. In order to set $x = 0$ in $\vec{U}(t,x) = \mathbb{G}(t,\cdot)*T(\cdot)\vec{u}_o$ we want $\vec{U}(t,\cdot)$ to be a function of x, say continuous, and to satisfy (2.16) we must have $D_t\vec{U}(t,0) = [D_t\vec{U}(t,x)]_{x=0} = [\mathbb{G}_t(t,\cdot)*T(\cdot)\vec{u}_o]_{x=0}$. Now for $\vec{u}_o \in D(A^M)$, $T(\cdot)\vec{u}_o \in C^M(F^m)$ by assumptions and thus $\mathbb{G}(t,\cdot)*T(\cdot)\vec{u}_o$ and $\mathbb{G}_t(t,\cdot)*T(\cdot)\vec{u}_o = \mathbb{P}(t,A)\mathbb{G}(t,\cdot)*T(\cdot)\vec{u}_o$ involves the convolution of elements $g_{ij}(t,\cdot)$ in $\mathbb{G}(t,\cdot)$ with terms $T(\cdot)f$ for $f \in F$. Hence evaluation E at $x = 0$ will be possible for example when, under convolution, $\mathbb{G}(t,\cdot)$ maps $C_\alpha^o(F^m) \to C^o(F^m)$ where $h(\cdot) \in C_\alpha^o(F)$ means $h(\cdot) \in C^o(F)$ and $p(h(x)) \leq k_p \exp \alpha|x|$ for any seminorm p in F (recall $p(T(x)f) \leq q(f) \exp \alpha|x|$ for quasiequicontinuous groups). Further requirements of the type $u_{ok} \in D(A^N)$ $(N \leq \infty)$ played off against the order of the distributions $g_{ij}(t,\cdot)$ in $\mathbb{G}(t,\cdot)$ will give other situations where evaluation would be possible (cf. A. Friedman [1], Gelfand-Šilov [3]). Under the present hypotheses one expects a suitable evaluation to exist for parabolic and hyperbolic systems for example and this will be illustrated later. For now we will simply assume $u_{ok} \in D(A^N)$ for some N such that evaluation E at $x = 0$ is possible and $E[\mathbb{G}(t,\cdot) * T(\cdot)\vec{u}_o] = D_t E[\mathbb{G}(t,\cdot) * T(\cdot)\vec{u}_o]$. Then we can state

Theorem 2.6 Let Φ be a space of C^∞ test functions where $\mathcal{D} \subset \Phi \subset \mathcal{S}'$ with $F : \Phi \to \hat{\Phi}$ a topological isomorphism; Φ and $\hat{\Phi}$ are also assumed to be barreled and complete in general. Let $T(x)$ be a quasiequicontinuous group generated by a closed densely defined linear operator A in the complete locally convex space F with the components u_{ok} of $\vec{u}_o$ belonging to $D(A^N)$ for suitable N and $T(\cdot)u_{ok} \in L_s(\Phi; F)$. Suppose $t \to \hat{\mathbb{G}}(t,s) \in C^1(L_s(\hat{\Phi}^m))$ as a "multiplier" satisfying (2.20). Assume $\mathbb{G}(t,\cdot)$ maps $C^N_\alpha(F^m) \to C^o(F^m)$ under convolution and that evaluation E at $x = 0$ satisfies $E[\mathbb{G}_t(t,\cdot) * T(\cdot)\vec{u}_o] = D_t E[\mathbb{G}(t,\cdot) * T(\cdot)\vec{u}_o]$. Then $\vec{u}(t)$ given by (2.22) is a solution of (2.16).

Remark 2.7 It was mentioned earlier that if the entries $\hat{g}_{ij}(t,s)$ of $\hat{\mathbb{G}}(t,s)$ are continuous multiplier functions of t and polynomials are multipliers in $\hat{\Phi}$ then $t \to \hat{g}_{ij}(t,s) \in C^1(L_s(\hat{\Phi}))$ as a multiplier. To see this write from (2.20)

$$(2.27) \qquad \frac{\Delta\hat{g}_{jk}}{\Delta t}\hat{\phi} = \frac{1}{\Delta t}\int_t^{t+\Delta t} \sum_\ell P_{j\ell}(\tau,-is)\hat{g}_{\ell k}(\tau,s)\hat{\phi}(s)d\tau.$$

The integrands $\tau \to p_{j\ell}(\tau,-is)\hat{g}_{\ell k}(\tau,s)\hat{\phi}(s)$ are continuous functions of τ with values in $\hat{\Phi}$ since $\tau \to \hat{g}_{\ell k}(\tau,s) \in C^o(L_s(\hat{\Phi}))$ as multipliers and $p_{j\ell}(\tau,-is) = \Sigma c_{j\ell k}(\tau)(-is)^k$ is a polynomial (cf. (2.17)). Hence by standard facts about vector valued integration (cf. Bourbaki [3; 4; 5]) $(\Delta\hat{g}_{jk}/\Delta t)\hat{\phi} \to D_t\hat{g}_{jk}(t,s)\hat{\phi}$ in $\hat{\Phi}$ and $\hat{g}_{jk} \in C^1(L_s(\hat{\Phi}))$ as a multiplier.

Now for uniqueness we will give two proofs and show them to be equivalent in the present context. The first is based on the technique of Carroll [12; 13; 14] whereas the second is the author's version of the approach in Donaldson [4; 5; 7] modeled on Carroll [16; 17]. First we consider the ordinary differential equations (cf. (2.20))

$$(2.28) \qquad \frac{d}{dt}\hat{\mathbb{G}}(t,\tau,s) = \mathbb{P}(t,-is)\hat{\mathbb{G}}(t,\tau,s); \quad \hat{\mathbb{G}}(\tau,\tau,s) = I$$

and as with (2.12) one knows that

(2.29) $\frac{d}{d\tau}\hat{\mathbb{G}}(t,\tau,s) = -\hat{\mathbb{G}}(t,\tau,s)\,\mathbb{P}(\tau,-is)$

Theorem 2.8 Assume the hypotheses of Theorem 2.6 and suppose further that $\tau \to \hat{\mathbb{G}}(t,\tau,s) \in C^0(L_s(\hat{\Phi}^m))$ as a multiplier while polynomials are multipliers in $\hat{\Phi}$. Then the solution $\vec{u}(t)$ of (2.16) given by (2.22) is unique in the class of solutions satisfying $T(\cdot)\vec{u}(t) \in L_s^m(\Phi;\ F)$.

Proof: Following Remark 2.7 we know from (2.29) that $\tau \to \hat{\mathbb{G}}(t,\tau,s) \in C^1(L_s(\hat{\Phi}^m))$. Let now $\hat{\vec{U}}(t,s)$ be any solution of (2.19) in $L_s^m(\hat{\Phi};\ F)$ with $\hat{\vec{U}}(0,s) = 0$ and we will show that $\hat{\vec{U}}(t,s) \equiv 0$. First consider $\vec{\psi}(t,\tau,s) = \hat{\mathbb{G}}(t,\tau,s)\hat{\vec{U}}(\tau,s) \in L_s^m(\hat{\Phi};\ F)$ for t fixed and $0 \le \tau \le t$ so that $\vec{\psi}(t,t,s) = \hat{\vec{U}}(t,s)$ while $\vec{\psi}(t,0,s) = 0$. Let $\hat{\vec{\phi}} \in \hat{\Phi}^m$ and look at $\tau \to \psi(t,\tau) = \langle\vec{\psi}(t,\tau,s),\ \hat{\vec{\phi}}(s)\rangle \in F$. To check continuity and differentiability it suffices to consider sequential $\Delta\tau = \tau_n - \tau$ (cf. Carroll [4]) and in an obvious notation we write

(2.30) $\frac{\Delta\psi}{\Delta\tau} = \langle\hat{\vec{U}}(\tau + \Delta\tau,s),\ \frac{\Delta\hat{\mathbb{G}}^T}{\Delta\tau}\hat{\vec{\phi}}(s)\rangle + \langle\frac{\Delta\hat{\vec{U}}}{\Delta\tau},\ \hat{\mathbb{G}}^T(t,\tau,s)\hat{\vec{\phi}}(s)\rangle.$

We recall that the bilinear map $(\hat{Q},\hat{\phi}) \to \langle\hat{Q},\hat{\phi}\rangle : L_s(\hat{\Phi};\ F) \times \hat{\Phi} \to F$ is separately continuous and since $\hat{\Phi}$ is barreled it follows that this map is hypocontinuous relative to bounded sets in $L_s(\hat{\Phi};\ F)$ (cf Horváth [1]). Now the sequence $\hat{\vec{U}}(\tau + \Delta\tau,s)$ is Cauchy in $L_s^m(\hat{\Phi};\ F)$ and hence bounded while $(\Delta\hat{\mathbb{G}}^T/\Delta\tau)\hat{\vec{\phi}} \to \mathbb{G}_\tau^T(t,\tau,s)\hat{\vec{\phi}}(s)$ in $\hat{\Phi}^m$ fo the first bracket in (2.30) converges to $\langle\hat{\vec{U}}(\tau,s),\ \hat{\mathbb{G}}_\tau^T(t,\tau,s)\hat{\vec{\phi}}(s)\rangle$ in F. Similarly the sequence $\Delta\hat{\vec{U}}/\Delta\tau$ is Cauchy in $L_s^m(\hat{\Phi};\ F)$ and hence bounded so the second bracket in (2.30) converges to $\langle\hat{\vec{U}}_\tau(\tau,s),\ \hat{\mathbb{G}}^T(t,\tau,s)\hat{\vec{\phi}}(s)\rangle$ in F. Hence we obtain

(2.31) $\psi_\tau(t,\tau) = \langle\hat{\mathbb{G}}_\tau(t,\tau,s)\hat{\vec{U}}(\tau,s),\ \hat{\vec{\phi}}(s)\rangle + \langle\hat{\mathbb{G}}(t,\tau,s)\hat{\vec{U}}_\tau(t,\tau,s),\ \hat{\vec{\phi}}(s)\rangle$

This shows that ψ is continuously differentiable and from (2.19) and (2.29) yields $\psi_\tau(t,\tau) = 0$. Hence the continuous function $\tau \to \psi(t,\tau)$ is constant so that $\psi(t,t) = \langle\hat{\vec{U}}(t,s),\ \hat{\vec{\phi}}(s)\rangle = \psi(t,0) = 0$ for any fixed t. Since $\hat{\vec{\phi}} \in \hat{\Phi}^m$ is arbitrary this implies that $\hat{\vec{U}}(t,s) \equiv 0$ in $L_s^m(\hat{\Phi};\ F)$. This entails uniqueness for (2.16) since if $\vec{u}(t)$ were any solution of (2.16) with $\vec{u}(0) = 0$ and

$T(\cdot)\vec{u}(t) = \vec{U}(t,\cdot) \in L_s^m(\Phi; F)$ then the corresponding $\hat{\vec{U}}(t,s)$ satisfies (2.19) in $L_s^m(\hat{\Phi}; F)$.

Second proof: Let $\hat{\vec{U}}(t,s)$ be any solution of (2.19) as above with $\hat{\vec{U}}(0,s) = 0$ and set $\hat{\vec{\phi}}(\tau,t,s) = \mathbb{G}^T(\tau,t,-is)\hat{\vec{\phi}}_o(s) \in \hat{\Phi}^m$ for $\hat{\vec{\phi}}_o \in \hat{\Phi}^m$ arbitrary and τ fixed ($0 \leq t \leq \tau$; see Remark 2.5 for the notation $\mathbb{G}^T(\tau,t,-is) = \hat{\mathbb{G}}^T(\tau,t,s)$). Then from (2.29) we have

$$\text{(2.32)} \qquad \frac{d}{dt}\hat{\mathbb{G}}^T(\tau,t,s) = -\mathbb{P}^T(t,-is)\hat{\mathbb{G}}^T(\tau,t,s)$$

so that in $\hat{\Phi}^m$, $\hat{\vec{\phi}}(\tau,t,s)$ satisfies

$$\text{(2.33)} \qquad \frac{d\hat{\vec{\phi}}}{dt} = -\mathbb{P}^T(t,-is)\hat{\vec{\phi}};\ \hat{\vec{\phi}}(\tau,\tau,s) = \hat{\vec{\phi}}_o(s).$$

Consider now $\theta(t,\tau) = \langle\hat{\vec{U}}(t,s), \hat{\vec{\phi}}(\tau,t,s)\rangle \in F$ which makes sense for t fixed. For $\Delta t = t_n - t$ sequential

$$\text{(2.34)} \qquad \frac{\Delta\theta}{\Delta t} = \langle\hat{\vec{U}}(t+\Delta t,s), \frac{\Delta\hat{\vec{\phi}}}{\Delta t}\rangle + \langle\frac{\Delta\hat{\vec{U}}}{\Delta t}, \hat{\vec{\phi}}(\tau,t,s)\rangle$$

Using again the hypocontinuity relative to bounded sets in $L_s^m(\hat{\Phi}; F)$ of the map $(\hat{Q},\hat{\phi}) \to \langle\hat{Q},\hat{\phi}\rangle : L_s(\hat{\Phi}; F) \times \hat{\Phi} \to F$ for $\hat{\Phi}$ barreled one sees that $\theta_t = \langle\hat{\vec{U}},\hat{\vec{\phi}}_t\rangle + \langle\hat{\vec{U}}_t,\hat{\vec{\phi}}\rangle = -\langle\hat{\vec{U}}, \mathbb{P}^T(t,-is)\hat{\vec{\phi}}\rangle + \langle\mathbb{P}(t,-is)\hat{\vec{U}},\hat{\vec{\phi}}\rangle = 0$ (from (2.33) and (2.19)). Thus $t \to \theta(t,\tau)$ is continuous and constant so $\theta(\tau,\tau) = \langle\hat{\vec{U}}(\tau,s), \hat{\vec{\phi}}_o(s)\rangle = \theta(0,\tau) = 0$ and since $\hat{\vec{\phi}}_o \in \hat{\Phi}^m$ is arbitrary it follows that $\hat{\vec{U}}(\tau,s) \equiv 0$ in $L_s^m(\hat{\Phi}; F)$ for any τ.

QED

<u>Remark 2.9</u> The equivalence of these two proofs in the present context is seen from the fact that $\psi(t,\tau) = \langle\mathbb{G}(t,\tau,-is)\hat{\vec{U}}(\tau,s), \hat{\vec{\phi}}(s)\rangle = \langle\hat{\vec{U}}(\tau,s), \mathbb{G}^T(t,\tau,-is)\hat{\vec{\phi}}(s)\rangle = \theta(\tau,t) = \langle\hat{\vec{U}}(\tau,s), \hat{\vec{\phi}}(t,\tau,s)\rangle$ where $\hat{\vec{\phi}}(s)$ is taken for $\hat{\vec{\phi}}_o(s)$. The technique of the first proof however, as used in Carroll [12; 13; 14] with singular or degenerat problems, involves no recourse to F-valued distributions and could be so applied here directly to (2.16). Thus the equivalence is a consequence of context.

<u>Remark 2.10</u> We note that the statement of Theorem 2.4(F) is somewhat stronger thar the theorem which would result as a corollary of Theorem 2.8. A strict corollary

would require $\tau \to D_t^k \hat{G}_n(t,\tau,s) \in C^1(L_s(\hat{\Phi}))$ for $0 \le k,n \le m-1$ whereas we only need this statement for $k = 0$ as indicated in the formal demonstration of Theorem 2.4(F). This can be made rigorous following the formulation of Theorems 2.6 and 2.8 but we will omit an explicit statement of this fact.

1.3. Examples. We will give first some results for singular and degenerate problems using the pairing of Definition 1.1 (cf. Carroll [12; 13; 14] and Carroll-Showalter [8]) and then turn to some examples of general parabolic and hyperbolic systems following Donaldson [4; 5; 7]. The singular and degenerate problems considered do not fall into the framework of Section 2 since they are not of the form (1.2) (either $A_j(t)$ is singular at $t = 0$ or $A_m(t)$ vanishes at $t = 0$).

Example 3.1. Euler-Poisson-Darboux (EPD) equations. As the basic equation we consider

$$(3.1) \qquad u_{tt} + \frac{2m+1}{t} u_t = A^2 u; \quad u(0) = u_o; \quad u_t(0) = 0$$

where A generates a locally equicontinuous group $T(x)$ in F. The case where $A^2 = \Delta$ in $\mathbb{R}^n$ has been of considerable interest and an extensive survey of such equations appears in Carroll-Showalter [8] (cf. also Carroll [33; 34] and Carroll-Silver [9; 10; 11] for group theoretic aspects). We will treat here only the case $m \ge -\frac{1}{2}$ and denote the corresponding solutions u by u^m; parallel results are easily proved for $\mathrm{Re}\, m > -\frac{1}{2}$ while for other values of m the theory changes a bit (cf. Carroll-Showalter [8]). In the study of such equations with $A^2 = \Delta$ various growth and convexity properties of the solutions were shown to be of interest and to handle this it was absolutely necessary to work in spaces F containing objects of unrestricted growth (e.g. $F = \mathcal{E} = C^\infty(\mathbb{R}^n)$). Thus our program of phrasing the theory of operator differential equations in general locally convex spaces F has a very realistic motivation. Banach or Hilbert spaces will simply not suffice to naturally encompass certain interesting types of solutions (using weighted Banach spaces suggests itself but this distorts the operator realization A^2 of Δ). Thus following the procedure of Section 1 we consider (cf. (1.2) - (1.3))

$$(3.2)\qquad R^m_{tt} + \frac{2m+1}{t} R^m_t = R^m_{xx};\quad R^m(0,x) = \delta(x);\quad R^m_t(0,x) = 0.$$

Here G has been replaced by R in conformity with the notation of Carroll-Showalter [8] where R^m was called a resolvant. One can work with $R^m(t,\cdot) \in \mathcal{S}'$ and Fourier transform (3.2) in x to obtain

$$(3.3)\qquad \hat{R}^m_{tt} + \frac{2m+1}{t} \hat{R}^m_t + s^2\hat{R}^m = 0;\quad \hat{R}^m(0,s) = 1;\quad \hat{R}^m_t(0,s) = 0$$

where $\hat{R}^m = \mathcal{F}R^m$. The solution of (3.3) is

$$(3.4)\qquad \hat{R}^m(t,s) = 2^m\Gamma(m+1)(st)^{-m}J_m(st) = \Gamma(m+1)\sum_{k=0}^{\infty} \frac{(-1)^k}{k!}\left(\frac{st}{2}\right)^{2k} \frac{1}{\Gamma(k+m+1)}.$$

Now a version of the Sonine integral formula can be employed here in writing for $m > -\frac{1}{2}$

$$(3.5)\qquad \hat{R}^m(t,s) = \frac{2\Gamma(m+1)}{\Gamma(\frac{1}{2})\Gamma(m+\frac{1}{2})}\int_0^1 \hat{R}^{-\frac{1}{2}}(\xi t,s)(1-\xi^2)^{m-\frac{1}{2}}d\xi$$

(cf. Carroll-Showalter [8]) where $\hat{R}^{-\frac{1}{2}}(t,s) = \mathrm{Cos}(st)$ with $R^{-\frac{1}{2}}(t,x) = \mu_x(t) = \frac{1}{2}[\delta(x+t) + \delta(x-t)]$. Consequently

$$(3.6)\qquad R^m(t,x) = \frac{2\Gamma(m+1)}{\Gamma(\frac{1}{2})\Gamma(m+\frac{1}{2})}\int_0^1 (1-\xi^2)^{m-\frac{1}{2}}\mu_x(\xi t)d\xi$$

where the integral is taken in $\mathcal{E}' = \mathcal{E}'_x$. We set $c_m = \Gamma(m+1)/\Gamma(\frac{1}{2})\Gamma(m+\frac{1}{2})$ so that (3.6) can be written

$$(3.7)\qquad R^m(t,x) = c_m\int_{-1}^1 (1-\xi^2)^{m-\frac{1}{2}}\delta(x-\xi t)d\xi$$

and express a solution of (3.1) as

$$(3.8)\qquad u^m(t) = \langle R^m(t,\cdot), T(\cdot)u_o\rangle = c_m\int_{-1}^1 (1-\xi^2)^{m-\frac{1}{2}}\langle\delta(x-\xi t), T(x)u_o\rangle d\xi$$

$$= c_m\int_{-1}^1 (1-\xi^2)^{m-\frac{1}{2}}T(\xi t)u_o d\xi = 2c_m\int_0^1 (1-\xi^2)^{m-\frac{1}{2}}\cosh(A\xi t)u_o d\xi$$

(bracket of Definition 1.1). For $m = -\frac{1}{2}$ (3.1) is a wave type equation and we have already expressed the unique solution as $u^{-\frac{1}{2}}(t) = \cosh At\, u_o$ (cf. Theorem 1.8). It is known that

$$(3.9) \qquad \hat{R}^m_t(t,s) = \frac{-s^2 t}{2(m+1)} \hat{R}^{m+1}(t,s)$$

$$(3.10) \qquad \hat{R}^m_t(t,s) = \frac{2m}{t} [\hat{R}^{m-1}(t,s) - \hat{R}^m(t,s)]$$

which correspond to recursion relations

$$(3.11) \qquad R^m_t(t,x) = \frac{t}{2(m+1)} \Delta R^{m+1}(t,x)$$

$$(3.12) \qquad R^m_t(t,x) = \frac{2m}{t} [R^{m-1}(t,x) - R^m(t,x)]$$

in $E'_x(\Delta = D^2_x)$. Consequently

$$(3.13) \qquad u^m_t(t) = \langle R^m_t(t,\cdot), T(\cdot)u_o\rangle =$$

$$\frac{t}{2(m+1)} \langle \Delta R^{m+1}(t,\cdot), T(\cdot)u_o\rangle = \frac{t}{2(m+1)} \langle R^{m+1}(t,\cdot), \Delta T(\cdot)u_o\rangle$$

$$= \frac{t}{2(m+1)} \langle R^{m+1}(t,\cdot), A^2 T(\cdot)u_o\rangle = \frac{t c_{m+1}}{2(m+1)} \int_{-1}^{1} (1 - \xi^2)^{m+\frac{1}{2}} T(\xi t) A^2 u_o \, d\xi$$

$$= \frac{A^2 t}{2(m+1)} u^{m+1}(t).$$

Similarly $u^m_t = \frac{2m}{t} [u^{m-1} - u^m]$ and combining this with (3.13) one obtains (3.1). The calculations can all be justified as in Example 1.7 and we can state

<u>Theorem 3.2</u> Let A generate a locally equicontinuous group in F with $u_o \in D(A^2)$. Then for $m > -\frac{1}{2}$ a solution of (3.1) is given by (3.8).

The uniqueness question for this problem $(m > -\frac{1}{2})$ is answered by the following argument (see Carroll [12], Carroll-Showalter [8]). For $0 < \tau \leq t$ and constructs resolvants $R^m(t,\tau,x)$ and $S^m(t,\tau,x)$ satisfying in E'

$$(3.14) \qquad y^m_{tt} + \frac{2m+1}{t} y^m_t = \Delta y^m = y^m_{xx}$$

with $R^m(\tau,\tau,x) = \delta(x)$, $R^m_t(\tau,\tau,x) = 0$, $S^m(\tau,\tau,x) = 0$, and $S^m_t(\tau,\tau,x) = \delta(x)$ (cf. Carroll-Showalter [8] for details). When $\tau = 0$, $S^m(t,0,x) \equiv 0$, while for $0 \leq \tau \leq t \leq b$ the distributions $R^m(t,\tau,\cdot)$ and $S^m(t,\tau,\cdot)$ have supports contained in a fixed compact set $K = \{x : |x| \leq x_o\}$ and, with their x derivatives in $\mathcal{D}'$, are of order ≤ 2. Furthermore one has

$$R^m_\tau(t,\tau,x) = -\Delta S^m(t,\tau,x) \tag{3.15}$$

$$S^m_\tau(t,\tau,x) = -R^m(t,\tau,x) + \frac{2m+1}{\tau} S^m(t,\tau,x) \tag{3.16}$$

and the constructions which follow all make sense as in Example 1.7. Thus let $u \in C^2(F)$ be any solution of (3.1) with $u(0) = u_t(0) = 0$ and set

$$R(t,\tau) = \langle R^m(t,\tau,\cdot),\ T(\cdot)u(\tau)\rangle \tag{3.17}$$

$$S(t,\tau) = \langle S^m(t,\tau,\cdot),\ T(\cdot)u_\tau(\tau)\rangle \tag{3.18}$$

(bracket of Definition 1.1). Then, referring to Example 1.7 to justify the calculations (cf. also Carroll-Showalter [8]), we can write (cf. (3.15) - (3.16))

$$R_\tau(t,\tau) = -\langle S^m(t,\tau,\cdot),\ T(\cdot)A^2u(\tau)\rangle + \langle R^m(t,\tau,\cdot),\ T(\cdot)u_\tau(\tau)\rangle \tag{3.19}$$

$$S_\tau(t,\tau) = \langle S^m(t,\tau,\cdot),\ T(\cdot)u_{\tau\tau}(\tau)\rangle \tag{3.20}$$

$$+ \langle -R^m(t,\tau,\cdot) + \frac{2m+1}{\tau} S^m(t,\tau,\cdot),\ T(\cdot)u_\tau(\tau)\rangle.$$

Setting $\phi(t,\tau) = R(t,\tau) + S(t,\tau)$ we have $\phi(t,0) = 0 = \phi(t,t) = u(t)$ since

$$\phi_\tau = \langle S^m(t,\tau,\cdot),\ T(\cdot)[u_{\tau\tau} + \frac{2m+1}{\tau} u_\tau - A^2u]\rangle = 0 \tag{3.21}$$

(i.e. $\tau \to \phi(t,\tau)$ is continuous with $\phi_\tau = 0$). This proves

<u>Theorem 3.3</u> Under the hypotheses of Theorem 3.2 the $u(t)$ of (3.8) is the unique solution of (3.1) in $C^2(F)$.

<u>Example 3.4</u> Degenerate Cauchy problems. We consider problems of the form $(a(0) = 0)$

$$u_{tt} + s(t)u_t + Ar(t)u - A^2a(t)u + b(t)u = f \tag{3.21}$$

$$u(0) = u_t(0) = 0 \tag{3.22}$$

where A is the generator of a locally equicontinuous group $T(x)$ in F. The discussion here follows Carroll [13; 14] to which we refer for bibliography. Such problems are an extension of the classical Cauchy problem for the Tricomi equation in the hyperbolic region and thus are of interest in transsonic gas dynamics for example. Various abstract versions have been treated by Carroll [26], Carroll-Wang [29], Krasnov [1], Lacomblez [1], Walker [1], and Wang [1; 2] while some classical results can be found in Berezin [1], Bers [1], and Protter [1] (cf. also Barantzev [1], Bitsadze [1], Brézis [2], Carroll-Showalter [8], Friedman-Schuss [4], Schuss [1], Showalter [2; 3; 4; 5], Smirnov [1; 2], Weinstein [1; 2]. Following Section 1 (cf. (1.2) - (1.3)) we replace A by $-d/dx$ in (3.21) and consider first

$$G_{tt} + S(t)G_t - r(t)G_x - a(t)G_{xx} + b(t)G = 0 \tag{3.23}$$

for $G(t) \in S'_x$. It is assumed that s, r, a, and b are continuous while $a(t) > 0$ with $a(0) = 0$. Taking Fourier transforms $x \to y$ (3.23) gives

$$\hat{G}_{tt} + s(t)\hat{G}_t + iyr(t)\hat{G} + a(t)y^2\hat{G} + b(t)\hat{G} = 0 \tag{3.24}$$

and it will be convenient to eliminate the $b(t)$ term as follows. Let $\hat{G}(t) = \hat{g}(t) \exp \int_0^t \gamma(\xi)d\xi$ where γ satisfies the Riccati equation

$$\gamma' + s\gamma + \gamma^2 + b = 0; \quad \gamma(0) = 0. \tag{3.25}$$

In order to produce a suitable γ we note that if one sets $\gamma = \alpha'/\alpha$ then α satisfies

$$\alpha'' + s(t)\alpha' + b(t)\alpha = 0 \tag{3.26}$$

and we choose α to be the unique solution of (3.26) satisfying $\alpha(0) = 1$ with $\alpha'(0) = 0$. Then $\gamma(0) = 0$ and the continuous function γ will remain finite on some interval $0 \leq t \leq T < t_o$ where t_o is the first zero of $\alpha(t)$. We can then restrict our attention to $[0,T]$ since for $t \geq T$ the problem is not degenerate and can be handled by the methods of Section 2. Thus on $[0,T]$ we consider

(3.27) $\hat{g}'' + (s + 2\gamma)\hat{g}' + (ay^2 + iry)\hat{g} = 0$

which we write as a system

(3.28) $\vec{g}' = \mathbb{P}(t,y)\vec{g};\quad \vec{g} = \begin{pmatrix} y\hat{g} \\ \hat{g}' \end{pmatrix};\quad \mathbb{P}(t,y) = \begin{pmatrix} 0 & y \\ -ir-ay & -s-2\gamma \end{pmatrix}.$

We look for solutions $\vec{Y}$ and $\vec{Z}$ of (3.28) satisfying

(3.29) $\vec{Y}(\tau) = \begin{bmatrix} y\hat{Y} \\ \hat{Y}_t \end{bmatrix}(\tau) = \begin{pmatrix} 0 \\ 1 \end{pmatrix};\quad \vec{Z}(\tau) = \begin{bmatrix} y\hat{Z} \\ \hat{Z}_t \end{bmatrix}(\tau) = \begin{pmatrix} 1 \\ 0 \end{pmatrix}$

where $0 \leq \tau \leq t \leq T$. By well known theorems there exist solutions $\hat{Y}(t,\tau,y)$ and $\hat{Z}(t,\tau,y)$ of (3.27) satisfying the prescribed intial conditions at $t = \tau$ and these functions are continuous in (t,τ,y) and analytic in y for $0 \leq \tau \leq t \leq T$ and $y \in \mathbb{C}$. It is easily shown also that

(3.30) $\hat{Z}_\tau = (ay^2 + iyr)(\tau)\hat{Y};\quad \hat{Y}_\tau = -Z + (s + 2\gamma)(\tau)\hat{Y}$

(cf. Carroll [4]). By an argument now in Gelfand-Šilov [3] if one writes the solution of (3.28) in the form $\vec{g}(t,\tau,y) = \mathbb{Q}(t,\tau,y)\vec{g}(\tau,\tau,y)$ where $\mathbb{Q}(\tau,\tau,y) = I$ then $\|\mathbb{Q}(t,\tau,y)\| \leq c \exp \tilde{c}|y|(t - \tau)$ where $\|\ \|$ denotes the matrix norm (so that in particular $|q_{ij}| \leq \|\mathbb{Q}\|$ when $\mathbb{Q} = (q_{ij})$). Hence one can state

<u>Lemma 3.5</u> The functions $\hat{Y}(t,\tau,y)$ and $\hat{Z}(t,\tau,y)$ are continuous in (t,τ,y) for $0 \leq \tau \leq t \leq T$ and $y \in \mathbb{C}$ while for (t,τ) fixed, $y\hat{Y}$, $y\hat{Z}$, $\hat{Y}_t$, and $\hat{Z}_t$ are entire analytic functions of y of exponential type $\leq \tilde{c}T$.

For various reasons we need to establish growth estimates on $\hat{Y}$, $\hat{Z}$, etc. and writing first $\hat{Y} = \phi + i\psi$ one has $\phi'' + (2\gamma + s)\phi' + ay^2\phi - yr\psi = 0$ with $\psi'' + (2\gamma + s)\psi' + ay^2\psi + yr\phi = 0$. Multiply the first (resp. second) equation by ϕ' (resp ψ') and add to obtain

(3.31) $\frac{d}{dt}|\hat{Y}_t|^2 + 2(2\gamma + s)|\hat{Y}_t|^2 + ay^2\frac{d}{dt}|\hat{Y}|^2 \leq (y^2r^2|\hat{Y}|^2 + |\hat{Y}_t|^2).$

Assuming now $a \in C^1$ we obtain for $0 < \tau \leq t \leq T$

(3.32)
$$|\hat{Y}_t|^2 + 2\int_\tau^t (2\gamma + s)|\hat{Y}_t|^2 d\xi + a(t)y^2|\hat{Y}|^2 \leq 1 + \int_\tau^t [(a' + r^2)y^2|\hat{Y}|^2 + |\hat{Y}_t|^2]d\xi.$$

This type of inequality can be handled by the use of Gronwall type lemmas (cf Carroll [4]). Thus set $P = (a' + r^2)y^2$ and $Q = 1 - 2(2\gamma + s)$ so that $|Q| \leq \tilde{c}$ on $[0,T]$. We add $\tilde{c}\int_\tau^t ay^2|\hat{Y}|^2 d\xi$ to the right side of (3.32) and set $\Xi = |\hat{Y}_t|^2 + ay^2|\hat{Y}|^2$ so that

(3.33) $$\Xi \leq 1 + \int_\tau^t P|\hat{Y}|^2 d\xi + \tilde{c}\int_\tau^t \Xi\, d\xi.$$

By Gronwall's lemma

(3.34) $$\Xi \leq E(t,\tau) + \int_\tau^t P|\hat{Y}|^2 E(t,\xi)d\xi$$

where $E(t,\xi) = \exp \tilde{c}(t-\xi)$. Forgetting the $|\hat{Y}_t|^2$ term in Ξ for the moment and following a Gronwall type procedure we obtain for $P \geq 0$

(3.35) $$ay^2|\hat{Y}|^2 \leq E(t,\tau)\exp\int_\tau^t \hat{P}d\xi$$

where $\hat{P} = (a' + r^2)/a$. This yields

<u>Lemma 3.6</u> Given $a \in C^1$, $b,r,s \in C^0$, $\hat{P} \geq 0$, and $\hat{Y}$ the solution of (3.27) satisfying $\hat{Y}(\tau,\tau,y) = 0$ with $\hat{Y}_t(\tau,\tau,y) = 1$ it follows that for y real and $0 < \tau \leq t \leq T$

(3.36) $$a(\tau)y^2|\hat{Y}(t,\tau,y)|^2 \leq E(t,\tau)\exp\int_\tau^t (r^2/a)d\xi.$$

Let now $F(t,\tau) = \exp(-\int_\tau^t (r^2/a)d\xi)$ with $F(\tau) = F(T,\tau)$ so that $F(\tau) \leq F(t,\tau)$. Then since $E(t,\tau) \leq \exp \tilde{c}\, T = k$ we have from (3.36)

(3.37) $$a(\tau)F(\tau)y^2|\hat{Y}(t,\tau,y)|^2 \leq k\ .$$

Note that $F(\tau)$ may tend to zero as $\tau \to 0$ while $a(\tau) \to 0$ by assumption but for $\tau > 0$ both $F(\tau)$ and $a(\tau)$ are positive. Similarly one obtains from the above inequalities

(3.38) $$|\hat{Y}_t(t,\tau,y)|^2 a(\tau)F(\tau) \leq \tilde{k}$$

where $\tilde{k} = k \max a(t)$ on $[0,T]$. Setting $\tilde{Q}(\tau) = (a(\tau)F(\tau))^{\frac{1}{2}}$ one can bound $\tilde{Q}(\tau)|\hat{Y}_{tt}(t,\tau,y)|$ from the differential equation. Now writing $\hat{W}(t,\tau,y) =$

$\tilde{Q}(\tau)\hat{Y}(t,\tau,y)$ it follows that $y \to y\hat{W}(t,\tau,y)$ is entire of exponential type $\leq \tilde{c}T$ and is bounded uniformly for y real and $0 \leq \tau \leq t \leq T$. A simple argument shows that $\hat{Y}(t,\tau,\cdot)$ is also of exponential type $\leq \tilde{c}T$. Also for y real with $|y| \leq R$ say, $|\hat{W}(t,\tau,y)|$ is bounded by continuity in (t,τ,y) while from the above $|\hat{W}(t,\tau,y)| \leq k^{\frac{1}{2}}/|y|$ is bounded for $|y| > R_o$. From the Paley-Wiener-Schwartz theorem it follows then that $W(t,\tau,\cdot) = F^{-1}\hat{W}(t,\tau,y) \in E'_x$ with supp W contained in a fixed compact set for $0 \leq \tau \leq t \leq T$. Similar conclusions apply to W_t and W_{tt} which will indeed represent the derivatives of W in E'_x (cf Carroll [4]). There results

<u>Lemma 3.7</u> Under the hypotheses indicated W, W_t, and W_{tt} belong to E'_x and have supports contained in a fixed compact set for $0 \leq \tau \leq t \leq T$. Moreover $(t,\tau) \to W(t,\tau,\cdot)$, $W_t(t,\tau,\cdot)$, and $W_{tt}(t,\tau,\cdot)$ are continuous with values in E'_x for $0 \leq \tau \leq t \leq T$ and $t \to W(t,\tau,\cdot) \in C^2(E'_x)$.

Going back to (3.21) - (3.22) now we omit the $b(t)$ term in view of (3.25) and replace $s(t)$ by $\tilde{s}(t) = s(t) + 2\gamma(t)$. Let $h(t) = f(t)/\tilde{Q}(t)$ and assume $h(\cdot) \in C^o(F)$ with $f(t) \in D(A^2)$ while $Ah(\cdot)$ and $A^2h(\cdot)$ belong to $C^o(F)$. We define a bracket $\langle W(t,\xi,\cdot), T(\cdot)h(\xi)\rangle$ as in Section 1 (cf. Lemmas 1.3 and 1.4, Definition 1.1, and Remark 1.2) and the arguments proceed as in Example 1.7. Here $W(t,\tau,\cdot)$ corresponds to $S \in E'_x$ of order ≤ 2 with supp $S \subset \hat{K}$ compact and for $\theta \in C^2(F)$ the map $(S,\theta) \to \langle S,\theta\rangle : E' \times C^2(F) \to F$ will again be hypocontinuous relative to bounded sets in $C^2(F)$. Consider for $\tau > 0$

$$(3.39) \qquad u(t) = \int_\tau^t \langle W(t,\xi,\cdot), \ T(\cdot)h(\xi)\rangle d\xi$$

Then the following calculations can be justified as in Section 1.

$$(3.40) \qquad u'(t) = \int_\tau^t \langle W_t(t,\xi,\cdot), \ T(\cdot)h(\xi)\rangle d\xi$$

$$(3.41) \qquad u''(t) = f(t) + \int_\tau^t \langle W_{tt}(t,\xi,\cdot), \ T(\cdot)h(\xi)\rangle d\xi$$

$$(3.42)\qquad Au(t) = \int_\tau^t \langle W(t,\xi,\cdot),\ AT(\cdot)h(\xi)\rangle d\xi$$

$$= \int_\tau^t \langle W(t,\xi,\cdot),\ D_xT(\cdot)h(\xi)\rangle d\xi = -\int_\tau^t \langle D_xW(t,\xi,\cdot),\ T(\cdot)h(\xi)\rangle d\xi.$$

Similarly $A^2u(t) = \int_\tau^t \langle D_x^2W(t,\xi,\cdot),\ T(\cdot)h(\xi)\rangle d\xi$ and we obtain $u'' + \tilde{s}u' + Aru - A^2au = f$. There is no trouble in passing to the limit as $\tau \to 0$ and using γ to transform back to the original equation. Hence

<u>Theorem 3.8</u> Let $a(t) > 0$ for $t > 0$ with $a(0) = 0$ and $a \in C^1$; let b, r, and s belong to C^o with $\hat{P} = (a' + r^2)/a \geq 0$; and choose T as indicated. Let $\tilde{Q} = (aF)^{\frac{1}{2}}$ with $F(\tau) = \exp(-\int_\tau^t (r^2/a)d\xi)$ and assume $h = f/\tilde{Q} \in C^o(F)$ with $Ah(\cdot)$ and $A^2h(\cdot) \in C^o(F)$ where A generates a locally equicontinuous group $T(x)$ in F. Then, after modification by a factor $\exp\int_o^t \gamma(\xi)d\xi$, $u(t)$ given by (3.39) with $\tau = 0$ is a solution of (3.21) - (3.22) on $[0,T]$.

To treat uniqueness we need some estimates for $\hat{Z}(t,\tau,y)$ and arguing as before with $\hat{Y}$ we obtain

$$(3.43)\qquad \frac{d}{dt}|\hat{Z}_t|^2 + 2\tilde{s}(t)|\hat{Z}_t|^2 + a(t)y^2\frac{d}{dt}|\hat{Z}|^2 \leq y^2r^2(t)|\hat{Z}|^2 + |\hat{Z}_t|^2$$

$$(3.44)\qquad |\hat{Z}_t|^2 + 2\int_\tau^t \tilde{s}(\xi)|\hat{Z}_t|^2 d\xi + a(t)y^2|\hat{Z}|^2 \leq$$

$$a(\tau)y^2 + \int_\tau^t [(a' + r^2)y^2|\hat{Z}|^2 + |\hat{Z}_t|^2]d\xi.$$

Set $P = (a' + r^2)y^2$ and $Q = 1 - 2\tilde{s}$ as before with $\hat{\Xi} = |\hat{Z}_t|^2 + ay^2|\hat{Z}|^2$ so that adding the term $\tilde{c}\int_\tau^t ay^2|\hat{Z}|^2 d\xi$ to the right side of (3.44) we obtain

$$(3.45)\qquad \hat{\Xi} \leq a(\tau)y^2 + \int_\tau^t P|\hat{Z}|^2 d\xi + \tilde{c}\int_\tau^t \hat{\Xi}\, d\xi.$$

There results

$$(3.46)\qquad \hat{\Xi} \leq a(\tau)y^2E(t,\tau) + \int_\tau^t P|\hat{Z}|^2E(t,\xi)d\xi$$

$$(3.47)\qquad a(t)y^2|\hat{Z}|^2 \leq a(\tau)y^2E(t,\tau)\exp\int_\tau^t \hat{P}d\xi.$$

Lemma 3.9 Given the hypotheses of Lemma 3.6 and $\hat{Z}$ the solution of (3.27) satisfying $\hat{Z}(\tau,\tau,y) = 1$ with $\hat{Z}_t(\tau,\tau,y) = 0$ it follows that for y real and $0 \le \tau \le t \le T$

$$(3.48) \qquad |\hat{Z}(t,\tau,y)|^2 \le E(t,\tau)\exp\int_\tau^t (r^2/a)d\xi$$

which can be written as $F(\tau)|\hat{Z}(t,\tau,y)|^2 \le E(t,\tau)$.

Next one can establish estimates on $\hat{Z}_\tau$ and $\hat{Y}_\tau$ from (3.30) using the estimates already obtained for $\hat{Y}$ and $\hat{Z}$. Reasoning as before we also conclude that $\hat{Y}_\tau$ and $\hat{Z}_\tau$ are entire functions in y of exponential type $\le \tilde{c}T$ and one can easily prove then

Lemma 3.10 Under the hypotheses indicated $F^{\frac{1}{2}}(\tau)Z = F^{\frac{1}{2}}(\tau)F^{-1}\hat{Z}$, $\tilde{Q}(\tau)Z_\tau$ (and $\tilde{Q}(\tau)Z)$, and $\tilde{Q}(\tau)Y_\tau$ belong to E'_x with supports contained in a fixed compact set for $0 \le \tau \le t \le T$. The τ derivatives can be taken in E'_x for $\tau > 0$ and $(t,\tau) \to F^{\frac{1}{2}}Z$ or $\tilde{Q}Z$, $\tilde{Q}Z_\tau$, and $\tilde{Q}Y_\tau$ are continuous with values in E'_x.

Now for uniqueness we extend the technique of Section 1 and define

$$(3.49) \qquad R(t,\xi) = \langle Z(t,\xi,\cdot),\ T(\cdot)u(\xi)\rangle$$

$$(3.50) \qquad S(t,\xi) = \langle Y(t,\xi,\cdot),\ T(\cdot)u'(\xi)\rangle$$

where u is any solution of the modified equation (3.21) (i.e. $s(t)$ is replaced by $\tilde{s}(t) = s(t) + 2\gamma(t)$ and $b(t) = 0$) with $f = 0$. By our lemma above (3.49) and (3.50) make sense as do the calculations which follow (cf. Example 1.7). Thus

$$(3.51) \qquad R_\xi = \langle Z_\xi,\ Tu\rangle + \langle Z,\ Tu'\rangle = \langle Z,\ Tu'\rangle - \langle a(\xi)D_x^2Y,\ Tu\rangle -$$

$$\langle r(\xi)D_xY,\ Tu\rangle = \langle Z,\ Tu'\rangle + \langle Y,\ r(\xi)ATu\rangle - \langle Y,\ a(\xi)A^2Tu\rangle$$

$$(3.52) \qquad S_\xi = \langle Y_\xi,\ Tu'\rangle + \langle Y,\ Tu''\rangle = \langle Y,\ Tu''\rangle - \langle Z,\ Tu'\rangle + \langle\tilde{s}(\xi)Y,\ Tu'\rangle.$$

Letting now $\phi(t,\xi) = R(t,\xi) + S(t,\xi)$ we obtain from (3.51) - (3.52)

$$(3.53) \qquad \phi_\xi = \langle Y,\ T(u'' + \tilde{s}u' + rAu - aA^2u\rangle = 0.$$

Consequently $\phi(t,t) = \phi(t,\tau)$ which implies

$$(3.54)\qquad u(t) = \langle F^{\frac{1}{2}}(\tau)Z(t,\tau,\cdot),\ T(\cdot)F^{-\frac{1}{2}}(\tau)u(\tau)\rangle + \langle \tilde{Q}(\tau)Y(t,\tau,\cdot),\ T(\cdot)\tilde{Q}^{-1}(\tau)u'(\tau)\rangle.$$

Theorem 3.11 Let u be any solution of the modified equation (3.21) such that $F^{-\frac{1}{2}}(\tau)u(\tau) \to 0$ and $\tilde{Q}^{-1}(\tau)u'(\tau) \to 0$ as $\tau \to 0$ while the other hypotheses on a, b, r, s hold. Then u is unique.

Remark 3.12 The condition $\hat{P} \geq 0$ or $a' \geq -r^2$ allows $a(t)$ not to be monotone (which had been assumed in some earlier work on such equations). Some remarks about this and about the behavior of $F(\tau) = \exp(-\int_\tau^t (r^2/a)d\xi)$ as $\tau \to 0$ are given in Carroll-Wang [29].

In general the requirements of Theorem 3.11 regarding the growth of $u(\tau)$ and $u'(\tau)$ as $\tau \to 0$ are somewhat strong although possible for suitable f in (3.21). It is therefore of interest to consider the case when $F(\tau) \to 0$ as $\tau \to 0$ and the relation of this to results of Krasnov [1] and Protter [1] has been discussed briefly in Carroll-Wang [29]. In this event the requirements of Theorem 3.11 are only that $u(0) = 0$ and $a^{-\frac{1}{2}}(\tau)u'(\tau) \to 0$ as $\tau \to 0$. To examine the feasibility of this let u satisfy the modified equation (3.21) with $f = 0$, $u(0) = 0$, and $u'(0) = 0$. Multiply this equation by the term $\exp(\int_0^t \tilde{s}(\xi)d\xi)$ and integrate to obtain

$$(3.55)\qquad u'(t) = -\int_0^t [Ar(\xi)u - A^2a(\xi)u]\exp(-\int_\xi^t \tilde{s}(\eta)d\eta)d\xi.$$

Since $\exp(-\int_\xi^t \tilde{s}(\eta)d\eta) \leq M$ on $[0,T]$, for any continuous seminorm p on F we have

$$(3.56)\qquad p(u'(t)) \leq \int_0^t [|r(\xi)|p(Au) + a(\xi)p(A^2u)]Md\xi.$$

Now $p(Au)$ and $p(A^2u)$ will be bounded for a solution $u \in C^2(F)$ of (3.21) (Au and A^2u belong to $C^0(F)$) and

$$(3.57)\qquad \int_0^t |r|d\xi = \int_0^t a^{\frac12}\left(\frac{|r|}{a^{\frac12}}\right)d\xi \le \left(\int_0^t ad\xi\right)^{\frac12}\left(\int_0^t (r^2/a)d\xi\right)^{\frac12}$$

whereas $\int_0^t ad\xi = ((\int_0^t ad\xi)^{\frac12})^2$. Hence for $\int_0^t (r^2/a)d\xi$ bounded

$$(3.58)\qquad p(a^{-\frac12}(t)u'(t)) \le \hat{M}a^{-\frac12}(t)\left(\int_0^t a(\xi)d\xi\right)^{\frac12}$$

and $a^{-\frac12}(t)u'(t) \to 0$ if $a^{-\frac12}(t)(\int_0^t a(\xi)d\xi)^{\frac12} \to 0$. This condition is examined in Carroll-Wang [29] and Wang [1; 2] and since oscillations in $a(t)$ are permitted by the stipulation $\hat{P} \ge 0$ it is not automatically satisfied. If a is monotone near $t = 0$ it is obviously valid. Consequently

<u>Theorem 3.13</u> Assume the hypotheses of Theorem 3.11 with $F(\tau) > 0$ on $[0,T]$ and suppose $a^{-\frac12}(t)(\int_0^t a(\xi)d\xi)^{\frac12} \to 0$ as $t \to 0$. Then if u satisfies (3.21) - (3.22) with $f = 0$ it follows that $u \equiv 0$ on $[0,T]$.

We go next to some examples illustrating Section 2 and recall briefly some constructions of Gelfand-Šilov [1; 2; 3] (cf also A. Friedman [1]).

<u>Example 3.14</u> The space $S^{\beta,B}_{\alpha,A}$ consists of C^∞ functions ϕ satisfying

$$(3.59)\qquad |x^kD^p\phi(x)| \le \bar{c}(\bar{A},\bar{B},\phi)\bar{A}^k\bar{B}^p k^{k\alpha}p^{p\beta}$$

for $\bar{A} > A$ and $\bar{B} > B$. For $\delta,\rho = 1,\frac12,\frac13,\dots$ the topology is defined in terms of norms

$$(3.60)\qquad \|\phi\|_{\delta,\rho} = \sup_{x,k,p} \frac{|x^kD^p\phi(x)|}{(A+\delta)^k(B+\rho)^p k^{k\alpha}p^{p\beta}}\,.$$

With this topology $S^{\beta,B}_{\alpha,A}$ is a Montel space and $S^\beta_\alpha = \cup S^{\beta,B_m}_{\alpha,A_m}$ where $A_m, B_m \to \infty$. If $\alpha,\beta > 0$ with $\alpha + \beta \ge 1$ then $FS^\beta_\alpha = \tilde{S}^\beta_\alpha = S^\alpha_\beta$ while S^β_α is nontrivial. For $p > 1$ the space Z^p_p is defined as $S^{1-1/p}_{1/p}$ and it consists of functions extendable to $\mathbb{C}$ as entire functions satisfying

$$(3.61)\qquad |\phi(z)| \le c\ \exp\ b|z|^p;\quad |\phi(x)| \le \hat{c}\ \exp(-a|x|^p).$$

Since $FS^\beta_\alpha = S^\alpha_\beta$ we have $FZ^p_p = \tilde{Z}^p_p = Z^q_q$ where $\frac1p + \frac1q = 1$. Multipliers

$f : S_\alpha^\beta \to S_\alpha^\beta$ are determined by functions f such that the map $\phi \to f\phi : S_\alpha^\beta \to S_\alpha^\beta$ is continuous and if $|f(z)| \le c\exp(a|z|^p)$ then f is a multiplier $Z_q^q \to Z_q^q$ for any $q > p$. Evidently polynomials are multipliers in any Z_p^p. We mention also that the spaces Z_p^p are sufficiently rich in the sense that $\int g\phi dx = 0$, with $\int |g\phi| dx < \infty$, for all $\phi \in Z_p^p$ implies $g \equiv 0$.

Remark 3.15 The space $\mathcal{D}$ is denoted by K in Gelfand-Šilov [1; 2; 3] and $F\mathcal{D} = Z$ is the space of entire functions of exponential type satisfying $|s^k\psi(\sigma + i\tau)| \le c\exp a|\tau|$ for $s = \sigma + i\tau$.

Now to work with equations of the form (2.16) let $\mathbb{P}(t, A) = \mathbb{P}(A)$ be independent of t and let $\lambda_k(s)$ be the roots of $\det(\mathbb{P}(-is) - \lambda I) = 0$. Set $\Lambda(s) = \max \operatorname{Re} \lambda_k(s)$ for $1 \le k \le m$ and denote by $\|\mathbb{Q}\|$ the operator norm of a matrix $\mathbb{Q}$ so that $\max_k \sum_j |q_{jk}|^2 \le \|\mathbb{Q}\|^2 \le \Sigma\Sigma|q_{jk}|^2$. The entries in the matrix $\mathbb{P}(-is)$ will have the form $p_{ij}(-is) = \Sigma c_{ijk}(-is)^k$ and we let p be the highest index k so that $|p_{ij}(-is)| \le c|s|^p$ for large $|s|$. Then $\|\mathbb{P}(-is)\| \le c_1|s|^p + c_2$ and for $0 \le t \le T < \infty$

$$\|\exp t\,\mathbb{P}(-is)\| \le c(1 + |s|)^{p(m-1)}\exp bt|s|^{p_o} \tag{3.62}$$

where $p_o \le p$ is called the reduced order of the system and is determined as follows. If one writes

$$\det(\mathbb{P}(-is) - \lambda I) = (-1)^m\lambda^m + p_1(s)\lambda^{m-1} + \ldots + p_m(s)$$

with degree $p_m(s) = p_m$ then p_o is defined as $\max p_k/k$ for $1 \le k \le m$.

Definition 3.16 The system is called hyperbolic if for $s = \sigma + i\tau$

$$\Lambda(s) \le c|s| + c_1; \quad \Lambda(\sigma) \le c_2 \tag{3.63}$$

in which case $\|\exp t\,\mathbb{P}(-is)\| \le c(1 + |s|)^{p(m-1)}e^{bt|s|}$ with $\|\exp t\,\mathbb{P}(-i\sigma)\| \le \hat{c}(1 + |\sigma|)^{p(m-1)}$. The system is called parabolic if $\Lambda(\sigma) \le -c|\sigma|^h + c_1$ in which case (3.62) holds with $\|\exp t\,\mathbb{P}(-i\sigma)\| \le c(1 + |\sigma|)^{p(m-1)}\exp(-at|\sigma|^h)$.

In the case of a hyperbolic system the elements of $\hat{\mathbb{G}}(t,s) = \exp t\,\mathbb{P}(-is)$ will be multipliers in $\hat{\Phi} = Z$ so $\Phi = K = \mathcal{D}$ can be taken in Section 2. Actually the elements of $\mathbb{G}(t,x)$ will be distributions with supports in a fixed compact set for $0 \le t \le T < \infty$ and $\vec{U}(t,x) = \mathbb{G}(t,\cdot) * T(\cdot)\vec{u}_o$ in Section 2 is defined for locally equicontinuous groups $T(x)$. The entries $g_{ij}(t,\cdot)$ are of bounded order and $T(\cdot)\vec{u}_o$ will belong to $C^M(F)$ when the elements of $\vec{u}_o$ belong to $D(A^M)$. Hence for suitable assumptions on $\vec{u}_o$ the evaluation $E[\mathbb{G}(t,\cdot) * T(\cdot)\vec{u}_o]$ will be appropriate in Theorem 2.6 (see A. Friedman [1] or Gelfand-Šilov [3] for further details in this direction).

For parabolic systems one knowns $p_o > 1$ and $0 < h \le p_o$ are integers with h even while for so called Petrovskij parabolic systems $h = p_o = p$. If $p_1 > p_o$ then the elements of $\hat{\mathbb{G}}(t,s)$ will be multipliers in $\hat{\Phi} = Z_{p_1}^{p_1}$ so $\Phi = Z_{q_1}^{q_1}$ with $\frac{1}{p_1} + \frac{1}{q_1} = 1$ is indicated. Now there is another index associated with parabolic systems called the genus, μ, where $1 - (p_o - h) \le \mu \le 1$. For constant coefficient systems this can be defined directly from $\Lambda(s)$ as the largest μ such that in the domain $|\tau| \le k(1 + |\sigma|)^{\mu}$ one has $\Lambda(s) \le -c|\sigma|^h + c_1$. It can be shown that

$$(3.64) \qquad \| \hat{\mathbb{G}}(t,x) \| \le \hat{c} \exp(-at|\sigma|^h + \hat{b}t|\tau|^{p_o/\mu})$$

and various useful properties of the elements of $\mathbb{G}(t,x)$ can be established from this. If $\mu > 0$ for example and $\hat{p} = p_o/p_o - \mu$ with $\bar{\theta} > \theta$ where $\theta^{p_o/\mu} = \hat{b}Tp_o/\mu$ then

$$(3.65) \qquad \| \mathbb{G}(t,x) \| \le c \exp\left(-\frac{1}{\hat{p}} \left|\frac{x}{\bar{\theta}}\right|^{\hat{p}}\right)$$

for $t \in [0,T]$. Thus the entries in $\mathbb{G}(t,x)$ are functions with exponential growth as indicated and convolution $\mathbb{G}(t,x) * T(\cdot)\vec{u}_o$ can be taken in the classical sense for continuous function elements in $T(\cdot)\vec{u}_o$ of growth bounded by $\exp b_o|x|^{\hat{p}}$ for suitable b_o. Similarly there will be estimates

$$(3.66) \qquad \| D_x^q D_t^j \mathbb{G}(t,x) \| \le c_{qj} \exp(-\rho|x|^{\hat{p}})$$

with c_{qj} and ρ depending on t (becoming ∞ as $t \to 0$). Thus for $t > 0$

one can differentiate $\mathbb{C}(t,\cdot) * T(\cdot)\vec{u}_o$ in the classical sense for continuous $T(\cdot)\vec{u}_o$ of suitable growth so that $E[\mathbb{C}(t,\cdot) * T(\cdot)\vec{u}_o]$ will be appropriate in Theorem 2.6. To insure continuity of $\mathbb{C}(t,\cdot) * T(\cdot)\vec{u}_o$ as $t \to 0$ suitable hypotheses of differentiability on $T(\cdot)\vec{u}_o$ have to be imposed (not so for Petrovskij parabolic systems however). Analogous remarks hold for $\mu \leq 0$ where $\|\mathbb{C}(t,x)\| \leq c \exp(-a|x|^{\hat{p}})$ with $\hat{p} = h/h-\mu$ and estimates of the form (3.66) are valid.

Remark 3.17 Using the notions just discussed of hyperbolic, parabolic, etc. and the formalism of Section 1.1 (cf Theorem 1.6(F)) Hersh [1] gives versions of the following theorems where F is a Banach space. Here the passage from a single equation to a system transpires as in (2.8) - (2.10) and the results will also follow from the technique of Section 1.2 as indicated in the discussion just concluded following Remark 3.15. Recall also that the Hersh technique involves $P(D_t, -D_x)$ or $P(D_t, is)$ in (2.8); one can retain the notation $\hat{\mathbb{C}}(t,\tau,s)$ etc. however and simply use $\mathbb{P}(is)$, $P_j(is)$, etc. instead of $\mathbb{P}(-is)$, $P_j(-is)$, etc. in the discussion concerning the roots $\lambda_k(s)$. An equation is said to be hyperbolic, parabolic, etc. if the corresponding system has these properties. The $\hat{G}_n$ are constructed then as in (2.3) (with $P(D_t, is)$ - cf. (1.3)) and the solution is represented in the form (1.4).

Theorem 3.18 Let F be a Banach space and $P(D_t, -D_x)$ be hyperbolic with $T(\cdot)$ a strongly continuous group and $u_n \in D^\infty(A) = \cap D^k(A)$ for example. Then the problem (1.2) has a solution for $t \in (-\infty, \infty)$ given by (1.4).

Theorem 3.19 Under the other hypotheses of Theorem 3.18 let $P(D_t, -D_x)$ be parabolic of positive genus. Then (1.4) represents a solution of (1.2) for $t < T$.

One notes here that $\mathbb{C}(t,x)$ for fixed $t < T$ decays like $\exp(-k|x|^\gamma)$ where $\gamma = \hat{p} > 1$ (cf. (3.65)). On the other hand it is well known that $\|T(x)u_n\| \leq M \exp \omega|x| \; \|u_n\|$ (cf Hille-Phillips [1]) so everything fits together in (1.4). We mention also that a system of the form (2.20) where $\hat{\mathbb{C}}(t,s) = \exp t\,\mathbb{P}(-is)$ is called Petrovskij correct if $\Lambda(\sigma) \leq c$ for σ real in which case one has

$\| \hat{G}(t,\sigma) \| \leq c(1 + |\sigma|)^h$ where $h \leq p(m-1)$. The genus is defined as the largest μ such that in the domain $|\tau| \leq k(1 + |\sigma|)^\mu$ one has $\Lambda(s) \leq c$. For $\mu > 0$ for example one has then for $t < T$

$$(3.67) \qquad \| \hat{\mathfrak{G}}(t,s) \| \leq c(1 + |\sigma|)^h \exp \theta |\tau|^{p_0/\mu}$$

and this leads to estimates on $\| \mathfrak{G}(t,x) \|$ permitting compositions in (1.4) of the $G_n(t,\cdot)$ with suitably restricted $T(\cdot)u_n$ (cf. A. Friedman [1], Gelfand-Šilov [3]).

Remark 3.20 It is pointed out in Hersh [1] (cf. also Bragg [1]) that the techniques of Section 1.1 can be extended to situations involving several commuting operators A_i generating groups $T_i(x) = \exp A_i x$. (The case of several commuting operators has also been treated in Hilbert space by spectral methods in Carroll [4; 20; 21; 22] and we refer here to section 1.4.) Thus let us take two commuting operators A and B to illustrate the matter and set

$$(3.68) \qquad P(D_t, A, B) = \sum c_{ijk} D_t^i A^j B^k$$

$(0 \leq i \leq m)$. As in (1.3) let the distributions G_n $(0 \leq n \leq m-1)$ satisfy

$$(3.69) \qquad P(D_t, -D_x, -D_y)G_n = 0; \quad D_t^k G_n|_{t=0} = \delta_{nk}\delta(x)\delta(y).$$

Then for suitable u_n the function

$$(3.70) \qquad u(t) = \sum_{n=0}^{m-1} \langle G_n(t,x,y), T_A(x)T_B(y)u_n \rangle$$

will formally satisfy

$$(3.71) \qquad P(D_t, A, B)u = 0; \quad D_t^n u(0) = u_n.$$

The bracket in (3.70) is a suitable pairing as before between $G_n(t,\cdot,\cdot) \in \mathcal{D}'_{x,y}$ and vector functions $T_A(x)T_B(y)u_n$ which will be as smooth as necessary upon appropriate selection of the u_n. The proof follows the lines of (1.6) exactly.

A slightly different approach to the same kind of problems is indicated in Bragg

[1] where the formula

$$(3.72) \qquad \exp t\,P(a) = \exp[tP(D)] \cdot \exp \sum_1^n a_i\xi_i \Big|_{\xi=0}$$

is exploited. Let for example $P(D) = P(D_1,\dots,D_n)$ be an elliptic operator of order $2k$ with real constant coefficients where the highest order term has the form $(-1)^{k-1}P_o(D)$ with $P_o(\zeta)$ a positive definite homogeneous polynomial of degree $2k$. One considers the Cauchy problem

$$(3.73) \qquad v_t(\xi,t) = P(D)v; \quad v(\xi,0) = f(\xi)$$

and following Rosenblum [1] a solution to (3.73) is given by

$$(3.74) \qquad v(\xi,t) = \int_{\mathbb{R}^n} K(\xi - \sigma,\ t)f(\sigma)d\sigma$$

where K is a Green's function for which growth estimates are obtained. Symbolically (3.74) says

$$(3.75) \qquad v(\xi,t) = \exp[tP(D)]f(\xi).$$

Let then $\phi \in \cap D(A_i^{2k})$ with $T_i(x) = \exp A_i x$ as above and consider

$$(3.76) \qquad u(t) = \exp[tP(A)]\phi = \int_{\mathbb{R}^n} K(-\sigma,t) \prod_1^n T_i(\sigma_i)\phi d\sigma.$$

This is a formal generalization of (3.72) with $f(\sigma) = \Pi T_i(\sigma_i)\phi$ ($\exp a_i\xi_i$ is replaced by $T_i(\xi_i)$ in (3.72)) and will agree with the corresponding result obtained via the Hersh technique.

Remark 3.21 A variation on the Hersh technique is presented in Donaldson [3]. One considers a Banach space F in which A generates a strongly continuous group. A solution $Y(t)$ of the equation

$$(3.77) \qquad \sum_{j=0}^n f_j(tA)A^jD_t^{n-j}Y = 0; \quad f_j(z) = \sum_{p=0}^{m_j} a_p z^p$$

is required. Set $m = \max m_j$ $(0 \le j \le n)$ and $\gamma = m + n$. Let $\phi \in D(A^\gamma)$ and consider

$$(3.78)\qquad Y(t) = \int_\alpha^\beta F(x)T(xt)\phi dx$$

over a suitable contour from α to β in $\mathbb{C}$ such that

$$(3.79)\qquad \{A^n P(F(x),\ T(xt)\phi\}\Big|_\alpha^\beta = 0$$

where F is a solution of

$$(3.80)\qquad Q(D)F = \sum_{j=0}^{n}\sum_{p=0}^{m_j} (-1)^p a_p D_x^p[x^{n-j}F] = 0$$

and P is a suitable bilinear concomitant defined below (cf. Ince [1]). Thus $Q(D)$ is the formal adjoint of $Q^*(D) = \Sigma\,\Sigma a_p x^{n-j}D^p$ and P is defined by

$$(3.81)\qquad \frac{d}{dx}P(u,v) = vQ(D)u - uQ^*(D)v$$

so that in the present situation one can write explicitly

$$(3.82)\qquad P(F(x),\ T(xt)\phi) =$$

$$\sum_{j=0}^{n}\sum_{p=0}^{m_j}\sum_{k=0}^{p-1} (-1)^k a_p\{D^{p-k-1}[T(xt)\phi]\}D^k[x^{n-j}F(x)]$$

<u>Theorem 3.22</u> Under the hypotheses above $Y(t)$ is a formal solution of (3.77).

Proof: We insert $Y(t)$ given by (3.78) into (3.77) to obtain

$$(3.83)\qquad \sum_{j=0}^{n} f_j(tA)A^jD_t^{n-j}Y =$$

$$\sum_{j=0}^{n}\sum_{p=0}^{m_j} a_p(tA)^pA^j\int_\alpha^\beta F(x)x^{n-j}A^{n-j}T(xt)\phi dx =$$

$$A^n\sum_{j=0}^{n}\sum_{p=0}^{m_j} a_p\int_\alpha^\beta x^{n-j}F(x)(tA)^pT(xt)\phi dx =$$

$$A^n\sum_{j=0}^{n}\sum_{p=0}^{m_j} a_p\int_\alpha^\beta x^{n-j}F(x)D_x^pT(xt)\phi dx = A^n\int_\alpha^\beta F(x)Q^*(D)T(xt)\phi dx =$$

$$A^n\int_\alpha^\beta [Q(D)F]T(xt)\phi dx - A^nP(F,T(xt)\phi)\Big|_\alpha^\beta = 0.$$

For finite intervals (α,β) of $\mathbb{R}$ there is no problem in justifying the calculation; for infinite intervals one needs some growth condition on $F(x)$; for general contours in $\mathbb{C}$ one requires that $T(x)$ have a suitable analytic extension. QED

Various applications of this formalism are given in Donaldson [3] but they are covered here by other techniques.

1.4 Other techniques. In this section we will sketch some results involving differential equations of the form

$$(4.1) \qquad L_n(D_t)v(t) = 0$$

$$(4.2) \qquad L_n(D_t) = D_t^n + \sum_{k=1}^{n} A_k D_t^{n-k}.$$

We will not deal with the case $n = 1$ as such since the equation $v_t + Av = 0$ is essentially the classical linear semigroup problem which is covered exhaustively for example in Carroll [4], Hille-Phillips [1], Yosida [1], etc. Such results as are needed for dealing with higher order equations will simply be cited. It does seem worthwhile to include here some results for the case $n = 2$ both for constant A_k and variable $A_k(t)$ since a lot of this has not yet appeared in book form and we will draw upon Baiocchi [3], Bruk [1], Carroll-State [25], DaPrato-Giusti [5], Gearhart [1], Goldstein [2; 3], Kato [6; 7], Kleiman [1], Krein [1], Levine [1; 2; 3], Levitan [4], Lions [4; 9], Mazumdar [4], Raskin-Sobolevskij [1], Sobolevskij [2; 3], Torelli [1], Treves [8; 9], Višik [1], Višik-Ladyženskaya [3], Yakubov [1; 2; 3; 4; 6; 7], Yosida [2], etc. (cf. also Lax-Phillips [1; 2]). For $n > 2$ there are methods due to Abdulkerimov [1], Bairamogly-Khalilova [1], Carroll [18; 19; 20; 21; 22], Chazarain [1], Djedour [1], Dubinskij [1; 2], Gasymov [1], Gorbačuk-Kočubei [3], Mazya-Plamenevskij [1], Raskin [2], Yakubov [5; 9], etc. which we will discuss and indicate results. Finally we mention some interesting new results on operator differential equations with noncommuting operator coefficients due to Hersh-Steinberg [7] (cf also Grabovskaya-Krein [1] and Krein-Šikhvatov [2]). In fact we will begin the section with this latter material since it seems to be of more current interest.

The Hersh-Steinberg [7] theory goes as follows. Consider for $0 \leq k \leq N-1$

$$(4.3)\qquad Lu = \sum_{i=0}^{N} P_i(A)D_t^{N-i}u;\quad D_t^k u(0) = u_k$$

where $P_i(A)$ is a polynomial sum of ordered products of generally noncommuting densely defined linear operators A_j in a Banach space F $(1 \leq j \leq n)$. One assumes the A_j are a basis for a finite dimensional Lie algebra of operators $\mathcal{A}$ over $\mathbb{R}$ which is "integrable". (Standard elementary facts about Lie theory are assumed here and we refer to Helgason [2] and Varadarajan [1] for details.) Explicitly

$$(4.4)\qquad [A_i, A_j] = \sum_{k=1}^{n} c_{ij}^k A_k$$

and to explain the word integrable we recall that if $\pi : G \to L(F)$ is a representation of a Lie group G on F (i.e. a continuous homomorphism into $L(F)$) then π induces a map $d\pi$ taking the Lie algebra $\tilde{g}$ into densely defined linear operators on F. Here if $g : \mathbb{R} \to G$ is a smooth curve with $g(0) = e$ and $g'(0) = a \in \tilde{g}$ then

$$(4.5)\qquad d\pi(a)\xi = \frac{d}{dt}\pi(g(t))\Big|_{t=0};\quad \xi \in F$$

is defined on a dense set of ξ. The algebra $\mathcal{A}$ is said to be integrable if there is a Lie group G and a representation π of G on F such that $A_j = d\pi(a_j)$ for all j where the a_j form a basis of $\tilde{g}$.

Now with $g : \mathbb{R} \to G$ as above and $a = g'(0) \in \tilde{g}$ one defines for $h \in G$

$$(4.6)\qquad \tilde{a}f(h) = \frac{d}{dt} f(g(t)h)\Big|_{t=0}$$

for smooth functions f on G. The right translation R_h is defined by $R_h f(g) = f(gh^{-1})$ and an operator L is said to be right invariant if $R_h^{-1}LR_h f(g) = Lf(g)$ so that $\tilde{a}$ is seen to be right invariant. Now if $s = (s_1,\ldots,s_n)$ is a local coordinate system at e so that the diffeomorphism $\phi(s)$ takes a nbh of 0 in $\mathbb{R}^n$ into a nbh of e then $\tilde{a}$ has an expression as a first order differential operator in these coordinates. Take the basis a_k of $\tilde{g}$ in the form

$a_k = \partial/\partial s_k \phi(s)|_{s=0}$ and one has then

(4.7) $$\tilde{a}_k f(\phi(s)h) = \sum \ell_k^i(s) \frac{\partial}{\partial s_i} f(\phi(s)h).$$

Indeed in a nbh of e one can write $f = f \circ \phi \circ \phi^{-1} = \tilde{f} \circ \phi^{-1}$ where $\tilde{f}$ is a function of euclidean coordinates and one writes $\partial f/\partial s_k$ for $\partial \tilde{f}/\partial s_k$ in standard notation. To evaluate $\partial/\partial r_k f(\phi(r)\phi(s)h)$ at $r = 0$ we consider $\tilde{f}(\phi^{-1}(\phi(r)\phi(s)h))$ so that $\ell_k^i(s) = \partial/\partial r_k \phi_i^{-1}(\phi(r)\phi(s)h)|_{r=0}$ where ϕ_i^{-1} denotes the ith coordinate of ϕ^{-1}. The matrix $(\ell_k^i(s))$ is clearly invertible.

Let next $C^\infty(\pi)$ be the set of $y \in F$ such that $\pi(g)y$ is C^∞ as an F valued function of g. One knows that $C^\infty(\pi)$ is dense in F and $\pi(g)y \in D(A_j)$ while clearly $\pi(g)y \in D(\tilde{a}_j)$. A simple calculation then yields $\tilde{a}_j\pi(g) = A_j\pi(g)$ so that on $\pi(g)C^\infty(\pi)$ $\mathcal{A}$ is isomorphic to the Lie algebra D_R of right invariant differential operators on $C^\infty(G)$ (note that $A_jA_i = \tilde{a}_i\tilde{a}_j$ in this correspondence). One has a standard notion of distribution on G so that $\mathcal{D}(G)$, $\mathcal{E}(G)$, $\mathcal{E}'(G)$, $\mathcal{D}'(G)$, etc are all well defined. Let $d\mu$ denote the right invariant Haar measure on G. Then if L is a linear operator $L : \mathcal{D}(G) \to \mathcal{D}(G)$ a (formal) adjoint is defined by $\langle L^*f, h\rangle = \langle f, Lh\rangle$ for $f,h \in \mathcal{D}(G)$ where $\langle f, h\rangle = \int f\bar{h}d\mu$. One shows then easily that $\tilde{a}^* = -\tilde{a} + c$ where c is a constant. Evidently $\tilde{a}$ and $\tilde{a}^*$ extend to be operators on $\mathcal{E}'(G)$.

Definition 4.1 If $T \in \mathcal{E}'(G)$ the quantization $Q(T)$ is defined as a linear operator with domain $C^\infty(\pi) \subset F$ by the rule

(4.8) $$Q(T)\xi = \int_G T(g)\pi(g)\xi d\mu.$$

This is an abuse of notation of course and means $\langle T(g), \pi(g)\xi\rangle$ for a vector valued $\mathcal{E}' - \mathcal{E}$ bracket relative to $d\mu$. This concept is very useful since it is easily shown that for $\xi \in C^\infty(\pi)$

(4.9) $$Q(\tilde{a}^*_{j_r} \dots \tilde{a}^*_{j_1} T)\xi = A_{j_r} \dots A_{j_1} Q(T)\xi.$$

To see this consider

$$(4.10) \qquad Q(\tilde{a}_2^*\tilde{a}_1^*T)\xi = \int_G (\tilde{a}_2^*\tilde{a}_1^*T(g))\pi(g)\xi d\mu = \int_G T(g)(\tilde{a}_1\tilde{a}_2\pi(g)\xi)d\mu$$

$$= \int_G T(g)A_2A_1\pi(g)\xi d\mu = A_2A_1 \int_G T(g)\pi(g)\xi d\mu = A_2A_1Q(T)\xi.$$

The property (4.9) allows one to transform (4.1) into a partial differential equation on G.

Now in Lu from (4.3) replace A_j by σ_j and D_t by τ to obtain a polynomial $P(\tau,\sigma) = \Sigma P_i(\sigma)\tau^{N-i}$. Let $P_N(\tau,\sigma)$ be the principal part of $P(\tau,\sigma)$ (i.e. the terms of order N) and define (4.3) to be of hyperbolic type if for all real $\sigma \neq 0$ the equation $P_N(\tau,\sigma) = 0$ has N distinct real roots $\tau_j(\sigma)$. Define further

$$(4.11) \qquad \tilde{L} = \sum_{i=0}^{N} P_i(\tilde{a}_j^*)D_t^{N-i}$$

Theorem 4.2 If (4.3) is hyperbolic then $\tilde{L}$ is hyperbolic as a partial differential operator in local coordinates $s \to \phi(s)$.

The proof is straightforward and we omit details. Next recall that for $T,S \in \mathcal{E}'(G)$ one defines (with abuse of notation)

$$(4.12) \qquad (T * S)(g) = \int_G T(gh^{-1})S(h)d\mu(h).$$

Then by patching together local solutions and using standard facts one can prove

Theorem 4.3 If (4.3) is of hyperbolic type the Cauchy problem $(0 \leq j,\ k \leq N-1)$

$$(4.13) \qquad \tilde{L}f_j(t,g) = 0; \quad D_t^k f_j(0,g) = \delta_{jk}\delta(g)$$

has a unique solution with $f_j(t,\cdot) \in \mathcal{E}'(G)$. If $\psi \in C^\infty(G)$ then $\int f_j(t,g)\psi(g)d\mu(g)$ is C^∞ in t.

Finally using (4.9) it is easily seen that for $\xi \in C^\infty(\pi)$

$$(4.14) \qquad LQ(f)\xi = Q(\tilde{L}f)\xi$$

and the next theorem follows immediately from the preceeding results.

Theorem 4.4 Let (4.3) be of hyperbolic type with the $P_i(A)$ in the enveloping algebra of an integrable algebra A. Then for $u_k \in C^\infty(\pi)$ the problem (4.3) has a solution given by

$$(4.15)\qquad u(t) = \sum_{j=0}^{N-1} Q(f_j(t,g))u_j .$$

Remark 4.5 It is asserted in Hersh-Steinberg [7] that if the algebra generated by the A_j^* is also integrable then the solution (4.15) of (4.3) in unique. It seems probable that a uniqueness theorem based on the techniques of Section 2 could also be devised and we will return to this later. The restriction to hyperbolic equations above is only to insure the existence of suitable f_j such that all brackets and integrals make sense; other situations can easily be envisioned.

Example 4.6 A number of interesting examples and applications are given in Hersh-Steinberg [7]. The Weyl-Heisenberg algebra has as a basis the skew adjoint operators

$$(4.16)\qquad A_j = \partial/\partial x_j ;\quad B_k = ix_k ;\quad iI$$

each of which generates a unitary group in $L^2(\mathbb{R}^n)$ of the form

$$(4.17)\qquad e^{tA_j}f(x) = f(x + te_j);\quad e^{tiI}f(x) = e^{it}f(x);\quad e^{tB_k}f(x) = e^{itx_k}f(x)$$

where e_k denotes an orthonormal basis in $\mathbb{R}^n$. Any linear partial differential equation with polynomial coefficients can be written as an operator equation of the form (4.3) using this algebra. However this is not always the best algebra to take. It is noted for example that the equation

$$(4.18)\qquad u_{tt} = -u_{xxxx} + x^2u_{xx} + 2xu_x - x^4u + \frac{1}{4}u$$

is not hyperbolic in the usual sense and would not lead to a hyperbolic operator equation in terms of the Weyl-Heisenberg algebra. However if one takes

$$(4.19)\qquad A = ix^2;\quad B = xD_x + \frac{1}{2};\quad C = iD_x^2$$

then (4.18) becomes $u_{tt} = (A^2 + B^2 + C^2)u$ which will be hyperbolic. Here one sees that $[A,B] = -2A$, $[A,C] = 4B$, and $[B,C] = -2C$ while A,B,C are formally skew adjoint on $C_o^\infty(\mathbb{R})$. Further they extend to unique skew adjoint operators on $L^2(\mathbb{R})$ generating unitary groups. The Hermite functions are analytic vectors for $\mathcal{A} = \{A,B,C\}$ and by Simon [1] $\mathcal{A}$ is integrable. In fact the associated group G is a twofold covering of $SL(2,\mathbb{R})$. Another aspect of choosing a "good" operator version of (4.18) is that one obtains naturally an energy

$$E(t) = \frac{1}{2}(\|u_t\|^2 + \|Au\|^2 + \|Bu\|^2 + \|Cu\|^2)$$

yielding good a priori estimates. A further application to equations of the form

$$(4.20) \qquad u_{tt} = L(x,D)u = \sum a_{i,j}^{k,\ell}\, x_i x_j\, D_k D_\ell u$$

with elliptic L is also given.

We go now to some second order equations and will sketch various methods. Consider first the equation (I.14) in a Banach space F with initial data $w(0) = u_o$ and $w'(0) = u_1$. Setting $w_1 = w$ and $w_2 = w'$ one has

$$(4.21) \qquad \vec{w}_t = \mathcal{A}\vec{w}; \quad \vec{w}(0) = \begin{pmatrix} u_o \\ u_1 \end{pmatrix}; \quad \mathcal{A} = \begin{pmatrix} 0 & 1 \\ A^2 & 0 \end{pmatrix}.$$

Assume A generates a strongly continuous group $T(\cdot)$ in F and assuming $0 \in \rho(A)$ = the resolvant set of A one expects $\mathcal{A}$ to generate a strongly continuous group given by

$$(4.22) \qquad \exp t\mathcal{A} = \cosh At \begin{pmatrix} I & 0 \\ 0 & I \end{pmatrix} + A^{-1}\sinh At\, \mathcal{A}.$$

In fact it is proved in Goldstein [3] that if $D(\mathcal{A})$ is taken to be $D(A^2) \times D(A)$ then this is indeed true in the space $D(A) \times F$ where $D(A)$ has the graph norm (but not in $F \times F$ since $A^{-1}\sinh At\,\mathcal{A}$ does not define a bounded operator in $F \times F$). For variable $A(t)$ with constant domain $D(A(t)) = D$ the following theorem is proved in Goldstein [2]. First for $B(t) \in L(E,F)$ one says $B(\cdot) \in Lip(E,F)$ if for any $T > 0$ there is a constant $m = m(T)$ such that for $|s|,|t| < T$ $\|B(t) - B(s)\| \le m|t-s|$. Also if $D(A) \subset D(B)$ with $A^{-1} \in L(F)$

then $BA^{-1} \in L(F)$ by the closed graph theorem.

Theorem 4.7 Let $F = H$ be a Hilbert space with $B(t)$ self adjoint satisfying $(B(t)x,x) \geq c(t)\|x\|^2$ for $x \in D(B(t)) = \hat{D}$ where $c(t) > 0$ is bounded away from 0 on bounded intervals. Assume $t \to B(t)B^{-1}(0) \in Lip(H)$. Then the problem

$$(4.23)\qquad w'' + B(t)w = 0;\quad w(0) = w_o;\quad w'(0) = w_1$$

with $w_o \in \hat{D}$ and $w_1 \in D$ has a unique solution $w \in C^2(H)$.

Here $D(B^{\frac{1}{2}}(t))$ $(= D(A(t)) = D$ is automatically constant with $\hat{D}$ and a perturbation term $P(t) \in L(D,F)$ with $P(\cdot) \in Lip(D,F)$ can be added to $B(t)$. We will sketch some of the steps of the proof. A preliminary lemma based on Kato [2; 8] (cf. also Mizohata [1]) states

Lemma 4.8 Let F be a Banach space with $C(t)$ the generator of a strongly continuous semigroup in $L(F)$ for $t \geq 0$. Let $Q(t) \in L(F)$ be such that $Q^{-1}(t) \in L(F)$ with $Q(t)C(t)Q^{-1}(t)$ generating a contraction semigroup in $L(F)$ and $Q(\cdot) \in Lip(F)$. Suppose there is a function $R(\cdot) : \mathbb{R}^+ \to L(F)$ with strong derivative $\dot{R}$ such that $\dot{R}(\cdot) \in Lip(F)$, $R^{-1}(t) \in L(F)$, and $E(t) = R(t)C(t)R^{-1}(t)$ has domain independent of t. Then

$$D(t) = (I - E(t))(I - E(0))^{-1} \in L(F)$$

and one supposes $D(\cdot) \in Lip(F)$. From these hypotheses it follows that there is a unique $U : \mathbb{R}^+ \to F$ satisfying

$$(4.24)\qquad \frac{dU}{dt} = C(t)U;\quad U(0) = f \in D(C(0)).$$

Proof of Theorem 4.7: (4.23) is equivalent to

$$(4.25)\qquad \frac{d\vec{w}}{dt} = A(t)\vec{w};\quad \vec{w}(0) = \begin{pmatrix} w_o \\ w_1 \end{pmatrix};\quad A(t) = \begin{pmatrix} 0 & 1 \\ -B(t) & 0 \end{pmatrix}$$

where $\vec{w}(t) \in F = D \times H$ with $D(A(t)) = \hat{D} \times D$. Let $F_t = F$ with the norm

$$\|(f,g)\|_t^2 = \|A(t)f\|^2 + \|g\|^2$$

and then for fixed t, $A(t)$ generates a unitary group in $L(F_t)$. Let $D_t = D$ with norm $\|f\| = \|A(t)f\|$ (D has the D_o norm). One shows by a routine argument that there is an operator $Q_o(t) \in L(D)$ such that $Q_o^{-1}(t) \in L(D)$, $Q_o(t) : D_t \to D = D_o$ is unitary, and $Q_o(t) : D \to D$ is positive self adjoint. Set $T_o(t) = Q_o^2(t)$ considered in D_o with domain $\hat{D}$ so that $T_o(t) = B^{-1}(0)B(t)$. The hypotheses of Theorem 4.7 will imply now that $T_o(\cdot) \in \mathrm{Lip}(D)$ (this takes a little calculation). Let $T > 0$ be fixed and then on $[0,T]$ $cI \leq T_o(t) \leq c^{-1}I$ with $T_o(t)$ self adjoint on D_o. A standard formula (cf. Carroll [4], Yosida [1]) allows one to write now for $f \in D = D_o$

$$(4.26) \qquad Q_o(t)f = \frac{1}{\pi}\int_0^\infty s^{-\frac{1}{2}}(sI + T_o(t))^{-1}T_o(t)f\,ds.$$

Using this formula one can show that $Q_o(\cdot) \in \mathrm{Lip}(D)$ (cf. Kato [4]). It is now possible to apply Lemma 4.8. First with $F = D \times H$ ($D = D_o$) define

$$(4.27) \qquad Q(t) = \begin{pmatrix} Q_o(t) & 0 \\ 0 & I \end{pmatrix}.$$

Then $Q(t)$ and $Q^{-1}(t)$ belong to $L(F)$, $Q(\cdot) \in \mathrm{Lip}(F)$, and $Q(t) : F_t \to F$ is unitary. Hence $Q(t)A(t)Q^{-1}(t)$ generates a unitary group in $L(F)$, namely $Q(t)\exp s\,A(t)Q^{-1}(t)$ ($s \in \mathbb{R}$). Next set $R(t) = I$. One knows $1 \in \rho(A(t))$ since $A(t)$ generates a uniformly bounded group in $L(F)$ with $D(A(t)) = \hat{D} \times D$. Hence in Lemma 4.8 one can write

$$D(t) = (I - A(0))^{-1} - A(t)A^{-1}(0)[A(0)(I - A(0))^{-1}]$$

and to show $D(\cdot) \in \mathrm{Lip}(F)$ it suffices to show $A(\cdot)A^{-1}(0) \in \mathrm{Lip}(F)$ which follows with a small calculation. Hence Lemma 4.8 applies and Theorem 4.7 is proved. QED

We consider next some variable domain problems following Carroll-State [25]. The basic problem will be phrased as

$$(4.28) \qquad w'' + B(t)w = f$$

in a separable Hilbert space H and we will allow the domains $D(B(t))$ to really vary in time. This is accomplished by using a technique of Carroll [4; 36; 37; 38] to describe the variation of subspaces in terms of operators (cf. also Carroll-Cooper [24], Cooper [1; 2], Mazumdar [1; 2; 3; 4]). Other abstract techniques for variable domain problems appear in Baiocchi [1; 2], Bernardi [1], Bernardi-Brezzi [2], Gearhart [1], Kato [6; 7], Lions [4; 5; 10; 11; 12] while some recent work with applications to scattering theory can be found in Bandos-Cooper [3], Cooper [3; 4; 5; 6], Cooper-Medeiros [7], Cooper-Strauss [8; 9], Inoue [1], Strauss [3].

We will deal with a weak version of (4.28) first. Thus let $V(t) \subset H$ be a family of Hilbert spaces dense in H with continuous injections. Let $V'(t)$ be the antidual so one can write $V(t) \subset H \subset V'(t)$. Let $R(t) : H \to V(t)$ (into) be determined by $(x,y) = ((R(t)x,y))_t$ for $x \in H$ and $y \in U(t)$. Then $R^{-1}(t)$ is a positive self adjoint operator in H and $S(t) = R^{-\frac{1}{2}}(t)$ maps its domain $V(t) = D(S(t))$ one to one onto H with $((x,y))_t = (Sx,Sy)$ (see Carroll [4] for details). The operators $S(t)$ were called "standard" operators at one time but perhaps a better name would be canonical $V(t)$ related operators. Let $W = L^2(V(t)) = \{u \in L^2(H)$ on $[0,T]$, $u(t) \in V(t)$ a.e., $Su \in L^2(H)\}$ with scalar product $((u,v))_W = \int_0^T (Su,Sv)dt$. One makes hypotheses which insure that $\| S^{-1}(t) \| \leq c$ with $S^{-1}(\cdot)h$ measurable in H for $h \in H$ (i.e. $(S^{-1}(\cdot)h,k)$ is to be measurable for $k \in H$). Then $W \subset L^2(H)$. Let $b(t,\cdot,\cdot)$ be a continuous, coercive, self adjoint sesquillinear form on $V(t) \times V(t)$ with $|b(t,x,y)| \leq c\| x\|_t \| y\|_t$ and $b(t,x,x) \geq \alpha \| x\|_t^2$. Such a form induces a topology on $V(t)$ equivalent to its original topology and we introduce a new Hilbert structure on $V(t)$ with scalar product $b(t,x,y)$. Clearly $b(t,x,y) = \langle B(t)x,y\rangle_t$ where $B(t) : V(t) \to V'(t)$ is linear and continuous ($\langle\ ,\ \rangle_t$ denotes $V(t) - V'(t)$ antiduality). This determines an unbounded operator in H, denoted again by $B(t)$, with domain

$$D(B(t)) = \{x \in V(t);\ B(t)x \in H\}$$

and for $x \in D(B(t))$ one has $b(t,x,y) = (B(t)x,y)$. Since $b(t,x,y) = \overline{b(t,y,x)}$

we see that $B(t)$ is a self adjoint operator in H, closed, and densely defined. Next we let $\theta(t) : V'(t) \to V(t)$ be the isometric isomorphism determined by $<x,y> = ((\theta(t)x,y))_t$ so that $B(t) = \theta^{-1}(t)$ and one shows easily that $\theta(t) = S^{-2}(t)$ when restricted to H. Thus $B(t) = S^2(t)$ is self adjoint and we will work with this $B(t)$ in (4.28). Suitable perturbations of $B(t)$ will of course be permissable (cf. Carroll-State [25]).

Now a weak version of (4.28) is to find $w \in W$, with $w' \in L^2(H)$ and $w(0) = 0$, so that

$$(4.29) \qquad -\int_0^T (w',v')dt + \int_0^T b(t,w,v)dt = \int_0^T (f,v)dt$$

for all $v \in W$ with $v' \in L^2(H)$ and $v(T) = 0$. In general a term $(w_1,v(0))$ will be added to the right side of (4.29) but we assume $w_1 = 0$ here for simplicity ($w_1 \sim w'(0)$). We assume that $S^{-1}(\cdot)$ is weakly C^1 (i.e. $(S^{-1}(\cdot)h,k)$ is C^1 for $h,k \in H$) in which case there are self adjoint operators $\dot{S}^{-1}(t)$ with $(S^{-1}(t)h,k)' = (\dot{S}^{-1}(t)h,k)$, $\| S^{-1}(t)\| \le c_1$, $\| \dot{S}^{-1}(t)\| \le c_2$, and $S^{-1}(\cdot)$ is Lipschitz continuous (cf. Carroll [4]). Now in making (4.28) or (4.29) into a first order system an additional term appears in order to take into account the variation of domain. Thus set $w = S^{-1}w_1$ and observe that in $\mathcal{D}'(H)$

$$w' = (S^{-1}w_1)' = \dot{S}^{-1}w_1 + S^{-1}w_1'$$

when w_1' makes sense. Let $\vec{w} = \begin{pmatrix} w_1 \\ w_2 \end{pmatrix} \in \mathcal{H} = L^2(H) \times L^2(H)$, $\mathcal{W} = W \times W$, $\vec{f} = \begin{pmatrix} 0 \\ f \end{pmatrix}$, and consider the problem of finding $\vec{w} \in \mathcal{H}$ such that

$$(4.30) \qquad -(\vec{w},\vec{v}') + \lambda(\vec{w},\vec{v}) + (A\vec{w},S\vec{v}) = (\vec{f},\vec{v})$$

$$A = \begin{pmatrix} \dot{S}^{-1} & -1 \\ 1 & 0 \end{pmatrix}$$

for all $\vec{v} \in \mathcal{W}$ with $\vec{v}' \in \mathcal{H}$ and $\vec{v}(T) = 0$. Here a λ term is added on for convenience in the argument; it does not affect existence or uniqueness. To see that (4.30) is a correct transposition of (4.29) we write out the equations as

(4.31) $$-\int_0^T (w_1,v_1')dt + \lambda\int_0^T (w_1,v_1)dt - \int_0^T (w_2,Sv_1)dt + \int_0^T (\dot{S}^{-1}w_1,Sv_1)dt = 0$$

(4.32) $$-\int_0^T (w_2,v_2')dt + \lambda\int_0^T (w_2,v_2)dt + \int_0^T (w_1,Sv_2)dt = \int_0^T (f,v_2)dt.$$

Thus taking $v_1 = S^{-1}\phi$ for $\phi \in \mathcal{D}(H)$ on $(0,T)$ we have $v_1' = \dot{S}^{-1}\phi + S^{-1}\phi'$ and setting $\lambda = 0$ (4.31) says

(4.33) $$-\int_0^T (S^{-1}w_1,\phi')dt = \int_0^T (w_2,\phi)dt$$

which means $w_2 = (S^{-1}w_1)'$ in $\mathcal{D}'(H)$. Hence given a solution $\vec{w}$ of (4.30) with $\lambda = 0$ define $w = S^{-1}w_1$ so that $w_2 = w'$ and $(w_1,Sv_2) = (Sw,Sv_2) = b(t,w,v_2)$. Then (4.32) says

(4.34) $$-\int_0^T (w',v_2')dt + \int_0^T b(t,w,v_2)dt = \int_0^T (f,v_2)dt$$

which is (4.29) with $v_2 = v$.

As in Carroll-Cooper [24] $W' = L^2(V'(t)) = \theta^{-1}W$ and one defines $L : W \to W'$ by $Lu = u'$ with $D(L) = \{u \in W;\ u' \in L^2(H);\ u(0) = 0\}$. Similarly $L' : W \to W'$ is defined as $L'u = -u'$ with $D(L') = \{u \in W;\ u' \in L^2(H);\ u(T) = 0\}$. Both L and L' are densely defined and as a consequence of $S^{-1}(\cdot)$ being weakly C^1 it is proved that $L_s = \bar{L} = (L')^* = L_w$. Actually a stronger result is proved in Carroll-Cooper [24], namely if $S^{-2}(\cdot)$ is weakly C^1 then $L_s = L_w$. This is a very useful criterion and the proof is nontrivial. We recall next that a map $Q : F \to F'$ is monotone if for $u,v \in D(Q)$,

$$\mathrm{Re} < Q(u) - Q(v),\ u - v> \ \geq 0$$

(take here F to be a reflexive Banach space for example). When Q is linear and monotone with $D(Q)$ dense we say Q is maximal monotone if it is not the proper restriction of another linear monotone operator. A (nonlinear) $Q : F \to F'$ with $D(Q) = F$ is bounded if it takes bounded sets into bounded sets and hemicontinuous if it is continuous from lines in F to the weak topology of F'; Q is coercive if

$$\mathrm{Re}\langle Q(x),x\rangle \geq \phi(\|x\|)\ \|x\|$$

where $\phi(x) \to \infty$ as $x \to \infty$ ($\phi(x)$ can be negative for x small). Now, following Carroll-Cooper [24], by a result of Brézis [2] L_s is maximal monotone since $L_s = L_w$. The following special case of a result of Browder [8] is then needed for our development.

Lemma 4.9 Assume F is a reflexive Banach space. Let $L : F \to F'$ be a closed densely defined linear maximal monotone map and $Q : F \to F'$ a monotone hemicontinuous bounded coercive map. Then $L + Q$ maps $D(L) \subset F$ onto F'.

Consider now the regularized strong parabolic problem associated with (4.30) (cf Lions [5] for parabolic regularization)

Problem 4.10 Find $\vec{w}^{\varepsilon} \in W$ such that

$$\mathcal{L}\vec{w}^{\varepsilon} + \lambda\vec{w}^{\varepsilon} + \mathcal{P}\vec{w}^{\varepsilon} + \varepsilon\mathcal{K}\vec{w}^{\varepsilon} = \vec{f} \tag{4.35}$$

$$\mathcal{L} = \begin{pmatrix} L_s & 0 \\ 0 & L_s \end{pmatrix}; \quad \mathcal{K} = \begin{pmatrix} \theta^{-1} & 0 \\ 0 & \theta^{-1} \end{pmatrix}; \quad \mathcal{P} = \begin{pmatrix} \theta^{-1}S^{-1}\dot{S}^{-1} & -S \\ S & 0 \end{pmatrix}. \tag{4.36}$$

Let us check that a solution of (4.35) satisfies (4.30) with a suitable ε term added. Let $\vec{v} \in \mathcal{W}$ with $\vec{v}' \in \mathcal{H}$ and $\vec{v}(T) = 0$. Then $\vec{v} \in D(\mathcal{L}')$ in an obvious notation and taking $\mathcal{W} - \mathcal{W}'$ brackets in (4.35) with $\vec{v}$ we obtain

$$-(\vec{w}^{\varepsilon},\vec{v}')_{\mathcal{H}} + \lambda(\vec{w}^{\varepsilon},\vec{v})_{\mathcal{H}} + (A\vec{w}^{\varepsilon},S\vec{v})_{\mathcal{H}} + \varepsilon(S\vec{w}^{\varepsilon},S\vec{v})_{\mathcal{H}} = (\vec{f},\vec{v})_{\mathcal{H}}. \tag{4.37}$$

Indeed the individual equations in (4.35) are

$$L_s w_1^{\varepsilon} + \lambda w_1^{\varepsilon} + \theta^{-1}S^{-1}\dot{S}^{-1}w_1^{\varepsilon} - Sw_2^{\varepsilon} + \varepsilon\theta^{-1}w_1^{\varepsilon} = 0 \tag{4.38}$$

$$L_s w_2^{\varepsilon} + \lambda w_2^{\varepsilon} + Sw_1^{\varepsilon} + \varepsilon\theta^{-1}w_2^{\varepsilon} = f \tag{4.39}$$

and one notes that, for $x \in H$ and $v \in V(t)$, $\langle x,v\rangle = (x,v)$ while, for $x \in V'(t)$ and $v \in V(t)$, $\langle x,v\rangle = ((\theta x,v)) = (S\theta x,Sv)$. Hence in appropriate spaces

$$\langle\theta^{-1}S^{-1}\dot{S}^{-1}w_1^{\varepsilon},v_1\rangle = (\dot{S}^{-1}w_1^{\varepsilon},Sv_1), \quad -\langle Sw_2^{\varepsilon}, v_1\rangle = -(Sw_2^{\varepsilon}, v_1)$$
$$= -(w_2^{\varepsilon}, Sv_1), \quad \langle\theta^{-1}w_1^{\varepsilon},v_1\rangle = (Sw_1^{\varepsilon}, Sv_1),$$

etc. so that $\langle P\vec{w}^{\varepsilon}, \vec{v}\rangle = (A\vec{w}^{\varepsilon}, S\vec{v})$, $\langle K\vec{w}^{\varepsilon}, \vec{v}\rangle = (S\vec{w}^{\varepsilon}, S\vec{v})$, etc. The program now is to find solutions $\vec{w}^{\varepsilon}$ of (4.35) with $|\vec{w}^{\varepsilon}|$ and $\varepsilon^{\frac{1}{2}}\|\vec{w}^{\varepsilon}\|_{\mathcal{W}} = \varepsilon^{\frac{1}{2}}|S\vec{w}^{\varepsilon}|_{H}$ bounded. Then by weak compactness we can take weak limits in (4.37) which leads to a solution of (4.30). This is accomplished by taking $F = \mathcal{W}$ in Lemma 4.9 with $Q = \lambda I + P + \varepsilon K$ and making the hypotheses

$$\text{(4.40)} \qquad \mathrm{Re}(x, \dot{S}^{-1}(t)S(t)x)_H \geq -\beta|x|_H^2$$

for $x \in V(t)$. Then taking $\lambda > \beta$ one obtains $\vec{w}^{\varepsilon}$ solving (4.35) (after some routine calculation to verify coercivity) and the estimates $|\vec{w}^{\varepsilon}|_H \leq c$ and $\varepsilon^{\frac{1}{2}}|S\vec{w}^{\varepsilon}|_H \leq \hat{c}$ follow easily. Consequently one has

<u>Theorem 4.11</u> Let $b(t,\cdot,\cdot)$ be a family of continuous self adjoint coercive sesquilinear forms on $V(t) \times V(t)$ and put on $V(t)$ the corresponding Hilbert structure. Assume the canonical $V(t)$ related operators $S(t)$ satisfy $S^{-1}(\cdot)$ weakly C' with (4.40). Then there exists a solution of (4.30).

<u>Remark 4.12</u> It is shown in Carroll-State !251 that (4.40) is equivalent to

$$\text{(4.41)} \qquad \frac{d}{dt}|S^{-1}(t)h|^2 \geq -2\beta|S^{-1}(t)h|^2$$

for $h \in H$ and this leads to the requirement that $V(s) \subset V(t)$ for $t \geq s$. Examples are provided to show that $V(t)$ need not be constant.

<u>Remark 4.13</u> Mazumdar [4] works with (4.35) directly for $\varepsilon = 0$ using further hypotheses and some results of Mazumdar [1; 2; 3]. In particular he assumes S^{-1} is weakly C^2, (4.40) holds, and $S^{-1}\dot{S}^{-1}S$, or equivalently $\dot{S}^{-2}S$, extends to be a bounded operator in H. Then if $f \in H$ with $f' \in H$ and $f(0) = 0$ there exists a solution $w \in W$ of $w'' + 2\lambda w' + \lambda^2 w + S^2 w = f$ satisfying $Sw \in W$, $(Sw)' \in W'$, $w' \in W$, and $w'' \in H$ while $w(0) = w'(0) = 0$. Note here that if $u = e^{\lambda t}w$ then $u'' + S^2 u = e^{\lambda t}[w'' + 2\lambda w' + \lambda^2 w + S^2 w]$ with $u(0) = w(0)$ and $u'(0) = w'(0) + \lambda w(0)$. Examples show that this is also a true variable domain situation.

<u>Reamrk 4.14</u> In Kato [6; 7] evolution equations of the type $w' + A(t)w = f$ are

studied where the $-A(t)$ generate C^o semigroups in a Banach space F but not necessarily analytic semigroups. Such equations are then said to be of hyperbolic type and have some applications to symmeyric hyperbolic systems of differential equations (cf. Crandall [4], Friedrichs [1], Phillips [1; 2]). The domain $D(A(t))$ is allowed to vary but it is assumed there is a dense linear subspace $D \subset D(A(t))$ on which $A(t)$ acts "smoothly".

The Russian school has treated second order problems of "parabolic" type extensively and extended the techniques to higher order equations. Thus going to Krein [1] we let $A(t)$ and $B(t)$ be closed densely defined linear operators in a Banach space E and consider

$$(4.42) \qquad u'' = A(t)u' + B(t)u; \quad u(0) = u_o; \quad u'(0) = u_1.$$

Suppose $D(A(t)) = D$ independent of t with $A^{-1}(t) \in L(E)$. If $u(0) \in D$ from $A(\cdot)u'$ continuous and $A(0)A^{-1}(t) \in L(E)$ it follows that $A(0)u'$ is continuous. Since $A(0)$ is closed

$$A(0)[u(t) - u(0)] = A(0)\int_0^t u'(\tau)d\tau = \int_0^t A(0)u'(\tau)d\tau$$

and hence $u(t) \in D$ with $A(0)u(\cdot) \in C^1$ $(A(0)u(t))' = A(0)u'(t))$. Let now $u' = v$ and $A(0)u(t) = w(t)$ so that

$$(4.43) \qquad v' = A(t)v + B(t)A^{-1}(0)w; \quad w' = A(0)v$$

and $\vec{x}' = \mathcal{Q}(t)\vec{x}$ where

$$(4.44) \qquad \mathcal{Q}(t) = \begin{pmatrix} A(t) & B(t)A^{-1}(0) \\ A(0) & 0 \end{pmatrix}; \quad \vec{x} = \begin{pmatrix} v \\ w \end{pmatrix}.$$

Solutions of (4.42) correspond to solutions of $\vec{x}' = \mathcal{Q}(t)\vec{x}$. The first technique assumes $A(\cdot)y \in C^1(E)$ for $y \in D$ with $A(t)$ suitably approximable by Yosida approximations

$$A_n(t) = -nA(t)R_{A(t)}(n) = A(t)(I - \frac{1}{n} A)^{-1}$$

(recall $R_A(\lambda) = (A - \lambda I)^{-1}$) while $B(t)A^{-1}(0) \in L(E)$ with $B(\cdot)A^{-1}(0)$ strongly differentiable. Then $\mathcal{Q}(t)$ is thought of as $\mathcal{A}(t) + \mathcal{B}(t)$ where $\mathcal{A}(t)$ (resp. $\mathcal{B}(t)$) has zeros in the second (resp. first) column. The problem is viewed as a perturbation of $\vec{y}' = \mathcal{A}(t)\vec{y}$ and a unique solution is obtained for $u_o, u_1 \in D$, using results for first order equations. If $A(\cdot)y \in C^1(E)$ for $y \in C^1(E)$ for $y \in D$ while $\| R_{A(t)}(\lambda) \| \leq 1/1+\lambda$ for $\lambda \geq 0$ (instead of the approximation property) and the other hypotheses are unchanged then the same conclusion follows by invoking other results on first order equations. The thrust of the matter in terms of operator strength is displayed for constant operators in the following theorem where we say that an operator C is subordinate to A if $D(C) \supset D(A)$ and $\| Cx \| \leq c \| Ax \|$ for $x \in D(A)$. If in addition for small η $\| Cx \| \leq \phi_\eta(x) + \eta \| Ax \|$ for a continuous convex functional ϕ_η then B is said to be completely subordinate.

Theorem 4.15 Let A and $A_o = -BA^{-1}$ generate analytic semigroups with A_o completely subordinate to A. Then for $u_o \in D(B) \cap D(A)$ and $u_1 \in D(A)$ there exists a unique solution of (4.42).

One also speaks of a weak solution of (4.42) in requiring $u \in C^2$, $A(\cdot)u' \in C^o$, and $B(\cdot)u \in C^o$ only on $(0,T]$ instead of $[0,T]$. In the siutation of Theorem 4.15 such solutions exist uniquely for $u_o \in D(A)$ and $u_1 \in E$ and $u(t)$ will be analytic in some sector containing $(0,\infty)$. Similarly one has

Theorem 4.16 Let $A(t)$ and $A_o(t) = B(t)A^{-1}(t)$ have domains independent of t and satisfy $\| R_{A(t)}(\lambda) \| \leq M/|\lambda - w|$ and $\| R_{A_o(t)}(\lambda) \| \leq M_o/|\lambda - w_o|$ for $\mathrm{Re}\,\lambda > w$ and $\mathrm{Re}\,\lambda > w_o$. Let $A'(t)A^{-1}(t)$, $A_o(t)A_o^{-1}(0)$ and $A_o(t)A^{-\alpha}(0)$ for some $\alpha \in [0,1]$ be bounded and satisfy a Hölder condition. Then for $u_o \in D(A) \cap D(B)$ and $u_1 \in D(A^\rho)$ for $\rho > \alpha$ there exists a unique weak solution of (4.42).

It is in this spirit that one finds certain results for higher order equations. For example following Yakubov [5] we consider for closed unbounded densely defined A_k

$$\sum_{k=0}^{n} A_k(t) D_t^{n-k} u = f(t, u, \ldots, D_t^{n-1} u) \tag{4.45}$$

with $D_t^k u(0) = u_k$ $(0 \le k \le n-1)$ where $A_o = 1$. Solutions of type 1 satisfy (4.45) on $[0,T]$ with $u \in C^n(E)$ (E is a Banach space here) whereas solutions of type 2 satisfy $u \in C^{n-1}(E)$ on $[0,T]$, $u \in C^n(E)$ on $(0,T]$, and $A_k(\cdot)D_t^{n-k}u \in C^o(E)$ on $[0,T]$ for $n \ge 2$ while $A_1(\cdot)D_t^{n-1}u \in C^o(E)$ on $(0,T]$ and is L^1 on $[0,T]$. The following theorems are proved by writing (4.45) as a first order system and invoking results on first order equations.

<u>Theorem 4.17</u> Let $D(A_1(t)) = D$ dense with $\| R_{A_1(t)}(\lambda) \| \le 1/(\lambda+\delta)$ for $\lambda > -\delta$ $(\delta > 0)$. Suppose $A_1(\cdot)A_1^{-1}(0) \in C^2(L_s(E))$ and $A_k(\cdot)A_1^{-1}(0) \in C^1(L_s(E))$ on $[0,T]$. Assume $u_k \in D$ with $f = f(t)$ a function in $C^1(E)$. Then (4.45) has a unqiue type 1 solution.

<u>Theorem 4.18</u> Let $D(A_1(t)) = D = D(A)$ dense with $\| R_{A_1(t)}(\lambda) \| \le c/(1+|\lambda|)$ for $\mathrm{Re}\,\lambda > 0$. Let $A_1(\cdot)A^{-1}(0) \in C^1(L_s(E))$ with $[A_1(\cdot)A^{-1}(0)]'$ and $A_k(t)A^{-1}(0)$ $(k = 2,\ldots,n)$ satisfying a Hölder condition on $[0,T]$. Let $u_k \in D$ for $0 \le k \le n-2$ with $u_{n-1} \in D(A^\alpha)$ for some $\alpha > 0$ and let $f = f(t)$ satisfy a Hölder condition. Then (4.45) has a unique type 2 solution.

Theorems for more general f are obtained in the framework

$$(4.46) \qquad D_t^n u + A(t)D_t^{n-1}u = F(t,u,\ldots,D_t^{n-1}u)$$

but we will not be more specific here. Let us mention in passing the paper of Kharitonenko-Yurčuk [1] where the abstract Love equation

$$(4.47) \qquad u'' + A(t)[u + \nu u''] = f$$

with suitable data $u(0)$ and $u'(0)$ is treated in a Hilbert space. We go next to a technique of Raskin [2] (cf. also Dubinskij [1; 2], Gasymov [1], Grisvard [2]) for equations of the form

$$(4.48) \qquad L_n(D_t)v = \sum_{k=0}^{n} A_k D_t^{n-k} v = 0 \qquad (A_o = I)$$

where the A_k are closed densely defined operators in a Banach space E. Initial conditions $D_t^k v(0) = v_k$ $(0 \le k \le n-1)$ are required for suitable v_k. Assume

$A_k^{-1} \in L(E)$ and set $-B_k = -A_k A_{k-1}^{-1}$ $(1 \le k \le n)$ with resolvant $R_k(\lambda) = (\lambda I + B_k)^{-1}$ satisfying

$$(4.49) \qquad \| R_k(\lambda) \| \le M|\lambda|^{-\alpha_k}; \quad 0 < \alpha_k \le 1; \quad \text{Re}\,\lambda \ge \sigma_o$$

where $n-1 \le \delta = \sum_i^n \alpha_k \le n$ (note we are taking $R_A(\lambda) = (\lambda I - A)^{-1}$ here instead of $(A - \lambda I)^{-1}$ as before). Let $L_n(\lambda) = \Sigma\lambda^{n-k}A_k$ and the main tool in this work is the following lemma.

<u>Lemma 4.19</u> For some $c > 0$ and $\beta < \beta_o = \delta - (n-1)$ let $\| B_{k+1}x\| \le c\| B_k x\|^{\beta} \| x\|^{1-\beta}$ for $x \in D(B_k) \cap D(B_{k+1})$ and $k = 1,\ldots,n-1$. Assume $B_1 = A_1$ commutes with B_k for $k = 2,\ldots,n$. Then $L_n(\lambda)$ has an inverse $L_n^{-1}(\lambda)$ for $\text{Re}\,\lambda \ge \sigma_1$ and $\| L_n^{-1}(\lambda)\| \le c_1|\lambda|^{-\delta}$.

Proof: One writes first

$$(4.50) \qquad (\lambda + B_n) \cdots (\lambda + B_1) = \lambda^n + \lambda^{n-1}\sum_{i=1}^{n} B_i + $$
$$+ \lambda^{n-2}\sum_{i_2 \le i_1} B_{i_1}B_{i_2} + \cdots + B_n \cdots B_1.$$

Since $B_k B_{k-1} \cdots B_1 = A_k$ it follows that

$$(4.51) \qquad L_n(\lambda) = (\lambda + B_n) \cdots (\lambda + B_1) - \sum_{m=1}^{n-1} \lambda^{n-m} \sum_{i_1 > m} B_{i_1} \cdots B_{i_m}$$

where $i_1 > i_2 > \cdots > i_m$. Hence writing

$$(4.52) \qquad K(\lambda) = \sum_{m=1}^{n-1} \lambda^{n-m} \sum_{i_1 > m} B_{i_1} \cdots B_{i_m} R_1(\lambda) \cdots R_n(\lambda)$$

we can express (4.51) in the form

$$(4.53) \qquad L_n(\lambda) = [I - K(\lambda)](\lambda + B_n) \cdots (\lambda + B_1).$$

One can establish the estimate $\| K(\lambda)\| \le c_2|\lambda|^{\beta-\beta_o}$ for $\text{Re}\,\lambda \ge \sigma_1$ (see below) and hence $\| K(\lambda)\| \le 1$ for sufficiently large $|\lambda|$ (since $\beta < \beta_o$). This guarantees the existence of $L_n^{-1}(\lambda)$ from (4.53) which we can write in the form

$$(4.54) \qquad L_n^{-1}(\lambda) = R_1(\lambda) \cdots R_n(\lambda)[I - K(\lambda)]^{-1}$$

with a norm estimate as indicated. To obtain the estimate on $\| K(\lambda) \|$ observe that for $|\lambda|$ large

$$(4.55)\qquad \| B_1 R_1(\lambda) \cdots R_n(\lambda) \| = \| (I - \lambda R_1(\lambda)) R_2(\lambda) \cdots R_n(\lambda) \| \le M^n |\lambda|^{1-\delta} k.$$

Hence for any $i_1 > 1$ it follows that (note $\beta < 1$)

$$(4.56)\qquad \| \lambda^{n-1} B_{i_1} R_1(\lambda) \cdots R_n(\lambda) \| \le$$
$$c^{i_1 - 1} |\lambda|^{n-1} \| B_1 R_1(\lambda) \cdots R_n(\lambda) \|^{\beta^{i_1-1}} \| R_1(\lambda) \cdots R_n(\lambda) \|^{1-\beta^{i_1-1}}$$
$$\le c^{i_1-1} \tilde{k} M^n |\lambda|^{\beta - \beta_o}.$$

Similarly $B_{i_1} B_{i_2} R_1(\lambda) \cdots R_n(\lambda)$ for $i_1 > 2$ can be estimated in terms of $B_2 B_1 R_1(\lambda) \cdots R_n(\lambda) = R_3(\lambda) \cdots R_n(\lambda) - \lambda R_2(\lambda) \cdots R_n(\lambda) - \lambda B_2 R_1(\lambda) \cdots R_n(\lambda)$ and using (4.56) it follows that

$$(4.57)\qquad \| \lambda^{n-2} B_{i_1} B_{i_2} R_1(\lambda) \cdots R_n(\lambda) \| \le c(i_1, i_2) |\lambda|^{\beta - \beta_o}.$$

All terms in $K(\lambda)$ can be so estimated and one obtains the desired bound on $\| K(\lambda) \|$. QED

Let now $v_i \in \bigcap_{s=i}^{n-i-1} D(A_s)$, $i = 0, \ldots, n-1$, and write

$$(4.58)\qquad \psi_o(\lambda) = \sum_{j=0}^{n-1} \Big(\sum_{i=j+1}^{n} \lambda^{n-i} A_{i-1} \Big) v_j.$$

Then there results

<u>Theorem 4.20</u> Under the hypotheses of Lemma 4.19 the function

$$(4.59)\qquad v(t) = \frac{1}{2\pi i} \int_{\sigma - i\infty}^{\sigma + i\infty} \exp \lambda t L_n^{-1}(\lambda) \psi_o(\lambda) d\lambda$$

is a C^∞ solution of (4.48) for $t > 0$.

<u>Theorem 4.21</u> Let $v_i \in \bigcap_1^n D(A_s)$ for $i = 0, \ldots, k$ with $v_{k+1} \in (\bigcap_1^{n-1} D(A_s)) \cup (\bigcap_1^n D(B_s))$ while $v_{k+2}, \ldots, v_{n-1}$ satisfy the condition used in constructing $\psi_o(\lambda)$. Then $v(t)$ given by (4.59) is $k+1$ times continuously differentiable for $t \ge 0$ with $D_t^j v(0) = v_j$ for $0 \le j \le k+1$.

Somewhat more abstractly and generally Dubinskij [2] considers the matter as follows (cf also Atkinson [1; 2], Browne [1], Carroll [4], Chazarain [1], Cordes [2], DaPrato-Grisvard [3], Dubinskij [1], B. Friedman [1], Gasymov [1], Grisvard [2], Ichinose [1; 2; 3; 4], Källström-Sleeman [1], Mohammed [1], Roach-Sleeman [1]). Let

$$(4.60)\qquad P(D_t) = \sum_{j=0}^{s} A_j D_t^j; \quad P(\lambda) = \sum_{j=0}^{s} A_j \lambda^j.$$

The idea is to consider first a general operator equation

$$(4.61)\qquad P(B)u = \sum_{j=0}^{s} A_j B^j u = h$$

with closed densely defined operators A_j and B in a Banach space E. Suppose that the spectra $\sigma(B)$ and $\sigma(P)$ are disjoint with Γ a suitable contour in $\mathbb{C}$ separating them. $\lambda = \sigma(P)$ if $\exists\, x \in D(P) = \cap D(A_j)$ such that $P(\lambda)x = 0$; $\lambda \in \sigma(B)$ if $(B - \lambda)x = 0$ for some $x \in D(B)$ - only point spectra are being considered. Then formally a solution to (4.61) will be

$$(4.62)\qquad u = -\frac{1}{2\pi i}\int_\Gamma \lambda^{-k} P^{-1}(\lambda)(B - \lambda I)^{-1} B^k h\, d\lambda$$

where k is an integer characterizing the growth of $P^{-1}(\lambda)$ at ∞. This is an extension of methods of Grisvard [2] (cf Carroll [4] and see also Chazarain [1], DaPrato-Grisvard [3], Ichinose [1; 2; 3; 4]). More precisely let Γ be a smooth curve in the halfplane $\text{Re}\,\lambda > 0$ going out to ∞ with arc length proportional to $|\lambda|$; let S_+ (resp. S_-) be the region lying to the right (resp. left) of Γ. Then I: Assume $D(A_j B) = D(BA_j)$ with $A_j B = BA_j$; II: Let $\sigma(B) \subset S_+$ and $\sigma(P) \subset S_-$; III: Assume there are Banach spaces $E_1 \subset E$ and $E_2 \subset E$ with continuous injections such that for any $\lambda \notin S_+$ $(B - \lambda I)^{-1} : E_1 \to D(B)$ is 1-1 with $(B - \lambda I)^{-1} : E \to E_2$ satisfying $\|(B - \lambda I)^{-1}\| \leq c/(1 + |\lambda|)$ in both norms $E_1 \to E$ and $F \to E_2$ where $F = E_1 \cap E_2$; further for $\lambda \notin S_-$ let $P^{-1}(\lambda) : E_2 \to D(P)$ be 1-1 ans analytic with $A_j P^{-1}(\lambda) : F \to E_1$ and $\|A_j P^{-1}(\lambda)\| \leq c(1 + |\lambda|)^{k-j}$ in both norms $E_2 \to E$ and $F \to E_1$. Then it is immediate to verify

Theorem 4.22 Let the above hypotheses hold. Then for any $h \in D(B^{k+1})$ with

$B^{k+1}h \in F$ there exists a solution u of (4.61) given by (4.62) with k replaced by $k+1$. Further one has $\Sigma \| A_j B^j u \|_E \le c \| B^{k+1} h \|_F$. If in addition $B^k h \in F$ then (4.62) gives the solution with k unchanged.

Set now $W_{k+1} = \{u;\ u \in D(B^{k+1}P(B));\ B^{k+1}P(B)u \in F\}$ and $\tilde{W}_{k+1} = \{u;\ A_j B^j u \in D(B^{k+1});\ B^{k+1}A_j B^j u \in F\}$. Then it follows that

<u>Theorem 4.23</u> The solution u of $P(B)u = h$ as above is unique provided $\tilde{W}_{k+1}$ is dense in W_{k+1} in the sense that for any $u \in W_{k+1}$ there is a sequence $u_n \in \tilde{W}_{k+1}$ such that $u_n \to u$ and $B^{k+1}P(B)u_n \to B^{k+1}P(B)u$ in E.

These theorems can be refined somewhat by making further abstract hypotheses but we omit this. In the case when B is a self adjoint operator in a Hilbert space $E = H$ with spectral measures E_λ one can reformulate the matter as follows. If $G \subset H$ with a scalar product $(\ ,\ )_F$ one defines $F = \{h \in H;\ (E_\lambda - E_\mu)h \in G$ $(\lambda \subset \mu);\ \|h\|_F^2 = \int (dE_\lambda h,\ dE_\lambda h)_F < \infty\}$ where the integral is over $\sigma(B)$. Assume I holds with $A_s = I$ and for $\lambda \in \sigma(B)$ assume the map $P^{-1}(\lambda) : F \to D(P)$ is 1-1 with $\| A_j P^{-1}(\lambda) \| \le C(1 + |\lambda|)^{k-j}$ in the $F \to H$ norm. Then from standard spectral considerations there results

<u>Theorem 4.24</u> Under the above hypotheses if $h \in F$ with $B^k h \in F$ then a solution $u \in D(P(B))$ of $P(B)u = h$ is determined by the formula

$$(4.63) \qquad u = \int_{\sigma(B)} P^{-1}(\lambda)\, dE_\lambda h$$

and there is an estimate $\Sigma \| A_j B^j u \|_H \le c(\|h\|_F + \|B^k h\|_F)$.

For uniqueness let $W_k = \{u \in D(P(B));\ B^k P(B)u \in F\}$ and $\tilde{W}_k = \{u \in D(P(B));\ B^k A_j B^j u \in F\}$. Assume for any $u \in W_k$ there is a sequence $u_n \in \tilde{W}_k$ such that $u_n \to u$ and $B^k A_j B^j u_n \to B^k A_j B^j u$ in H $(j = 0,\dots,s)$. Then solutions of $P(B)u = h$ as above are unique. To apply these considerations to differential equations let X be a Hilbert or Banach space and define

$$H_q(s,P)(X) = \{u : I \to D(P) = \cap D(A_j); \quad I \subset \mathbb{R} \quad \text{an interval};$$

$$\text{such that} \quad \| u\|^q_{s,P,q} = \int_I (\Sigma \| A_j D^j_t u(t)\|^q + \| u(t)\|^q)dt < \infty\}.$$

Let $H_q(s,P,\gamma)(X)$ for γ a real number be the space of u such that $u(t)\exp(-\gamma t) \in H_q(s,P)(X)$. Further let $H(q,k,I,\gamma)(X) = \{u; \sum_0^k \| D^j(u(t)\exp(-\gamma t))\|^q < \infty\}$ (note $H(q,0,I,\gamma)(X) = L_{q,\gamma}(X) = L_q(X)$ with weight function $\exp(-\gamma t)$). Consider now (4.60) and equations $P(D_t)u = h$ with $A_s = I$.

<u>Definition 4.25</u> $P(D_t)$ is called γ-regular if $\sigma(P)$ lies outside the line $\text{Re}\,\lambda = \gamma$ and there is a Banach space $Y \subset X$ such that for $\text{Re}\,\lambda = \gamma$, $P^{-1}(\lambda)$: $Y \to D(P)$ is 1-1 with $\| A_j P^{-1}(\lambda)\| \leq c(\gamma)(1 + |\lambda|)^{k-j}$ $(j = 0,\ldots,s)$ where the norm denotes $Y \to X$.

It is straightforward then to verify the following theorem.

<u>Theorem 4.26</u> Let $P(D_t)$ be γ-regular with Y and X Hilbert spaces. Let E_λ be the spectral measures for $-iD_t$ in $L^2(\mathbb{R}, X)$. Then for any $h \in H(2,k,I,\gamma)(Y)$ the equation $P(D_t)u = h$ has a solution $u \in H_2(s,P,\gamma)(X)$ given by

$$(4.64) \qquad u(t) = e^{\gamma t}\int_{-\infty}^{\infty} P^{-1}(\gamma + i\lambda)dE_\lambda(he^{-\gamma t}).$$

If $\tilde{F} = \{u \in D(P); A_j u \in Y\}$ is dense in $D(P)$ then this solution is unique.

<u>Definition 4.27</u> The operator $P(D_t)$ is called parabolic if there is a γ_0 such that $\sigma(P) \subset \{\lambda; \text{Re}\,\lambda \leq \gamma_0\}$ and on some Banach space $Y \subset X$, $P^{-1}(\lambda) : Y \to D(P)$ is 1-1 and analytic for $\text{Re}\,\lambda > \gamma_0$ while if $\text{Re}\,\lambda \geq a > \gamma_0$, $\| A_j P^{-1}(\lambda)\| \leq c(a)(1 + |\lambda|)^{k-j}$ $(0 \leq j \leq s)$ with norm $Y \to X$. $P(D_t)$ is called backward parabolic if $P(-D_t)$ is parabolic.

<u>Definition 4.28</u> $P(D_t)$ is called hyperbolic if it is simultaneously parabolic and backward parabolic. Thus there exists $\gamma_0 \leq \gamma_1$ such that $\sigma(P)$ lies in the strip $\gamma_0 \leq \text{Re}\,\lambda \leq \gamma_1$ while for $\text{Re}\,\lambda \leq a < \gamma_0$ and $\text{Re}\,\lambda \geq a_1 > \gamma_1$ the estimate of Definition 4.27 holds.

Definition 4.29 $P(D_t)$ is called quasielliptic if $\sigma(P)$ lies outside some (open or closed) strip $\gamma_o \le \text{Re}\,\lambda \le \gamma_1$ while for $\gamma_o < a_o \le \text{Re}\,\lambda \le a_1 < \gamma_1$ $\|A_j P^{-1}(\lambda)\| \le c(a_o,a_1)(1 + |\lambda|)^{k-j}$ $(0 \le j \le s)$ with norm $Y \to X$. $P(D_t)$ is called quasi-hyperbolic if $\sigma(P)$ lies outside of a finite number of (open or closed) strips $\gamma_{2p} \le \text{Re}\,\lambda \le \gamma_{2p+1}$ $(p = 0,\ldots,N;\ N \ge 1)$ while in each of these "resolvant" strips the estimate above holds.

There are a number of theorems for these types of operators the proofs of which are all short and straightforward. First

Theorem 4.30 Let $h(t) \in H(q,k+1, I, \gamma)(Y)$ for $\gamma > \gamma_o$ and $P(D_t)$ be parabolic; then the equation $P(D_t)u = h$ with $D_t^k u(0) = 0$ for $0 \le k \le s-1$ has a solution $u(t) \in H_q(s,P,\gamma)(X)$ given by the formula

$$(4.65)\qquad u(t) = \frac{1}{2\pi i}\int_{a-i\infty}^{a+i\infty} \lambda^{-k} e^{\lambda t} * \theta(t) P^{-1}(\lambda) h^{(k)}(t)\,d\lambda$$

where $\gamma_o < a < \gamma$ and θ is the Heavyside function (* denotes convolution in t and one may assume $\gamma_o \ge 0$ without loss of generality).

Proof: Let $E = L_{q,\gamma}(\mathbb{R}^+,X)$ with $Bu = D_t u$ and $D(B) = \{u;\ u(t) \in H(q,1,I,\gamma)(X):\ u(0) = 0\}$. Then Theorem 4.22 can be applied ($E_1 = E$ and $E_2 = L_{q,\gamma}(\mathbb{R}^+,Y)$).

Similarly for the backward parabolic $P(D_t)$ let $h(t) \in H(q,k+1,I,\gamma)(Y)$ and $\gamma < -\gamma_o$; then for $\gamma < a < -\gamma_o$ a solution $u(t) \in H_q(s,P,\gamma)(X)$ of $P(D_t)u = h$ is given by

$$(4.66)\qquad u(t) = -\frac{1}{2\pi i}\int_{a-i\infty}^{a+i\infty} \lambda^{-k} P^{-1}(\lambda)\int_t^{\infty} e^{\lambda(t-\tau)} D_\tau^k h(\tau)\,d\tau\,d\lambda.$$

Theorem 4.31 Let $P(D_t)$ be hyperbolic and take $\gamma_o \le 0 \le \gamma_1$ with no loss of generality. Let $h(t) \in H(q,k+1,I,\gamma)(Y)$ with $\gamma > \gamma_1$. Then a solution of $P(D_t)u = h$ with $D_t^k u(0) = 0$ for $0 \le k \le s-1$ is given by (4.65). If $\gamma < \gamma_o$ then a solution of $P(D_t)u = h$ (same h) with no initial conditions is given by (4.66).

Results for quasielliptic and quasihyperbolic $P(D_t)$ are also given in Dubinskij

[2] where correct problems are indicated. For the study of operator "pencils" such as $L_n(\lambda)$ in Lemma 4.19 or $P(\lambda)$ in the Dubinskij work and some related topics we mention for example Abdukadyrov [1], Allakhverdiev-Gasanov [1], Burk [2; 3], Gasymov [1; 2; 3], Gasymov-Maksudov [4], Gohberg-Krein [1; 2], Gorbačuk [1], Gorbačuk-Gorbačuk [2], Gorbačuk-Vainerman [4], Markus-Mereutsa [1], Virozub-Matsaev [1], Voropaeva-Maslov [1], Yakubov-Balaev [8].

Remark 4.32 The spectral theory of Carroll [18; 19; 20; 21; 22] has already been sketched in book form in Carroll [4] so we will not develop it in detail here. However a few comments are appropriate here for completeness and some motivation was provided in the introduction. Recall that in $F = L^2(\mathbb{R}^n)$ one can write $iD_k = (-\Delta)^{\frac{1}{2}}R_k$ where the Riesz operators R_k are continuous so that in particular $\Sigma a_{j\alpha}(t)D_x^\alpha = \Sigma\tilde{a}_{jk\alpha}(t)R^\alpha A^k$ where $A = 1 + (-\Delta)^{\frac{1}{2}}$ is positive and self adjoint while the R_k and A all commute. For simplicity we treat the corresponding abstract problem with $a_{j\alpha}$ independent of t. Thus let H be a separable Hilbert space and Λ_k $(1 \leq k \leq N)$ a family of positive self adjoint operators with $a_k = \Lambda_k^{-1} \in L(H)$. Let $b_j \in L(H)$ $(1 \leq j \leq M)$ be operators commuting among themselves and with the a_k and Λ_k. Let A be the commutative Banach algebra generated by the a_k, b_i, and the identity e. We will use a few facts from Banach algebra theory here and refer to Carroll [4] for the necessary details. Let $\sigma_A \subset \mathbb{C}^{N+M}$ be the joint spectrum of the generators a_k and b_i: thus σ_A is the image of the carrier space Φ_A under the map $\gamma = \Phi_A \to \mathbb{C}^{N+M}$ defined by $\gamma(\phi) = (\hat{a}_1(\phi),\ldots,\hat{a}_N(\phi), \hat{b}_1(\phi),\ldots,\hat{b}_M(\phi))$ where $\phi(x) = \hat{x}(\phi)$; then σ_A is compact and homeomorphic to Φ_A. The spectral variables z_k, $1 \leq k \leq N$ (resp. z_i, $N+1 \leq i \leq N+M$) correspond to a_k (resp. b_i) and $\sigma_A(b)$ denotes the joint spectrum of the b_i in $\mathbb{C}^M$ ($\sigma_A(b) = pr_M(\sigma_A)$ where pr_M denotes the projection of $\mathbb{C}^{N+M}$ on $\mathbb{C}^M$). Clearly $z_k = \hat{a}_k(\phi) \geq 0$ in σ_A and we set $\lambda_k = z_k^{-1}$. Associated to the equation

$$(4.67)\qquad D_t^m u + \sum_{j=0}^{m-1} P_j(\Lambda_k,b_i)D_t^j u = 0$$

is the characteristic polynomial $(z' = (z_{N+1},\ldots,z_{N+M}))$

(4.68) $$Q(\zeta,\lambda,z') = \zeta^m + \sum_{j=0}^{m-1} P_j(\lambda_k,z_i)\zeta^j$$

(cf. (1.2.10) and the associated framework). Here one thinks of (4.67) as a first order system as in Section 1.2 of the form $\vec{u}' + \mathbb{P}(\Lambda,b)u = 0$ where $\mathbb{P}$ is a matrix like (1.2.10) with bottom row of the form $(P_o(\Lambda,b),\ldots,P_{m-1}(\Lambda,b))$ and 1 replaced by -1. Now one can develop the theory here in terms of σ_A but more generally let us take a compact set $\Xi \subset \mathbb{C}^{N+M}$ on which z_k is real and $z_k \geq 0$; usually $\sigma_A \subset \Xi$ but this is not assumed. Let $\zeta_s(\lambda,z')$, $s = 1,\ldots,m$ be the roots of $Q(\zeta,\lambda,z')$ and write $\psi(\lambda,z') = \max \operatorname{Re} \zeta_s(\lambda,z')$.

Definition 4.33 The equation (4.67) will be called Ξ hyperbolic if for $z' \in \mathrm{pr}_M\Xi$, $\psi(\lambda,z') \leq a|\lambda| + b$, while $\psi(\lambda,z') \leq c$ for $\lambda_k \geq c_k$ real. The equation is Ξ parabolic if for $z' \in \mathrm{pr}_M\,\Xi$, $\psi(\lambda,z') \leq -a|\lambda|^h + b$ for real $\lambda_k \geq c_k$; it is Ξ correct if for $z' \in \mathrm{pr}_M\Xi$, $\psi(\lambda,z') \leq c$ for real $\lambda_k \geq c_k$.

Now the spectral form of the Green's operator for the system $\vec{u}' + \mathbb{P}(\Lambda,b)\vec{u} = 0$ is $\mathbb{G}(t,\tau,\lambda,z') = \exp[-(t-\tau)\mathbb{P}(\lambda,z')]$ and one wants to show that $\mathbb{G}$ is in the domain of a suitable functional calculus relative to Ξ with suitable (t,τ) dependence. Then a matrix $G(t,\tau)$ with entries in $L(H)$ will arise yielding a solution of (4.67) in the form $\vec{u}(t) = G(t,\tau)\vec{u}(\tau)$. This involves considerable analysis which we omit here (cf Carroll [4; 18; 19; 20; 21; 22]). A typical theorem is cited for completeness however. Thus we will say that Ξ is p convex if for any $\zeta \notin \Xi$ there is a polynomial p such that $p(\zeta) = 1$ and $|p(z)| < 1$ for $z \in \Xi$. We say Ξ is p spectral for A if for any polynomial p one has $\|p(a,b)\| \leq c \sup|p(z,z')|$ over Ξ. Then

Theorem 4.34 Let Ξ be a p convex p spectral set for A with (4.67) a Ξ parabolic equation. Then there exists a weak Green's operator $G(t,\tau)$ with $(t,\tau) \to G(t,\tau) \in C^o(L_w(H^m))$, $G(t,s)G(s,\tau) = G(t,\tau)$, $G(\tau,\tau) = I$, and $t \to G(t,\tau) \in C^1(L_s(H^m))$, while $G_t + \mathbb{P}(\Lambda,b)G = 0$.

Remark 4.35 Taking $P(\lambda) = A + \lambda$ in (4.62) one obtains the solution of $(A + B)u = h$ in the form

(4.69) $$u = -\frac{1}{2\pi i}\int_\Gamma (A+\lambda)^{-1}(B-\lambda)^{-1}h\,d\lambda$$

for $k = 0$ (with suitable traverse of a suitable Γ). This is the formula of Grisvard [2] and has been systematically exploited in DaPrato-Grisvard [3]. They distinugish two main types of problems as follows.

Definition 4.36 The operator $L = A + B$ is hyperbolic if $(A - \lambda)$ and $(B - \lambda)$ are invertible for $\lambda > 0$ with

(4.70) $$\|(A-\lambda)^{-m}\| \leq \frac{M}{\lambda^m}; \quad \|(B-\lambda)^{-m}\| \leq \frac{M}{\lambda^m}$$

for $\lambda > 0$, $m = 1,2,\ldots$, with M independent of λ and m. The operator is parabolic if $(A - \lambda)$ and $(B - \lambda)$ are invertible in sectors Σ_A and Σ_B where $\Sigma_A = \{\lambda;\ |\arg\lambda| \leq \pi - \theta_A\}$ and $\Sigma_B = \{\lambda:\ |\arg\lambda| \leq \pi - \theta_B\}$ with $\theta_A + \theta_B < \pi$, while in these respective sectors

(4.71) $$\|(A-\lambda)^{-1}\| = O(|\lambda|^{-1}); \quad \|(B-\lambda)^{-1}\| = O(|\lambda|^{-1}).$$

Thus for L hyperbolic $(A - \lambda)$ and $(B - \lambda)$ are invertible in sectors as above with $\theta_A, \theta_B < \pi/2$ and $\|(A-\lambda)^{-1}\| \leq M/\mathrm{Re}\,\lambda$ with $\|(B-\lambda)^{-1}\| \leq M/\mathrm{Re}\,\lambda$. Thus the hyperbolic case is a limit of parabolic cases where $\theta_A + \theta_B = \pi$ in the limit. Time dependent evolution equations of the form $-u' + \Lambda(t)u(t) = f(t)$ and $u'' + \Lambda(t)u(t) - \lambda u = f(t)$ are studied also in the context of noncommuting A and B and results analogous to those of Kato [2; 6; 7], Kato-Tanabe [1], and Tanabe [1] are obtained. For noncommutative A and B a formula similar to (4.69) can also be utilized under suitable hypotheses. One can work in either case most effectively with a solution of $Lu - \zeta u = h$ $(L = A + B)$ expressed by (4.69) with B replaced by $B - \zeta$ (for technical reasons). Thus set for $\zeta > 0$

(4.72) $$S_\zeta = -\frac{1}{2\pi i}\int_\Gamma (B-\lambda-\zeta)^{-1}(A+\lambda)^{-1}d\lambda.$$

In the commutative case $S_\zeta = (\bar{L} - \zeta)^{-1}$ whereas in the noncommutative case if one assumes for example that $(B - \lambda)^{-1} D(A) \subset D(A)$ for $\lambda \in \rho(B)$ with

(4.73) $$\|[A,(B-\lambda)^{-1}](A-z)^{-1}\| \leq c(\theta,\theta')\phi(|\lambda|,|z|)$$

for suitable ϕ ($\theta = \arg\lambda$, $\theta' = \arg z$) then $(\bar{L} - \zeta)S_\zeta = 1 + R_\zeta$ where R_ζ will be small for ζ large and $(\bar{L} - \zeta)^{-1} = S_\zeta(1 + R_\zeta)^{-1}$. Hypotheses of stability as in Kato [6; 7] are also utilized.

CHAPTER 2

TRANSMUTATION

2.1. Preliminary remarks and examples. The idea of transmutation of differential operators will be useful in dealing with related abstract differential equations and we will discuss this from various points of view following at first Delsarte [1], Delsarte-Lions [2; 3], Levitan [1; 3; 5], and Lions [2; 3; 4] (cf. also Androščuk [1], Braaksma [1], Braaksma-deSnoo [2], Carroll [35], Carroll-Showalter [8], Chebli [1], Gasymov-Magerramov [5], Helgason [1], Hersh [3], Hutson-Pym [1; 2], LeBlanc [1], Löfström-Peetre [1], Marčenko [1; 2; 3], Povzner [1], Thyssen [1; 2]). Let $D = \partial/\partial x$ or $\partial/\partial y$ and consider (linear) differential polynomials $P(D)$ and $Q(D)$ acting in suitable spaces X and Y of functions or distributions. P and Q need not have constant coefficients nor be of the same order. We say that an operator B, usually an integral operator, transmutes $P(D)$ into $Q(D)$ if formally $Q(D)B = BP(D)$. Then for example if $P(D)u = 0$ we have $Q(D)Bu = 0$ so $v = Bu$ satisfies $Q(D)v = 0$. If B is an isomorphism from X onto Y then from $Q(D)v = 0$ we get $P(D)B^{-1}v = B^{-1}Q(D)v = 0$ so $u = B^{-1}v$ satisfies $P(D)u = 0$. More generally if u is a vector function with $u(x) \in F$ and $P(D)u(x) = g(A)u(x)$ for a suitable operator function $g(A)$ acting in F then setting $v(y) = (Bu)(y)$ we have

$$Q(D)v = BP(D)u = Bg(A)u = g(A)Bu = g(A)v$$

so that in F $Q(D)v(y) = g(A)v(y)$. It is in this spirit that one can connect solutions of ODE with operator coefficients and one speaks then of related differential equations.

Let us cite a few results on transmutation first to gain some perspective. Let $X = Y$ be the space H of analytic functions of one complex variable z with the topology of uniform convergence on compact subsets of $\mathbb{C}$. Let $P(D) =$

$\sum_{j=0}^{m} a_j(z)D^j$ where $a_j(\cdot) \in H$, $D = d/dz$, and $a_m(z) = 1$. Then Delsarte-Lions [2] prove

Theorem 1.1 There exists a continuous isomorphism $B : H \to H$ such that $D^m B = BP(D)$. Explicitly B has the form

$$(1.1) \qquad (Bf)(z) = \sum_{k=0}^{\infty} \sum_{p=0}^{m-1} D^p P(D)^k f(0) \frac{z^{p+km}}{(p+km)!} .$$

As a corollary it follows immediately that if $Q(D) = \sum_{j=0}^{m} b_j(z)D^j$ with $b_j(\cdot) \in H$ and $b_m(z) = 1$ then there exists an isomorphism $\tilde{B} : H \to H$ such that $Q(D)\tilde{B} = \tilde{B}P(D)$. The formal verification of (1.1) is straightforward and one need only confirm the convergence in order to prove Theorem 1.1. Such results are however special for analytic situations and the general question of just what $P(D)$ can be transmuted into what $Q(D)$ is unsolved; it cannot always be done (cf. Levitan [1]).

Remark 1.2 Delsarte [1] was also concerned with generalizing the idea of translation and this is developed further in Levitan [1; 3]. Thus let L be a second order differential operator for example of the form $Lf = f'' - \rho f$. Let $L\phi = \lambda\phi$ with $\phi(0,\lambda) = 1$ and $\phi'(0,\lambda) = 0$. Write formally $\phi = \sum_0^{\infty} \lambda^k \phi_k$ with $L\phi_k = \phi_{k-1}$ and $L\phi_o = 0$ so that

$$L\phi = \sum_0^{\infty} \lambda^k L\phi_k = \sum_1^{\infty} \lambda^k \phi_{k-1} = \sum_0^{\infty} \lambda^{n+1}\phi_n = \lambda\phi.$$

Then the generalized translation operator associated with L is defined by

$$(1.2) \qquad T_x^y f(x) = \sum_{k=0}^{\infty} \phi_k(y) L^k f(x)$$

for $f \in C^{\infty}$. Since $\phi_k(0) = 0$ for $k \geq 1$ with $\phi_o(0) = 1$ we have $T_x^0 f(x) = f(x)$ while $\phi_k'(0) = 0$ for $k \geq 0$ implies $D_y T_x^y f(x)|_{y=0} = 0$. The analogy with translation follows upon taking $L = D$, $\phi(x,\lambda) = \exp \lambda x = \Sigma\lambda^n x^n/n!$, and $\phi_n(x) = x^n/n!$ so that $D\phi_n = \phi_{n-1}$ and $T_x^y f(x) = \sum_0^{\infty} D^k f(x) y^k/k! = f(x + y)$. We note also that $L_x T_x^y f(x) = L_y T_x^y f(x)$ since $L_x T_x^y = \Sigma_k \phi_k(y) L_x^{k+1} f(x)$ while

$$L_y T_x^y f(x) = \sum L_y \phi_k(y) L_x^k f(x) = \sum \phi_{k-1}(y) L_x^k f(x).$$

For further comments in this direction see Remark 3.18.

Now following Lions [2] we consider the Cauchy problem, for $Q(D)$ of order m

$$(1.3) \qquad P(D_x)\phi(x,y) = Q(D_y)\phi(x,y)$$

$$(1.4) \qquad \phi(x,0) = f(x); \quad D_y^i\phi(x,0) = 0, \quad 1 \le i \le m-1$$

for suitable f. Assume that (1.3) - (1.4) has a unique solution in some appropriate space and define B by the rule $(Bf)(y) = \phi(0,y)$. Setting $\psi(x,y) = P(D_x)\phi(x,y)$ we have

$$(1.5) \qquad [P(D_x) - Q(D_y)]\psi = P(D_x)[P(D_x) - Q(D_y)]\phi = 0$$

$$(1.6) \qquad D_y^i\psi(x,0) = P(D_x)D_y^i\phi(x,0) = \begin{cases} P(D_x)f & 0 \\ 0 & 1 \le i \le m-1 \end{cases}$$

Consequently, given B defined as above and the uniqueness of solutions of (1.3) - (1.4) and (1.5) - (1.6), it follows that

$$(1.7) \qquad \psi(0,y) = (BP(D_x)f)(y) = P(D_x)\phi(x,y)|_{x=0}$$

$$= Q(D_y)\phi(0,y) = (Q(D_y)Bf)(y)$$

so that B transmutes $P(D)$ into $Q(D)$. We will utilize precise versions of this later and for now simply state (cf also Remark 2.18)

Theorem 1.3(F). Assume (1.3) - (1.4) and (1.5) - (1.6) have unique solutions and define B as indicated. Then $Q(D)B = BP(D)$.

Example 1.4 Let us apply this technique to (I.1.3) - (I.1.4) to produce (I.1.6) as in Example 1.1.10. Thus let $P(D) = D^2$ and $Q(D) = D$ with $g(A) = A^2$. To find B we look at (cf. (1.3) - (1.4))

$$(1.8) \qquad \phi_{xx} = \phi_y; \quad \phi(x,0) = f(x)$$

where f is a function in say $\mathcal{S}'$ which we take (or extend) to be even (cf. Remark 2.18). As with (1.1.12) one can take Fourier transforms in x to obtain

$\hat{\phi}_t + s^2\hat{\phi} = 0$, $\hat{\phi}(s,0) = \hat{f}(s)$, from which $\hat{\phi}(s,y) = \hat{f}(s)\exp(-s^2y)$ where $\hat{\phi} = F_x\phi$ amd recalling that $F\frac{1}{2}K(y,x) = \exp(-s^2y)$ we have

$$(1.9) \qquad \phi(x,y) = \frac{1}{2}\int_{-\infty}^{\infty} K(y,\xi)f(x-\xi)d\xi$$

$$1.10) \qquad \phi(0,y) = (Bf)(y) = \frac{1}{2}\int_{-\infty}^{\infty} K(y,\xi)f(-\xi)d\xi = \int_0^{\infty} K(y,\xi)f(\xi)d\xi$$

a formula which makes sense even if $f(\xi)$ is of say exponential growth as $\xi \to \infty$. Now apply this to the vector valued functions u and w of (I.1.3) - (I.1.4). Thus $P(D)w = A^2w$ and

$$Q(D)Bw = BP(D)w = BA^2w = A^2Bw;$$

the use of even functions f in (1.8) - (1.10) corresponds to the condition $w_t(0) = 0$. Consequently setting $u = Bw$ we obtain (I.1.6) from the form of (1.10) (cf. also (1.1.13)). One should emphasize that $w(\xi)$ must make sense as $\xi \to \infty$ and be integrable against $K(y,\xi)$ so we must take $T(\xi)$ to be a quasi-equicontinuous group for example (local equicontinuity is not enough).

Remark 1.5 With Example 1.4 in mind let us look at some other aspects of the transmutation question. Suppose we assume $DB = BD^2$ where B is an integral operator of the form

$$(Bf)(y) = \langle b(y,x), f(x)\rangle = \int_I b(y,x)f(x)dx$$

where I is $(-\infty,\infty)$ or $(0,\infty)$. For $I = (0,\infty)$ we obtain formally, assuming suitable behavior at ∞,

$$(1.11) \qquad (BD^2f)(y) = \int_0^{\infty} b(y,x)f''(x)dx =$$

$$\int_0^{\infty} b_{xx}(y,x)f(x)dx + b_x(y,0)f(0) - b(y,0)f'(0)$$

whereas $(DBf)(y) = \int_0^{\infty} b_y(y,x)f(x)dx$. Hence if $b_y = b_{xx}$ with $b_x(y,0) = 0$ and $f'(0) = 0$ we will have $DB = BD^2$ acting on such f. Clearly $b(y,x) = K(y,x) = (\pi y)^{-\frac{1}{2}}\exp(-x^2/4y)$ satisfies these requirements. For $I = (-\infty,\infty)$ the situation is similar except the condition $f'(0) = 0$ does not arise; the two intervals I are equivalent if we consider even functions f on $(-\infty,\infty)$ so that $f'(0) = 0$. On the other hand suppose w satisfies (I.1.4), namely $w_{xx} = A^2w$ with $w_x(0) = 0$, and we want to find u satisfying (I.1.3) connected to w by a formula $u(y) = \int_0^\infty b(y,x)w(x)dx$. Then $u_y = A^2u$ if $b_y = b_{xx}$ with $b_x(y,0) = 0$ since from (1.11)

$$A^2u(y) = \int_0^\infty b(y,x)A^2w(x)dx = \int_0^\infty b(y,x)w_{xx}(x)dx \tag{1.12}$$

$$= \int_0^\infty b_{xx}(y,x)w(x)dx = \int_0^\infty b_y(y,x)w(x)dx = u_y(y).$$

The initial condition $u(0) = w(0)$ is achieved when $b(0,x) = \delta(x)$ (one sided) as is the case with $b(y,x) = K(y,x)$.

Example 1.6 Let us consider (I.1.5) when A generates an equicontinuous group in F and follow the procedure of Section 1.1. Thus for $-\infty < x < \infty$ and $t \geq 0$ we are led intuitively to the half plane Dirichlet problem

$$G_{tt} + G_{xx} = 0; \quad G(0,x) = \delta(0) \tag{1.13}$$

rather than an improperly posed Cauchy problem. It is easily checked that $G(t,x) = \frac{1}{2}H(t,x) = t\pi^{-1}(t^2 + x^2)^{-1}$ is a solution of (1.13) and we write

$$z(t) = \langle \tfrac{1}{2}H(t,\cdot),\ T(\cdot)u_o\rangle = \frac{1}{2}\int_{-\infty}^{\infty} H(t,x)T(x)u_o dx \tag{1.14}$$

$$= \int_0^\infty H(t,x)\cosh Axu_o dx = \int_0^\infty H(t,x)w(x)dx$$

to obtain (I.1.8) formally. Since $p(T(x)u_o) \leq q(u_o)$ for seminorms p and q it follows that the bracket, or integral, in (1.14) makes sense for $t > 0$ (since $|H(t,x)| \leq \frac{2t}{\pi}(x^2 + t^2)^{-1}$). Similarly H_t, H_x, and $H_{tt} = -H_{xx}$ will be dominated by $c(t)/x^n$ $(n \geq 2)$ as $x \to \infty$ and one can differentiate under the integral sign in (1.14) and integrate by parts. Thus

$$z_{tt} = \frac{1}{2}\langle H_{tt}(t,x),\ T(x)u_o\rangle = -\frac{1}{2}\langle H_{xx}(t,x),\ T(x)u_o\rangle$$

$$= -\frac{1}{2}\langle H(t,x),\ D_x^2T(x)u_o\rangle = -A^2z$$

(when working on $(0,\infty)$ note that $H_x(t,0) = 0$ and $w_x(0) = 0$). Thus we can state (leaving open the question of uniqueness)

Theorem 1.7 Let A generate an equicontinuous group in F with $u_o \in D(A^2)$ while w is the unique solution of (I.1.4) from Theorem 1.1.8. Then the function z given by (1.14) (or equivalently (I.1.8)) is a solution of (I.1.5).

Remark 1.8 Let us look at Example 1.6 from the point of view of transmutation. Following Formal Theorem 1.3 we consider for $P(D) = D^2$ and $Q(D) = -D^2$

(1.15) $$\phi_{xx} + \phi_{yy} = 0; \quad \phi(x,0) = f(x).$$

Imposing a condition $\phi_y(x,0) = 0$ would lead to an improperly posed Cauchy problem so we want to find some other way to produce unique solutions of (1.15). Extend f to be an even function on $(-\infty,\infty)$ (cf. Remark 2.18) and Fourier transform (1.15) to obtain $\hat{\phi}_{yy} - s^2\hat{\phi} = 0$ with $\hat{\phi}(s,0) = \hat{f}(s)$. If one requires $\phi(x,y) \to 0$ as $y \to \infty$ in such a way that $\hat{\phi}(s,y) \to 0$ as well then a unique $\hat{\phi}$ is given by $\hat{\phi}(s,y) = \hat{f}(s)\exp(-sy)$. But $F\frac{1}{2}H(y,x) = \exp(-sy)$ where $\frac{1}{2}H(y,x) = y\pi^{-1}(x^2 + y^2)^{-1}$ and hence

(1.16) $$\phi(x,y) = \frac{1}{2}\int_{-\infty}^{\infty} H(y,\xi)f(x-\xi)d\xi$$

(1.17) $$\phi(0,y) = (Bf)(y) = \frac{1}{2}\int_{-\infty}^{\infty} H(y,\xi)f(-\xi)d\xi = \int_0^{\infty} H(y,\xi)f(\xi)d\xi.$$

Applying this to vector functions as before we obtain $z = Bw$ as in (1.14).

Remark 1.9 The specification of Cauchy data in (1.4) is one way of assuring a unique solution ϕ of (1.3) - (1.4) so as to determine as well defined operator B. Obviously any specification of data at $y = 0$ plus growth conditions which uniquely determine ϕ will serve to define a transmutation operator (cf Remark 2.18).

Example 1.10 Let us examine now the formula (I.1.7) connecting solutions u and z of (I.1.3) and (I.1.5). Setting $P(D) = D$ and $Q(D) = -D^2$ is natural and following Theorem 1.3(F) to construct a transmutation operator B one considers

(1.18) $\phi_x = -\phi_{yy}$; $\phi(x,0) = f(x)$

with some further stipulation necessary in order to insure a unique solution. We want to produce a solution involving as a kernel $I(y,x) = \frac{1}{2}\pi^{-\frac{1}{2}}yx^{-3/2}\exp(-y^2/4x)$ which itself satisfies $I_x = I_{yy}$ (clarified below). Now Fourier transforming (1.1.8) in x we have $\hat{\phi}_{yy} - is\hat{\phi} = 0$ with $\hat{\phi}(s,0) = \hat{f}(s)$ the solution of which has the form $\hat{\phi}(s,y) = A(s)\exp(is)^{\frac{1}{2}}y + B(s)\exp(-(is)^{\frac{1}{2}}y)$. Considering s real $(is)^{\frac{1}{2}} = \frac{1}{\sqrt{2}}(1+i)s^{\frac{1}{2}}$ and if we stipulate say that $\hat{\phi}(s,y) \to 0$ as $y \to \infty$ then $A(s) = 0$ is indicated. Hence we take

$$\hat{\phi}(s,y) = \hat{f}(s)\exp(-(is)^{\frac{1}{2}}y)$$

$$= \hat{f}(s)\exp(-s^{\frac{1}{2}}y/\sqrt{2})[\mathrm{Cos}\, s^{\frac{1}{2}}y/\sqrt{2} - i\,\mathrm{Sin}\, s^{\frac{1}{2}}y/\sqrt{2}].$$

But it is known that $\exp(-s^{\frac{1}{2}}y/\sqrt{2})\mathrm{Cos}\, s^{\frac{1}{2}}y/\sqrt{2} = F_c I(y,x)$, $\exp(-s^{\frac{1}{2}}y/\sqrt{2})\mathrm{Sin}\, s^{\frac{1}{2}}y/\sqrt{2} = F_s I(y,x)$ where F_c and F_s denote the Fourier cosine and sine transforms respectively (cf. Erdelyi [1]). Thus $\exp(-(is)^{\frac{1}{2}}y) = F_c I - iF_s I$ and if we take I^- to be the distribution $I(y,|x|)$ for $x \leq 0$ and 0 for $x > 0$ then $F_c I - iF_s I = \int_0^\infty I(y,x)\exp(-isx)dx = \int_{-\infty}^0 I(y,|x|)\exp isxdx = FI^-$. Consequently

(1.19) $$\phi(x,y) = \int_{-\infty}^0 I(y,|\xi|)f(x-\xi)d\xi$$

(1.20) $$\phi(0,y) = (Bf)(y) = \int_{-\infty}^0 I(y,|\xi|)f(-\xi)d\xi$$

$$= \int_0^\infty I(y,\xi)f(\xi)d\xi.$$

Applying this to vector functions we obtain $z = Bu$ which is (I.1.7).

Remark 1.11 It is instructive to construct B in Example 1.10 by assuming it is an integral operator with kernel $b(y,x)$. We can in fact go directly to u and z and assume $z(y) = \langle b(y,x), u(x)\rangle = \int_0^\infty b(y,x)u(x)dx$. To see what properties $b(y,x)$ should have we assume $u_x = A^2u$ and then if $b(y,x)$ has suitable growth as $x \to \infty$ while $b(y,0) = 0$ it follows that

$$A^2z(y) = \int_0^\infty b(y,x)A^2u(x)dx = \int_0^\infty b(y,x)u_x(x)dx = -\int_0^\infty b_x(y,x)u(x)dx.$$

Thus

$$(1.21)\qquad (D_y^2 + A^2)z(y) = \int_0^\infty (b_{yy}(y,x) - b_x(y,x))u(x)dx$$

and we are led to look for a distribution or function $b(y,x)$ such that

$$(1.22)\qquad b_{yy} - b_x = 0;\quad b(0,x) = \delta(x);\quad b(y,0) = 0$$

We know a solution of (1.22) is $b(y,x) = I(y,x)$ but to derive this we could Laplace transform (1.22) $(x \to s)$ to obtain

$$(1.23)\qquad \tilde{b}_{yy} = s\tilde{b};\quad \tilde{b}(0,s) = 1$$

where $\tilde{b}(y,s) = Lb(y,x) = \int_0^\infty b(y,x)\exp(-sx)dx$ (see Doetsch [1], Schwartz [1], Sneddon [1], Widder [1], Zemanian [1] for Laplace transforms). The solution of (1.23) is $\tilde{b}(y,s) = \exp(-ys^{\frac{1}{2}})$ if we require say $\tilde{b}(y,s) \to 0$ as $y \to \infty$ for s real (the function $\exp(ys^{\frac{1}{2}})$ can be excluded also by general growth requirements for Laplace transforms - cf. Schwartz [1]). But $LI(y,x) = \exp(-ys^{\frac{1}{2}})$ (cf. Erdelyi [1]) which yields $b(y,x) = I(y,x)$.

For completeness we will formulate a theorem relative to (I.1.7) as follows.

Theorem 1.12 Let A generate an equicontinuous group in F with $u_o \in D(A^2)$ and $u(t)$ be the unique solution of (I.1.3) given by Theorem 1.1.11. Then z given by (I.1.7) is a solution of (I.1.5).

Proof: One needs only to check that the various integrations, etc., make sense. We note first that with $c = 1/2\pi^{\frac{1}{2}}$

$$(1.24)\qquad I_y = cx^{-3/2}[1 - \frac{y^2}{2x}]\exp(-y^2/4x)$$

$$(1.25)\qquad I_x = \frac{c}{2}\, yx^{-5/2}\,[\frac{y^2}{2x} - 3]\exp(-y^2/4x)$$

while $I_{yy} = I_x$. If p is a seminorm in F then $p(w(t)) \le q(u_o)$ for some seminorm q and

$$(1.26)\qquad p(u(x)) \le (\pi x)^{-\frac{1}{2}}q(u_o)\int_0^\infty \exp(-t^2/4x)dt \le h(x)q(u_o)$$

where $h(x) = (\pi x)^{-\frac{1}{2}}\int_0^\infty \exp(-t^2/4x)dt$ by (I.1.6) (cf. Theorem 1.1.11). Similarly, since $u_x = A^2 u$, $p(u_x(x)) \le h(x)q(A^2 u_o)$. Consequently, from (I.1.7) for $t > 0$

$$(1.27) \qquad p(z(t)) \le \frac{t}{2\pi} q(u_o)\int_0^\infty \int_0^\infty \xi^{-2}\exp(-\frac{t^2+\tau^2}{4\xi})d\xi d\tau$$

$$= \frac{2t}{\pi} q(u_o)\int_0^\infty \frac{d\tau}{t^2+\tau^2} = q(u_o)$$

(cf. Erdelyi [1]) so that (I.1.7) makes sense. Similarly $\int_0^\infty I(y,x)u_x(x)dx$ is well defined and by (1.25) $\int_0^\infty I_x(y,x)u(x)dx$ can be checked by noting that for $t > 0$ and $n \ge 2$

$$(1.28) \qquad \int_0^\infty \int_0^\infty \xi^{-n}\exp(-\frac{t^2+\tau^2}{4\xi})d\xi d\tau = 4^{n-1}\Gamma(n-1)\int_0^\infty \frac{d\tau}{(t^2+\tau^2)^{n-1}} < \infty.$$

Similarly z_t and z_{tt} can be obtained by differentiating under the integral sign in (I.1.7). QED

Remark 1.13 The three formulas (I.1.6) - (I.1.8), with (I.1.9) - (I.1.11), are examined together in a concrete analytical setting by Dettman [3]. They are also discussed separately in various contexts in Bragg-Dettman [6; 7; 8] and Dettman [1]. Classical Laplace transform and complex variable methods are used. The matter is taken up again in Dettman [2] (cf. also Dettman [4]) in general spaces using the Hilbert transform and we will sketch some of this. For the classical Hilbert transform see Butzer-Trebels [1], Cotlar [1], Horváth [2], Okikiolu [1], Orton [1; 2], Stein [1], and Stein-Weiss [2]. The idea is to replace the classical Hilbert transform

$$(1.29) \qquad Hf = \lim_{y\to 0} \frac{1}{\pi}\int_{-\infty}^\infty \frac{tf(x-t)dt}{t^2+y^2}$$

by a formula (cf also Westphal [1])

$$(1.30) \qquad \tilde{H}f = \lim_{y\to 0} \frac{1}{\pi}\int_{-\infty}^\infty \frac{tT(t)fdt}{t^2+y^2}$$

where $T(t)$ is an equicontinuous group of operators in F generated by A. F is taken to be a complete, barreled, separated LCS over $\mathbb{R}$. Then $\tilde{H}^2 f = -f$ will be proved as in the classical case where $H^2 f = -f$ (suitable hypotheses on f

are indicated below). One wants here enough hypotheses on A so that the problems (I.1.3) - (I.1.5) are uniformly well posed in the sense that for u_o in a dense subspace D there exists a unique strong solution for $t \geq 0$, assuming the initial values in a strong sense, and such that if $u_o^\alpha \to 0$ in D the corresponding solutions converge uniformly to zero on compact subsets of $\mathbb{R}^+$. Also the problem

$$(1.31) \qquad v_{tt} - A^2 v = 0; \quad v(0) = 0; \quad v_t(0) = v_o$$

should be uniformly well posed for this theory. In particular if 0 is in the resolvant set of A and A generates an equicontinuous group $T(t)$ in F then setting $S(t) = \cosh At = \frac{1}{2}[T(t) + T(-t)]$ it follows that $U(t) = \int_0^t S(\eta)d\eta$ is also equicontinuous (note $v(t) = U(t)v_o$ for (1.31)). Under these assumptions all of the problems are uniformly well posed (cf Balakrishnan [2], Bragg-Dettman [6; 7], Dettman [1; 2], Fattorini [1], Goldstein [6] and our previous discussion of these problems). Now define

$$(1.32) \qquad \Phi(y)f = \frac{y}{\pi}\int_{-\infty}^{\infty} \frac{T(t)f}{t^2 + y^2}\, dt; \quad \Psi(y)f = \frac{1}{\pi}\int_{-\infty}^{\infty} \frac{tT(t)f}{t^2 + y^2}\, dt.$$

Then $\Phi(y)f$ is defined for $f \in F$ and $\Psi(y)f$ exists as a Cauchy principal value in the sense that

$$(1.33) \qquad \Psi(y)f = \lim_{N\to\infty} \frac{1}{\pi}\int_{-N}^{N} \frac{tT(t)f}{t^2 + y^2}\, dt$$

$$= \lim_{N\to\infty} \frac{1}{\pi}\int_{0}^{N} \frac{t[T(t) - T(-t)]f}{t^2 + y^2}\, dt$$

$$= \frac{1}{\pi}\int_0^\infty \frac{(t^2 - y^2)}{(t^2 + y^2)^2}\, \alpha(t)f dt$$

where

$$\alpha(t) = \int_0^t [T(s) - T(-s)]ds = 2\int_0^t \sinh Asds$$

is equicontinuous. Note here that $\alpha(t) = 2[T(t/2) - T(-t/2)]U(t/2)$ where $U(t) = \int_0^t S(\eta)d\eta$ and $R(A)$ is dense. On the other hand

(1.34) $$\int_0^N \frac{t[T(t) - T(-t)]f}{t^2 + y^2}\, dt = \frac{t\alpha(t)f}{t^2+y^2}\Big|_0^N + \int_0^N \frac{(t^2 - y^2)}{(t^2+y^2)^2}\alpha(t)f dt.$$

The first term on the right side in (1.34) tends to 0 as $N \to \infty$ since $\alpha(t)$ is equicontinuous. By a straightforward calculation one now obtains

Theorem 1.14 [Cauchy-Riemann equations). For $f \in D(A)$ and $y > 0$ one has $(\Phi(y)f)' = A\Psi(y)f$ and $(\Psi(y)f)' = -A\Phi(y)f$.

The classical Cauchy Reimann equations are exhibited in the case $A = -d/dx$ and $F = L^p(-\infty,\infty)$ for $1 \leq p \leq \infty$. However the classical theory is not a special case since 0 is not in the resolvant set of A. Now for $f \in D(A^2)$ one has evidently $(\Phi f)'' = -A^2\Phi f$ and $(\Psi f)'' = -A^2\Psi f$; in particular $z = \Phi f$ satisfies (I.1.5) and yields the formula (I.1.8) (for $f = u_o$). The Hilbert transform

(1.35) $$\tilde{H}f = \lim_{y\to 0} \Psi(y)f = \frac{1}{\pi}\int_{-\infty}^{\infty} \frac{[T(t) - T(-t)]f}{t}\, dt$$

is easily seen to be well defined for $f \in D(A)$. Further if A is a continuous operator then $\tilde{H}f$ is defined for all $f \in F$ and $\tilde{H}$ is continuous. To provide an inversion formula for $\tilde{H}$ let $g = \tilde{H}f$ and assume $\tilde{H}g$ exists. Then for $f' \in F'$

$$\langle f', T(-x)\Psi(y)g\rangle \to \langle f', T(-x)\tilde{H}^2 f\rangle$$

as $y \to 0$. If one shows that $\langle f', T(-x)\tilde{H}^2 f\rangle = -\langle f', T(-x)f\rangle$ then $\tilde{H}^2 f = -f$ will follow. To do this consider

(1.36) $$\phi(x,y) = \frac{y}{\pi}\int_{-\infty}^{\infty} \frac{\langle f',T(-t)f\rangle}{(x-t)^2 + y^2}\, dt;$$

$$\psi(x,y) = \frac{1}{\pi}\int_{-\infty}^{\infty} \frac{(x-t)\langle f',T(-t)f\rangle}{(x-t)^2 + y^2}\, dt$$

so that $\phi(x,y) = \langle f', T(-x)\Phi(y)f\rangle$ and $\psi(x,y) = \langle f', T(-x)\Psi(y)f\rangle$. As $y \to 0$ $\phi(x,y) \to \langle f', T(-x)f\rangle$ while $\psi(x,y) \to \langle f', T(-x)\tilde{H}f\rangle$; further ϕ and ψ are bounded harmonic functions in the half plane $y > 0$. Let now

(1.37) $$\Omega(z) = \phi(x,y) + i\psi(x,y) = \frac{i}{\pi}\int_{-\infty}^{\infty} \frac{\langle f',T(-t)f\rangle}{z - t}\, dt$$

which is a bounded analytic function for $y > 0$ and has therefore a Poisson integral representation

$$(1.38)\qquad \Omega(z) = \frac{y}{\pi}\int_{-\infty}^{\infty} \frac{<f',T(-t)f> + i<f',T(-t)\tilde{H}f>}{(x-t)^2 + y^2}\,dt.$$

Now let $\phi^*(x,y) = <f',T(-x)\Phi(y)g>$ and $\psi^*(x,y) = <f',T(-x)\Psi(y)g>$ so that $\Omega^*(z) = \phi^*(x,y) + i\psi^*(x,y)$ has a Poisson integral representation like (1.38) with f replaced by $g = \tilde{H}f$. Then $\mathrm{Re}\,\Omega^*(z) = \mathrm{Im}\Omega(z)$ so that $\mathrm{Im}\Omega^*(z) = -\mathrm{Re}\,\Omega(z)$ to within a constant which must be zero since

$$(1.39)\qquad \lim_{y\to\infty} \frac{y}{\pi}\int_{-\infty}^{\infty} \frac{<f',T(-t)f>}{(x-t)^2 + y^2}\,dt = 0;$$

$$\lim_{y\to\infty} \frac{1}{\pi}\int_{-\infty}^{\infty} \frac{(x-t)<f',T(-t)f>}{(x-t)^2 + y^2}\,dt = 0.$$

Hence $\psi^*(x,y) = -\phi(x,y)$ and consequently

$$(1.40)\qquad \lim_{y\to 0} \psi^*(x,y) = <f',T(-x)\tilde{H}g> = -\lim_{y\to 0} \phi(x,y) = -<f',T(-x)f>.$$

This proves the inversion formula

Theorem 1.15 If $\tilde{H}f$ has a Hilbert transform then $-f = \tilde{H}^2 f$.

Remark 1.16 These ideas are extended in Dettman [4] to generalized Hilbert transforms related to those of Heywood [1], Kober [1], and Okikiolu [1]. He considers the problems

$$(1.41)\qquad u'' + \frac{2\nu}{t}u' = A^2 u;\quad u(0) = f;\quad u'(0) = 0$$

$$(1.42)\qquad w'' + \frac{2\mu}{t}w' = -A^2 w;\quad w(0) = f\quad (\mu < 1/2)$$

again in some complete barreled separated LCS F over $\mathbb{R}$ where A generates an equicontinuous group $T(t)$ with 0 in the resolvant set of A. We have already discussed (1.41) in Example 1.3.1 (where $2\nu = 2m + 1$ with $\mathrm{Re}\, m > -1/2$ or $m = -1/2$) and it will come up again in Example 2.6, Example 3.11, and Example 3.14. It provides a nice example of a variable coefficient operator $D^2 + \frac{2m+1}{t}D$ susceptible to various techniques of interest. The problem (1.41) is a Cauchy problem for the EPD equation whereas (1.42) is an abstract Dirichlet type problem for the equation of generalized axially symmetric potential theory (cf. Weinstein [3]). The solution to (1.41) is given by (1.3.8) when A generates a locally equicontinuous group and this formula also holds of course for A equicontinuous in which case the problem (1.41) will be uniformally well posed. The problem (I.1.3) for the abstract heat equation serves as a link between (1.41) and (1.42) (cf. Theorem 1.1.11 for the solution of (I.1.3) given by (1.1.13) which also specializes to the case of an equicontinuous group); the problem (I.1.3) will be also uniformly well posed as in Remark 1.13. Now letting v be the solution of (I.1.3) one can write (cf. Bragg [3], Bragg-Dettman [6; 7; 8], Dettman [1])

$$(1.43)\qquad v(t) = \frac{4^{-\nu}t^{-\nu-\frac{1}{2}}}{\Gamma(\nu+\frac{1}{2})}\int_0^\infty e^{-\sigma^2/4t}\sigma^{2\nu}u(\sigma)\,d\sigma;$$

$$w(t) = \frac{4^{\mu-\frac{1}{2}}t^{1-2\mu}}{\Gamma(\frac{1}{2}-\mu)}\int_0^\infty e^{-t^2/4\sigma}\sigma^{\mu-3/2}\,v(\sigma)\,d\sigma$$

and composing these formulas one obtains

$$(1.44)\qquad w(t) = \frac{2t^{1-2\mu}}{\beta(\frac{1}{2}-\mu,\frac{1}{2}+\nu)}\int_0^\infty \frac{\sigma^{2\nu}u(\sigma)\,d\sigma}{(t^2+\sigma^2)^{\nu-\mu+1}}$$

where $\beta(\cdot,\cdot)$ is the beta function. In particular when $\nu = 0$ $u(\sigma) = \cosh A\sigma f$ and (1.44) can be written

$$(1.45)\qquad w(t) = \frac{t^{1-2\mu}}{\beta(\frac{1}{2}-\mu,\frac{1}{2})}\int_{-\infty}^\infty \frac{T(\sigma)f\,d\sigma}{(t^2+\sigma^2)^{1-\mu}}$$

When $F = L^p(-\infty,\infty)$, $1 < p < \infty$, and $T(\sigma)$ is the translation group this becomes

$$(1.46)\qquad w = \phi(x,t) = \frac{t^{1-2\mu}}{\beta(\frac{1}{2}-\mu,\frac{1}{2})}\int_{-\infty}^\infty \frac{f(\xi)\,d\xi}{[t^2+(x-\xi)^2]^{1-\mu}}$$

which is the "Poisson formula" for the solution of the generalized axially symmetric potential problem in the half plane $t > 0$ in terms of boundary values $f(x)$ (cf Weinstein [3]). The function $\phi(x,t)$ is the real part of a pseudo-analytic function in the upper half plane with conjugate function

$$(1.47)\qquad \psi(x,t) = \frac{1}{\beta(\frac{1}{2}-\mu,\frac{1}{2})}\int_{-\infty}^\infty \frac{(x-\xi)f(\xi)\,d\xi}{[t^2+(x-\xi)^2]^{1-\mu}}\ .$$

For $-\frac{1}{2} < \mu < \frac{1}{2}$ one has

(1.48) $$\lim_{t\to 0} \psi(x,t) = \frac{1}{\beta(\frac{1}{2}-\mu,\frac{1}{2})} \int_{-\infty}^{\infty} \frac{\operatorname{sgn}(x-\xi) f(\xi) d\xi}{|x-\xi|^{1-2\mu}}$$

which, except for a constant, is the generalized Hilbert transform of Heywood [1], Kober [1], and Okikiolu [2]. With this background Dettman [4] sets for $-1/2 < \mu < 1/2$

(1.49) $$\Phi_\mu(t)f = \frac{t^{1-2\mu}}{\beta(\frac{1}{2}-\mu,\frac{1}{2})} \int_{-\infty}^{\infty} \frac{T(\sigma) f d\sigma}{(t^2+\sigma^2)^{1-\mu}}$$

(1.50) $$\Psi_\mu(t)f = \frac{1}{\beta(\frac{1}{2}-\mu,\frac{1}{2})} \int_{-\infty}^{\infty} \frac{\sigma T(\sigma) f d\sigma}{(t^2+\sigma^2)^{1-\mu}}$$

and defines a generalized Hilbert transform as the strong limit (when it exists)

(1.51) $$H_\mu f = \lim_{t\to 0} \Psi_\mu(t) f.$$

Using techniques similar to those of Remark 1.13 it is proved that

Theorem 1.17 If $f \in D(A)$ then for $t > 0$ $(\Psi_\mu(t)f)' = -t^{2\mu} A\Phi_\mu(t)f$ and $A\Psi_\mu(t)f = t^{2\mu}(\Phi_\mu(t)f)'$.

Theorem 1.18 If $f \in D(A)$ then $H_\mu f$ exists and

(1.52) $$H_\mu f = \frac{1}{\beta(\frac{1}{2}-\mu,\frac{1}{2})} \int_0^{\infty} \frac{[T(\sigma) - T(-\sigma)]f}{\sigma^{1-2\mu}} d\sigma.$$

Theorem 1.19 If $H_\mu f$ has a generalized Hilbert transform then $H_{-\mu}(H_\mu f) = -f$.

Remark 1.20 Paškovskij [1; 2] examines an idea of related equations as follows. Two partial differential operators L_1 and L_2 of second order are said to be connected or related if there are smooth functions $a_i(x)$ and $b_i(x)$, $x = (x_1,\dots,x_n)$, such that for any $u,v \in C^2$ one has

(1.53) $$\ell_2 v L_1 u = \ell_1 u L_2 v + \sum_1^n D_i \phi_i(x,u,v,u_x,v_x)$$

(1.54) $$\ell_1 u = \sum b_i D_i u + b_0 u; \quad \ell_2 v = \sum_1^n a_i D_i v + a_0 v.$$

Boundary value problems for L_1 and L_2 in a region G with boundary ∂G are said to be related if L_1 and L_2 are related and whenever u and v satisfy appropriate homogeneous boundary conditions $\int_{\partial G} \Sigma\phi_i(x,u,v,u_x,v_x)\cos nx_i d\sigma = 0$. Taking $n = 2$ for convenience this means $\int_{\partial G} \phi_1 dx - \phi_2 dy = 0$. Then letting L_i^* be the formal adjoint it is easy to see that L_1 and L_2 are related if and only if $L_1^*\ell_2 = \ell_1^*L_2$. Thus if $L_2 v = 0$ then $w = \ell_2 v$ satisfies $L_1^* w = 0$. A number of straightforward calculational results are proved.

2.2. FURTHER EXAMPLES. The preceeding section has given an idea of the types of results one can obtain about transmutation and related equations and we will compile some further information here in this regard. First let us tabulate what has been learned about some simple operators and enlarge the list of examples. From Example 1.4 and Remark 1.5

$$DB = BD^2; \quad (Bf)(y) = \frac{1}{2}\int_{-\infty}^{\infty} K(y,\xi)f(-\xi)d\xi; \tag{2.1}$$

$$(Bf)(y) = \int_0^{\infty} K(y,\xi)f(\xi)d\xi \qquad (f'(0) = 0).$$

From Example 1.6 and Remark 1.8

$$-D^2B = BD^2; \quad (Bf)(y) = \frac{1}{2}\int_{-\infty}^{\infty} H(y,\xi)f(-\xi)d\xi; \tag{2.2}$$

$$(Bf)(y) = \int_0^{\infty} H(y,\xi)f(\xi)d\xi \qquad (f'(0) = 0).$$

The first expression for Bf in (2.1) and (2.2) apply to any vector function α satisfying $D^2\alpha = g(A)\alpha$ and $\beta = B\alpha$ then satisfies $D\beta = g(A)\beta$ or $-D^2\beta = g(A)\beta$ with $\beta(0) = \alpha(0)$; however $\alpha(\xi)$ must be defined for $-\infty < \xi < \infty$. Thus if $\tilde{w}$ is any solution of $\tilde{w}_{tt} = A^2\tilde{w}$ and $\tilde{u} = B\tilde{w}$ or $\tilde{z} = B\tilde{w}$ respectively we get $\tilde{u}_t = A^2\tilde{u}$ and $\tilde{z}_{tt} = -A^2\tilde{z}$. Similarly taking $g(A) = -A^2$ if $\tilde{z}_{tt} = -A^2\tilde{z}$ and $\tilde{v} = B\tilde{z}$ or $\tilde{w} = B\tilde{z}$ respectively then $\tilde{v}_t = -A^2\tilde{v}$ and $\tilde{w}_{tt} = A^2\tilde{w}$ respectively. The second expressions for B in (2.1) and (2.2) are refined versions which apply to α defined for $\xi \geq 0$ satisfying $\alpha_t(0) = 0$ and we get $u = Bw$ or $z = Bw$ as before for w, u, z as in the Introduction. Finally from Example 1.10 and Remark 1.11 one has

(2.3) $$-D^2B = BD; \quad (Bf)(y) = \int_0^\infty I(y,\xi)f(\xi)d\xi.$$

The formulas from special function theory corresponding to (2.1) - (2.3) are (setting $g(A) = A^2$ with $A \sim i\lambda$)

(2.4) $$e^{-\lambda^2 t} = (\pi t)^{-\frac{1}{2}} \int_0^\infty \exp(-\tau^2/4t)\cos\lambda\tau d\tau;$$

$$e^{-\lambda t} = 2t\pi^{-1} \int_0^\infty \frac{\cos\lambda\tau}{t^2+\tau^2} d\tau;$$

$$e^{-\lambda t} = \frac{t}{2\sqrt{\pi}} \int_0^\infty \tau^{-3/2} \exp(-t^2/4\tau)\exp(-\lambda^2\tau)d\tau.$$

The first two formulas are well known Fourier cosine transforms and the last formula can be obtained from a table of Laplace transforms $(L(x^{-3/2}\exp(-\zeta/x)) = 2(s/\zeta)^{\frac{1}{4}}K_{\frac{1}{2}}[2(s\zeta)^{\frac{1}{2}}]$ and $K_{\frac{1}{2}}(z) = (\pi/2z)^{\frac{1}{2}}\exp(-z)$ - cf. Magnus - Oberhettinger-Soni [1]).

Example 2.1 The kernel I will arise again if we take $P(D) = D$ and $Q(D) = D^2$. Thus one considers $\phi_x = \phi_{yy}$ with $\phi(x,0) = f(x)$ and after Fourier transformation this becomes $\hat{\phi}_{yy} + is\hat{\phi} = 0$ with $\hat{\phi}(s,0) = \hat{f}(s)$. We take as a solution $\hat{\phi}(s,y) = \hat{f}(s)\exp(-(-is)^{\frac{1}{2}}y)$ since $(-is)^{\frac{1}{2}} = \frac{1}{\sqrt{2}}(1 - i)s^{\frac{1}{2}}$ and this will be unique under the stipulation that $\hat{\phi}(s,y) \to 0$ as $y \to \infty$ (cf. Example 1.10). But $\exp(-(is)^{\frac{1}{2}}y) = F_cI(y,x) + iF_sI(y,x) = FI^+(y,x)$ where $I^+(y,x) = I(y,x)$ for $x \geq 0$ and $I^+(y,x) = 0$ for $x < 0$. Thus $\phi(x,y) = I^+(y,\cdot) * f(\cdot)$ which yields

(2.5) $$(Bf)(y) = \phi(0,y) = \int_0^\infty I(y,\xi)f(-\xi)d\xi.$$

Hence if $Du = A^2u$ then $Bu = \tilde{w}$ satisfies $D^2\tilde{w} = A^2\tilde{w}$ while if $Dv = -A^2v$ then $\tilde{z} = Bv$ satisfies $D^2\tilde{z} = -A^2\tilde{z}$. However the values $u(-\xi)$ or $v(-\xi)$ which arise from (2.5) may introduce complications.

A few other simple examples are worth mentioning here. First let $P(D) = D$ and $Q(D) = -D$ so one considers $\phi_x = -\phi_y$ with $\phi(x,0) = f(x)$. Clearly $\phi(x,y) = \delta(x - y) * f(x)$ so that

$$\phi(0,y) = (Bf)(y) = \int_{-\infty}^\infty \delta(\xi - y)f(-\xi)d\xi = f(-y).$$

Applied to vector functions if $D\alpha = A^2\alpha$ then $\beta = B\alpha$ satisfies $D\beta = -A^2\beta$ and $\beta(y) = \alpha(-y)$. A more interesting example is

Example 2.2 Let $P(D) = D^3$ and $Q(D) = D$ so we consider

$$(2.6) \qquad \phi_{xxx} = \phi_y; \quad \phi(x,0) = f(x); \quad \hat{\phi}_y - is^3\hat{\phi} = 0; \quad \hat{\phi}(s,0) = \hat{f}(s)$$

where $\hat{\phi} = F\phi$. Thus $\hat{\phi}(s,y) = \hat{f}(s)\exp is^3y$ and one knows that

$$(2.7) \qquad \frac{1}{2\pi}\int_{-\infty}^{\infty} e^{-isx}e^{is^3y}ds = K(3y,x) = (3y)^{-1/3}\,\mathrm{Ain}(x(3y)^{-1/3})$$

where $\mathrm{Ain}(z) = \mathrm{Ai}(-z)$ denotes the Airy function

$$\mathrm{Ai}(-z) = \frac{1}{3}z^{\frac{1}{2}}[J_{1/3}(\tfrac{2}{3}z^{3/2}) + J_{-1/3}(\tfrac{2}{3}z^{3/2})]$$

(cf. Roetman [1], Magnus-Oberhettinger-Soni [1]). Consequently $\phi(x,y) = K(3y,\cdot) * f(\cdot)$ and

$$(2.8) \qquad \phi(0,y) = (Bf)(y) = \int_{-\infty}^{\infty} K(3y,\xi)f(-\xi)d\xi.$$

By results of Roetman [1] this will make sense for continuous f with $|f(\xi)| \leq c\exp k|\xi|^{3/2-\varepsilon}$ as $\xi \to \infty$ and $\phi(x,y)$ will satisfy (2.6). In particular we can state

Theorem 2.3 Let $g(A)$ be a well defined operator function in F with $u_o \in D(g(A))$. Assume $u(t)$ satisfies $D^3u = g(A)u$ for $-\infty < t < \infty$ with $u(0) = u_o$ while $p(u(t)) \leq k_p \exp c_p|t|^{3/2-\varepsilon_p}$ for any seminorm p in F. Then $v(t) = \int_{-\infty}^{\infty} K(3t,\xi)u(-\xi)d\xi = (Bu)(t)$ satisfies $Dv = g(A)v$ with $v(0) = u_o$.

The formulas of Example 2.2 are perhaps better exploited in solving the problem $Du = -A^3u$ $(t \geq 0)$ with $u(0) = u_o \in D(A^3)$ by the technique of Section 1.1. Thus we assume A generates a group $T(\cdot)$ and replace A by $-D_x$ to obtain $G_t = G_{xxx}$ with $G(0,x) = \delta(x)$. By (2.6) - (2.7) $G(t,x) = K(3t,x)$ and a solution of the abstract Cauchy problem is

$$(2.9) \qquad u(t) = \langle G(t,\cdot),\, T(\cdot)u_o\rangle = \int_{-\infty}^{\infty} K(3t,\xi)T(\xi)u_o d\xi$$

which makes sense since $p(T(\xi)u_o) \leq q(u_o)e^{\alpha|\xi|}$. If we let $A = -D_x$ generate the group $T(\xi)u_o(x) = u_o(x - \xi)$ in a suitable space of functions F then $u(t) = u(t,x)$ solves $u_t = u_{xxx}$ with $u(0,x) = u_o(x)$ as in Roetman [1]. Summarizing

Theorem 2.4 Let A generate a quasiequicontinuous group $T(\cdot)$ in F with $u_o \in D(A^3)$. Then $u(t)$ defined by (2.9) satisfies $u_t = -A^3u$ with $u(0) = u_o$.

Remark 2.5 Evidently the transmutation method of relating solutions of abstract differential problems involving an operator A is quite independent of A generating a group. Groups arose in Example 1.4, Example 1.6, and Theorem 1.12 because of the fact that $w(t) = \cosh At\, u_o$ and the corresponding $u(t) = \int_0^\infty K(t,x)w(x)dx$ are known solutions of $w_{tt} = A^2w$ and $u_t = A^2u$ respectively when A generates a group.

Example 2.6 An important example of transmutation was furnished by Lions [2; 3; 4] in treating D^2 and $L_m = D^2 + \frac{2m+1}{t}D$ together with related generalizations. As in Section 1.3 we take $m > -\frac{1}{2}$ here for convenience and consider $P(D) = D^2$ and $Q(D) = L_m(D) = L_m$. Thus we look for ϕ satisfying

$$\phi_{xx} = \phi_{yy} + \frac{2m+1}{y}\phi_y;\quad \phi(x,0) = f(x);\quad \phi_y(x,0) = 0. \tag{2.10}$$

Referring to Example 1.3.1 we see that $\phi(x,y) = R^m(y,x) * f(x)$ so that (cf. (1.3.6))

$$\begin{aligned}(B_mf)(y) &= \phi(0,y) = \langle R^m(y,x),\, f(-x)\rangle \\ &= c_m\int_0^1 (1-\xi^2)^{m-\frac{1}{2}}[f(\xi t) + f(-\xi t)]d\xi\end{aligned} \tag{2.11}$$

where $c_m = \Gamma(m+1)/\Gamma(1/2)\Gamma(m+1/2)$. If $D^2w = A^2w$ with $w_t(0) = 0$ so that $w(t) = \cosh Atu_o$ then $u^m = B_mw$ satisfies $L_mu^m = A^2u^m$ and we recover the formula (1.3.8) from (2.11) (i.e. $u^m(t) = 2c_m\int_0^1 (1-\xi^2)^{m-\frac{1}{2}}\cosh A\xi tu_o d\xi$). Lions actually constructs an isomorphism B_m with inverse B_m^{-1} in suitable spaces for L_m and various generalizations but we will not spell this out here (cf.

Examples 3.11 and 3.14 for our version of this).

Remark 2.7 Various transmutation operators can be envisioned for higher order situations such as $P(D) = (-1)^{n+1}D^{2n}$ with $Q(D) = D$ so that one considers

$$(2.12) \qquad \phi_y = (-1)^{n+1}D_x^{2n}\phi; \quad \phi(x,0) = f(x).$$

After Fourier transmutation this becomes

$$(2.13) \qquad \hat{\phi}_y + s^{2n}\hat{\phi} = 0; \quad \hat{\phi}(s,0) = \hat{f}(s)$$

with solution $\hat{\phi}(s,y) = \hat{f}(s)\exp(-s^{2n}y)$. As in Section 1.3 we denote by Z_p^p the space of entire functions satisfying $|g(z)| \leq c \exp b|z|^p$ with $|f(x)| \leq c_1\exp(-a|x|^p)$ for x real so $Z_p^p = S_{1/p}^{1-1/p}$ in the notation of Gelfand-Silov [2; 3]. It is known that $FZ_p^p = \hat{Z}_p^p = Z_q^q = S_{1-1/p}^{1/p}$ where $\frac{1}{p} + \frac{1}{q} = 1$ and thus $\hat{\psi}(s,y) = \exp(-s^{2n}y) \in Z_{2n}^{2n}$ corresponds to a transmutation kernel $\phi(x,y) \in Z_q^q$ with $q = 2n/2n-1 = 1 + 1/2n-1$. Thus for suitable f, which can be specified, $\phi(x,y) = \psi(x,y) * f(x)$ will yield a transmutation operator $(Bf)(y) = \phi(0,y)$. This will be perfectly rigorous but an explicit description of $\psi(x,y)$ will generally not be available from a table of Fourier transforms. Thus since higher order differential problems do not usually arise in practice anyway we will not concern ourselves with the question of transmuting general $P(D)$ into $Q(D)$ (which is not always possible in any event as an example with $P(D) = D^3$ and $Q(D) = D^3 - q(x)$, q not analytic, illustrates - cf. Levitan [1]). Rather let us concentrate on second order differential polynomials where explicit formulas can be obtained. Some general results for this situation are developed in Levitan [1] and Lions [2; 3; 4] but since we are concerned primarily with operational calculus and related equations we will proceed rather differently via some examples of interest in applications and develop general theorems in that context.

Example 2.8 An example of realted non-homogeneous equations connected by a generalized Stieltjes transform is provided in Bragg [4] (cf. also Dettman [1] and see Remark 2.10 for further information on non-homogeneous related equations). The discussion in Bragg [4] takes place in a Banach space F where A generates an equibounded C^o group $T(t)$; suitable extensions to more general F are clearly possible but we omit this. Consider the problems $(\alpha \geq 0,\ \beta \geq 0)$

(2.14) $$u_{tt} - t^{\alpha}A^2u = f;\quad u(0) = u'(0) = 0$$

(2.15) $$v_{tt} + t^{\beta}A^2v = g;\quad v(0) = 0$$

where f and g are suitable continuous functions with values in $D(A^2)$. One makes a change of variables $t = [(\alpha + 2)\tau/2]^{2/\alpha+2}$ in (2.14) and $t = [(\beta+2)\tau/2]^{2/\beta+2}$ in (2.15) to obtain

(2.16) $$U_{\tau\tau} + \frac{\alpha}{(\alpha+2)\tau}U_{\tau} - A^2U = f^*$$

(2.17) $$V_{\tau\tau} + \frac{\beta}{(\beta+2)\tau}V_{\tau} + A^2V = g^*$$

where $U(\tau) = u([\frac{(\alpha+2)\tau}{2}]^{2/\alpha+2})$, $f^*(\tau) = (2/(\alpha+2)\tau)^{2\alpha/\alpha+2}f([\frac{(\alpha+2)\tau}{2}]^{2/\alpha+2})$, etc. The equation

(2.18) $$W_{\tau} - A^2W = h$$

is taken as an intermediate link and in keeping with a general formalism indicated below (see Bragg [3] and Remark 2.10) one takes $\xi = (1/4)\tau^2$ in (2.16) and $\tau = \xi$ in (2.18) to obtain

(2.19) $$[\xi D_{\xi}(\xi D_{\xi} + \frac{a+1}{2} - 1) - \xi A^2]U(2\xi^{\frac{1}{2}}) = \xi g^*(2\xi^{\frac{1}{2}});$$

$$(\xi D_{\xi} - \xi A^2)W = \xi h(\xi)$$

where $a = \frac{\alpha}{\alpha+2}$. One can then write

(2.20) $$U(2\xi^{\frac{1}{2}}) = \Gamma(\frac{a+1}{2})\xi^{1-\frac{a+1}{2}}L_s^{-1}\{s^{-(\frac{a+1}{2})}W(\frac{1}{s})\}_{s\to\xi}$$

provided that

(2.21) $$\xi f^*(2\xi^{\frac{1}{2}}) = \xi^{2-\frac{(a+1)}{2}}\Gamma(\frac{a+1}{2})L_s^{-1}\{s^{-[\frac{a+1}{2}-1]}s^{-1}h(\frac{1}{s})\}_{s\to\xi}.$$

Here $L_s^{-1}\{\ \}_{s\to\xi}$ means the inverse Laplace transform with $s \to \xi$. Similarly setting $\tau = \xi^{-\frac{1}{2}}$ in (2.17) and $\tau = 1/4\xi$ in (2.18) one obtains

$$\text{(2.22)}\qquad V(\xi^{-\frac{1}{2}}) = \frac{1}{\Gamma(\frac{1-b}{2})}\int_0^\infty e^{-\sigma}\sigma^{-\frac{b+1}{2}} W(1/4\sigma\xi)\,d\sigma$$

where $b = \beta/\beta+2$, provided

$$\text{(2.23)}\qquad g^*(\xi^{-\frac{1}{2}}) = \frac{-1}{\Gamma(\frac{1-b}{2})}\int_0^\infty e^{-\sigma}\sigma^{-\frac{b+1}{2}} h(1/4\sigma\xi)\,d\sigma.$$

Eliminating W between (2.20) and (2.22) there results

$$\text{(2.24)}\qquad v(t) = K(\alpha,\beta)\int_0^\infty \frac{t\sigma^{\alpha}u(\sigma)\,d\sigma}{[(\frac{2}{\beta+2})^2 t^{\beta+2} + (\frac{2}{\alpha+2})^2\sigma^{\alpha+2}]}$$

where $K(\alpha,\beta) = C(\alpha,\beta)\left[\frac{\alpha-\beta}{(\alpha+2)(\beta+2)} + 1\right]$ with

$$\text{(2.25)}\qquad C(\alpha,\beta) = \frac{2^{2[\frac{\alpha+1}{\beta+2}+\frac{1}{\beta+2}]}\,\Gamma(\frac{\alpha-\beta}{(\alpha+2)(\beta+2)}+1)}{\Gamma(\frac{\alpha+1}{\alpha+2})\Gamma(\frac{1}{\beta+2})(\alpha+2)^{\alpha/\alpha+2}(\beta+2)^{2/\beta+2}}\,.$$

The corresponding relation between f and g is

$$\text{(2.26)}\qquad g(t) = -t^2K(\alpha,\beta)\int_0^\infty \frac{f(\xi)\,d\xi}{[(\frac{2}{\beta+2})^2t^{\beta+2} + (\frac{2}{\alpha+2})^2\xi^{\alpha+2}]}\,.$$

Now to produce a satisfactory theorem from these calculations Bragg [4] first constructs explicitly a solution $u(t)$ to (2.14) when $f \in C^0(F)$ on $[0,\infty)$ with $f(t) \in D(A^2)$. One writes for $\phi \in D(A^2)$

$$\text{(2.27)}\qquad U_1(t)\phi = \frac{1}{\beta(\frac{1}{2},\frac{\alpha}{2\alpha+4})}\int_0^1 (1-\sigma)^{-\frac{\alpha+4}{2\alpha+4}}\sigma^{-\frac{1}{2}}\cosh Am(\sigma,t)\phi\,d\sigma$$

where $m(\sigma,t) = 2t^{\frac{\alpha+2}{\alpha}}\sigma^{\frac{1}{2}}/(\alpha+2)$ and

$$\text{(2.28)}\qquad U_2(t)\phi = K(\alpha)t\int_0^1 (1-\sigma)^{-\frac{\alpha}{\alpha+2}}U_1(t\sigma^{\frac{1}{\alpha+2}})\sigma^{-\frac{1}{\alpha+2}}\phi\,d\sigma$$

where $K^{-1}(\alpha) = \beta(\frac{2}{\alpha+2},\frac{\alpha+1}{\alpha+2})$. Then $U_1(t)\phi = \tilde{\phi}_1(t)$ satisfies (2.14) with $f = 0$, $\tilde{\phi}_1(0) = \phi$, and $\tilde{\phi}_1'(0) = 0$ while $U_2(t)\phi = \tilde{\phi}_2(t)$ satisfies (2.14) with $f = 0$,

$\tilde{\phi}_2(0) = 0$, and $\tilde{\phi}_2'(0) = \phi$. A solution to (2.14) is then given by

$$u(t) = \int_0^t [U_1(\xi)U_2(t) - U_2(\xi)U_1(t)]f(\xi)d\xi. \tag{2.29}$$

Now let G be the space of $f \in C^o(F)$ on $[0,\infty)$ with $f(t) \in D(A^2)$ and $\int_0^\infty \| f(t) \| dt < \infty$. There results

Theorem 2.9 Let $f \in G$ and $\alpha > \beta$. Then a solution v of (2.15) with g defined in terms of f by (2.26) is given by putting the $u(t)$ of (2.29) in (2.24).

Remark 2.10 A class of related differential equations connected to generalized hypergeometric functions are studied in Bragg [3] and this of course will yield corresponding transmutation formulas (which we mercifully will not spell out). Let

$$\Theta_1(q,\beta,t,D_t) = tD_t \prod_1^q (tD_t + \beta_j - 1) \tag{2.30}$$

for $\beta = (\beta_1,\ldots,\beta_q)$ and

$$\Theta_2(p,\alpha,t,D_t) = t\prod_1^p (tD_t + \alpha_i). \tag{2.31}$$

Let $P = P(x,D) = \Sigma a_\lambda(x)D^\lambda$ with $Q = Q(x,D)$ of the same form; suppose their orders are ℓ_1 and ℓ_2 respectively. Equations of the type

$$[Q\Theta_1(q,\beta,t,D_t) - P\Theta_2(p,\alpha,t,D_t)]U(x,t) = 0 \tag{2.32}$$

are considered and here it is seen that P and Q could be suitable more general abstract operators in a space F with no essential change in technique. Laplace transforms are used and some typical theorems are

Theorem 2.11 Let $u(x,t)$ be a solution of

$$[Q\Theta_1(q,\beta,t,D_t) - P\Theta_2(p-1,\alpha,t,D_t)]u = 0 \tag{2.33}$$

with $p \leq q$, α_i and β_j positive, $u \in C^r$ in (x,t) for $t > 0$ where $r = \max(p + \ell_1, q + \ell_2)$, and u and all derivatives of order $\leq r$ bounded. Let

$\alpha_p > 0$; then the function

$$(2.34) \qquad V(x,t) = \Gamma^{-1}(\alpha_p)\int_0^\infty e^{-\sigma}\sigma^{\alpha_p - 1}u(x,t\sigma)d\sigma$$

satisfies (2.32). If $u(x,t) \to \phi(x)$ as $t \to 0$ with ϕ bounded and continuous then $v(x,t) \to \phi(x)$ as $t \to 0$.

The proof is straightforward. Similarly one has

<u>Theorem 2.12</u> Let α_i and β_j be positive with $\beta_q > 1$. Let $U(x,t)$ be a solution of

$$(2.35) \qquad [Q\Theta_1(q-1,\beta,t,D_t) - P\Theta_2(p,\alpha,t,D_t)]U = 0$$

for $t > 0$ and set

$$(2.36) \qquad V(x,t) = t^{1-\beta_q}\Gamma(\beta_q)L_s^{-1}\{s^{-\beta_q}U(x,1/s)\}_{s\to t}\,.$$

If V exists, $V \in C^r$, and V with all derivatives or order $\leq r$ are bounded then V satisfies (2.32) for $t > 0$. If $U(x,1/s)$ is bounded and analytic in s for $\operatorname{Re} s > k > 0$ then as $t \to 0$ $\lim V(x,t) = \phi(x) = U(x,0)$.

<u>Theorem 2.13</u> Let $p \leq q+1$, $\beta_q > \alpha_p > 0$, and no $\beta_1,\ldots,\beta_{q-1}$ equal to zero or a negative integer. Let $U \in C^{r-1}$ satisfy

$$(2.37) \qquad [Q\Theta_1(q-1,\beta,t,D_t) - P\Theta_2(p-1,\alpha,t,D_t)]U = 0$$

for $t > 0$ such that $U(x,t) \to \phi(x)$ as $t \to 0$ with ϕ continuous. Define V by

$$(2.38) \qquad V(x,t) = \frac{\Gamma(\beta_q)}{\Gamma(\alpha_p)\Gamma(\beta_q - \alpha_p)}\int_0^1 \sigma^{\alpha_p - 1}(1-\sigma)^{\beta_q - \alpha_p - 1}U(x,t\sigma)d\sigma.$$

Then V satisfies (2.32) for $t > 0$ and, as $t \to 0$, $V(x,t) \to \phi(x)$.

Numerous applications of such formulas are given in Bragg [3] and this formalism is extended to nonhomegeneous problems in Bragg [2]. We cite a few typical results relative to Theorems 2.11 - 2.13 where by (2.32) = f for example we mean $[Q\Theta_1 - P\Theta_2]u = f$.

Theorem 2.14 Let U be a bounded solution of (2.33) = f and let V be given by (2.34). Then V satisfies (2.32) = g provided

$$(2.39)\qquad g(x,t) = \Gamma^{-1}(\alpha_p)\int_0^\infty e^{-\sigma}\sigma^{\alpha_p-1} f(x,t\sigma)d\sigma.$$

Theorem 2.15 Let U be a bounded solution of (2.35) = f such that $\lim U(x,t)$ exists as $t \to 0$. Let V be given by (2.36). The V satisfies (2.32) = g provided

$$(2.40)\qquad g(x,t) = t^{2-\beta_q}\Gamma(\beta_q)L_s^{-1}\{s^{1-\beta_q}f(x,\tfrac{1}{s})\}_{s\to t}.$$

Theorem 2.16 Let U satisfy (2.37) = f and define V by (2.38) with $\alpha_p > 0$ and $\beta_q - \alpha_p > 1$. Then V satisfies (2.32) = g provided

$$(2.41)\qquad g(x,t) = G_p^q\int_0^1 \sigma^{\alpha_p-1}(1-\sigma)^{\beta_q-\alpha_p-2} f(x,t\sigma)d\sigma$$

where $G_p^q = (\beta_q - \alpha_p - 1)\Gamma(\beta_q)/\Gamma(\alpha_p)\Gamma(\beta_q - \alpha_p)$.

Example 2.17 The Schrodinger equation $u_t = iu_{xx}$ viewed in the Gelfand-Šilov framework (cf the discussion around Definition 1.3.16 and Theorem 1.3.19) leads to $\lambda(s) = -is^2$ and $\Lambda(s) = 2\sigma\tau$ is zero for $\tau = 0$ so that the equation is Petrovskij correct. Under the Fourier transform $F_x u = \hat{u}$ we have in fact $\hat{u}_t = -is^2\hat{u}$ which leads to $\hat{G}(t,s) = \exp(-is^2t)$ for a fundamental solution ($\hat{G}(0,s) = 1$). Since $\Lambda(s)$ is bounded in the domain $|\tau| \leq k\ (1 + |\sigma|)^{-1}$ the genus $\mu = -1$. Further $|\hat{G}^{(q)}(t,\tau)| \leq c_q(1 + |\sigma|)^q$ and one can say that for any $\ell > 0$ the one dimensional Cauchy problem is well posed in the class of functions $u(x)$ having, together with their derivatives of order $\leq \ell + 2$, polynomial growth of order $\leq \ell - 2$ (while $u(x,t)$ has polynomial growth of order $\leq \ell$). Regarding the abstract Schrodinger equation we consider

$$(2.42)\qquad u_t = iA^2u;\quad u(0) = u_o$$

and the Hersh technique leads to $G_t = iG_{xx}$ with $G(0,x) = \delta(x)$. It is easily shown (cf. Treves [3]) that

$$(2.43)\qquad G(t,x) = \frac{e^{-i\pi/4}}{2\sqrt{\pi t}} \exp(-x^2/4it).$$

Consequently $u(t) = \langle G(t,x),\ T(x)u_o\rangle$ is given by

$$(2.44)\qquad u(t) = \frac{e^{-i\pi/4}}{2\sqrt{\pi t}} \int_{-\infty}^{\infty} e^{-x^2/4it} T(x)u_o dx$$

where A generates a suitable group $T(\cdot)$ in a space F. We note that if $\tilde{u}(t)$ is the solution of the abstract heat problem (I.1.3) then $\tilde{u}(it) = u(t)$ and formally (2.44) can be written

$$(2.45)\qquad u(t) = \int_0^{\infty} K(it,x)\cosh Axu_o dx.$$

We note that there are some errors in Polyanskij-Fedoryuk [1] where the formula $(\pi/p)^{\frac{1}{2}} \cdot \exp(-\alpha p^{\frac{1}{2}}) = \int_0^{\infty} \tau^{-\frac{1}{2}}\exp(-\frac{\alpha^2}{4\tau} - p\tau)d\tau$ is used (cf. Bragg-Dettman [6; 7; 8]). Their formulation is however easily modified to give a connection formula between the u of (2.42) and the z of (I.1.5) using spectral hypotheses on A^2.

Remark 2.18 The construction of transmutation operators by the method of Theorem 1.3(F) usually involves the extension of f to $(-\infty,0)$ in some manner (eg $f(-x) = f(x)$). The resulting transmutation operator will depend on the extension and the following discussion and remarks should clarify the matter. First, there are a number of results available concerning the solution of half plane or quarter plane differential problems (see eg. Dikopolov [1], Dikopolov-Šilov [2; 3], Hersh [4; 5; 6], Palamodov [2], Pavlov [1], Sarason [1; 2], Šilov [1], Zarnitskaya-Selezneva-Eidelman [1] and cf. also Fattorini [3], Fattorini-Radnitz [4], Gasymov [3]) and we will give some sample theorems to indicate natural frameworks. Let us say that Q is P-admissible if $P(D_x)\phi = Q(D_y)\phi$, with $\phi(x,0) = f(x) \in \Psi$ for some space Ψ, has a unique solution in some space Φ where Φ will usually involve some conditions $D_y^i\phi(x,0) = 0$ and/or growth conditions (see Section 2.3 also where this is subsumed under condition C). For such a problem to make sense one will usually have to extend $f(x)$, given for $x \geq 0$, to $(-\infty,0)$ and the space Φ deals with functions or distributions in the half space $-\infty < x < \infty$, $y \geq 0$. Then $Bf(y) = \phi(0,y)$ determines the transmutation operator B such that

$QB = BP$. Note that if Q is of order m say and data $D_y^k\phi(x,0) = E_k f(x)$, $k \in \{0,m-1\}$, is given, where $E_k \in L(\Psi,\Psi_k)$ for suitable Ψ and Ψ_k then conditions $\mathcal{C}$ would change of course, as would Φ, and the operator $B_E f(y) = \phi(0,y)$, $E = (E_k)$, would transmute $P \to Q$ as beofre provided $P(D_x)E_k = E_k P(D_x)$ (cf (1.6) and see Levitan [1; 3] for simple versions of this). Similarly working with the same equation for ϕ one can say P is Q-admissible if $P(D_x)\phi = Q(D_y)\phi$ with $\phi(0,y) = g(y) \in \mathcal{A}$ has a unique solution in say $\tilde{\Phi}$ where $\tilde{\Phi}$ is like Φ but defined over a region $-\infty < y < \infty$, $x \geq 0$; an extension of $g(y)$ to $y < 0$ is now indicated and one defines $\mathcal{B}g(x) = \phi(x,0)$ with $P\mathcal{B} = \mathcal{B}Q$. The half space problems for ϕ require global solutions in $t = y$ and the standard Gelfand-Šilov theory for example is not immediately applicable (cf Section 1.3 where $t \leq T < \infty$ often arises in fixing spaces). Some modifications of this theory are possible and we cite first a few typical results following Dikopolov-Šilov [2]. Consider for example

(2.46) $$D_t^m u = \sum_{k=0}^{m-1} P_k(iD_x)D_t^k u$$

where the P_k are polynomials and we work in the region $-\infty < x < \infty$ with $t \geq 0$. In order to relate this to our format let P_k be constant for $k > 0$ with $P_0(iD_x) = P(D_x)$. Setting $Q(D_t) = D_t^m - \sum_1^{m-1} P_k D_t^k$ we have $Q(D_t)u = P(D_x)u$. Let $\lambda_j(\sigma)$ be the roots of $\lambda^m - \sum_1^{m-1} P_k\lambda^k = P_0(\sigma)$, ordered by the rule

(2.47) $$\text{Re}\,\lambda_1(\sigma) \leq \ldots \leq \text{Re}\,\lambda_m(\sigma)$$

for σ real, and let G_j be the set of such σ where $\text{Re}\,\lambda_j(\sigma) \leq 0$ so that $G_1 \supset \ldots \supset G_m$. Let $\mathcal{H}$ be the space of L^2 functions on $\mathbb{R}$ together with all derivatives (computed in $\mathcal{D}'$) so that $H = \mathcal{F}\mathcal{H}$ consists of L^2 functions on $\mathbb{R}$ multiplied by arbitrary polynomials. These are countably normed spaces in the sense of Gelfand-Šilov [1; 2; 3] and one writes for example $H = \cup H^p$ where H^p is the Hilbert space with scalar product

$$(\hat{f},\hat{g})_p = \int \hat{f}(\sigma)\overline{\hat{g}(\sigma)}(1 + \sigma^2)^{-p}d\sigma.$$

A sequence $\hat{f}_n \to 0$ in H if for some p $\hat{f}_n \in H^p$ and $\|\hat{f}_n\|_p \to 0$ as $n \to \infty$. Similarly $f_n \to 0$ in $\mathcal{H}$ if for some differential polynomial $P(D)$, $f_n = P(D)F_n$, $F_n \in L^2$, and $F_n \to 0$ in L^2. A typical theorem relative to (2.46) is

<u>Theorem 2.19</u> Let functions $v_j(\sigma)$ be given on G_j where v_j is the product of a polynomial times a function in $L^2(G_j)$ (one writes $v_j \in H(G_j)$). The equation (2.46) has a solution $u(x,t)$ with $[D_t^{k-1}u(x,0)]^\wedge = v_k(\sigma)$ on G_k $(1 \le k \le m)$. For $t \ge 0$ $u(\cdot,t) \in \mathcal{H}$ and, as $t \to \infty$, $D_t^k u(\cdot,t)$ $(0 \le k \le m-1)$ grows in $\mathcal{H}$ no faster than some power of t. This u is unique in the class of such functions and depends continuously in $\mathcal{H}$ on the data $v_k \in H(G_k)$.

<u>Example 2.20</u> Consider $D_t^2 u = D_x^2 u$ so that $P_o(iD_x) = -(iD_x)^2$ and $\lambda^2 + \sigma^2 = 0$ gives $\lambda_1(\sigma) = i\sigma$ with $\lambda_2(\sigma) = -i\sigma$. Thus $G_1 = G_2 = \mathbb{R}$ and we have the familiar Cauchy problem for the wave equation. On the other hand for $D_t^2 u = -D_x^2 u = (iD_x)^2 u$ where $\lambda^2 - \sigma^2 = 0$ and $\lambda_1(\sigma) = -|\sigma|$ with $\lambda_2(\sigma) = |\sigma|$, one has $G_1 = \mathbb{R}$ with $G_2 = \Phi$. Thus $v_1(\sigma)$ can be prescirbed but not $v_2(\sigma)$ and we have a well posed Dirichlet problem for the Laplace equation.

<u>Remark 2.21</u> If more rapid growth is allowed $(O(t^p \exp ct))$ the G_j can be replaced by $G_j^c = \{\mathrm{Re}\, \lambda_j(\sigma) \le c\}$ and in the Laplace example above $v_2(s)$ can be specified for $|s| \le c$. On the other hand for any equation (2.46) the sets G_j^c are nonempty for some c and there will be nontrivial correct problems. If one works locally in t the Cauchy problem itself will be correct for suitable exponential growth and $0 \le t \le T$ (cf. the end of Section 1.3).

Various refinements of such results are possible but we will not enter into a discussion of this.

<u>Remark 2.22</u> Let us mention now for completeness some criteria of Šilov [1] for quarter plane problems and specialize his results to some second order equations for simplicity. Consider

$$(2.48) \qquad D_t^2 u = P_o(iD_x)u = \alpha u + \gamma(iD_x)^2 u$$

(2.49) $u(0,t) = w_o(t)$; $D_x u(0,t) = w_1(t)$

(2.50) $u(x,0) = u_o(x)$; $D_t u(x,0) = u_1(x)$

where the number of conditions imposed at $t = 0$ may vary here as indicated below. First order terms in (2.48) are omitted here for convenience. One looks at the roots $\lambda_j(s)$ $(j = 0,1)$ of

(2.51) $\lambda^2 = \alpha + \gamma s^2$

and there are critical points $s = \pm(-\alpha/\gamma)^{\frac{1}{2}}$ where the $\lambda_j(s)$ have algebraic branch points. Thus if α and γ have the same sign we have $s = \pm i(\alpha/\gamma)^{\frac{1}{2}}$ whereas if they have opposite signs $s = \pm(-\alpha/\gamma)^{\frac{1}{2}}$ is real. In either case $\text{Ims} \leq \beta = |\alpha/\gamma|^{\frac{1}{2}}$ for both roots and one works in the half plane $\text{Ims} > \beta$ where the roots are single valued and distinct. A number r $(r = 0,1,2)$ is then determined specifying how many conditions (2.50) are allowed, such that, for $\text{Ims} > \beta$, $\text{Re}\,\lambda_j(s) \leq 0$ for $0 \leq j \leq r-1$ and for $j \geq r$ $\text{Re}\,\lambda_j(s) > 0$ at least for some s with $\text{Ims} > \beta$. For $r = 2$ both conditions in (2.50) are permitted. One writes (2.48) in the form

(2.52) $D_t^2 u = \alpha u - \gamma[D_x^2 u - w_1\delta - w_o\delta']$

and the quantity

(2.53) $g(s,t) = -\gamma[w_1(t) - isw_o(t)]$

will be of importance in the theory. Indeed set

(2.54) $G_\nu(s) = \int_0^\infty e^{-\theta\lambda_\nu(s)} g(s,\theta)\,d\theta$

for $\nu \geq r$ $(r = 0,1)$. This integral is defined for $\text{Ims} > \beta$ and $\text{Re}\,\lambda_\nu(s) > 0$ provided for example that the $w_i(t)$ have at most polynomial growth as $t \to \infty$. One looks for solutions u in a space $\mathcal{H}_+^\beta$ delimited by conditions of the form $\int_0^\infty \exp(-2\beta x)|I^q u(x,t)|^2 dx < \infty$ where $I^q u$ denotes a suitable "primitive" of u. The space $H_+^\beta = F\mathcal{H}_+^\beta$ on the other hand consists of functions $\phi(s)$, analytic for $\text{Im}s > \beta$, such that $\int_{-\infty}^{\infty} |\phi(s)|^2/|s|^{2q} d\sigma < \infty$ for some q, uniformly in $\tau = \text{Im}s > \beta$ $(s = \sigma + i\tau)$.

Theorem 2.23 There exists a unique solution of (2.48) - (2.50) provided the $G_\nu(s)$ can be continued analytically in s to the half plane $\text{Im}s > \beta$ as functions in H_+^β.

Example 2.24 Let $\alpha = 0$ and $\gamma = -1$ so that (2.48) is the wave equation. One obtains $\lambda^2 = -s^2$, $\lambda_0 = is$, $\lambda_1 = -is$, $\beta = 0$, and for $\text{Im}s > 0$ $\text{Re}\,\lambda_0(s) = \text{Re}\, i(\sigma + i\tau) = -\tau < 0$ while $\text{Re}\,\lambda_1(s) = \text{Re}[-i(\sigma + i\tau)] = \tau > 0$. Hence $r = 1$ and w_0, w_1, and u_0 can be prescribed which is a standard mixed problem for the wave equation.

Example 2.25 Let $\alpha = 0$ and $\gamma = 1$ which makes (2.48) a Laplace equation where $\lambda^2 = s^2$, $\lambda_0 = s$, $\lambda_1 = -s$, $\beta = 0$, and for $\text{Im}s > 0$ $\text{Re}\,\lambda_0(s) = -\sigma$ while $\text{Re}\,\lambda_1(s) = \sigma$ (neither of which is always negative). Thus $r = 0$ and the $G_\nu(s)$, defined for $\text{Re}s > 0$ and $\text{Re}s < 0$ respectively, are

$$(2.55) \qquad G_0(s) = -\int_0^\infty e^{-\theta s}[w_1(\theta) - isw_0(\theta)]d\theta = -\int_0^\infty e^{-\theta s}[w_1(\theta) - iw_0'(\theta)]d\theta - iw_0(0)$$

$$(2.56) \qquad G_1(s) = -\int_0^\infty e^{\theta s}[w_1(\theta) - isw_0(\theta)]d\theta = -\int_0^\infty e^{\theta s}[w_1(\theta) + iw_0'(\theta)]d\theta + iw_0(0).$$

An analysis of the Cauchy Riemann equations in the present context (cf Šilov [1]) allows one to conclude that the $G_\nu(s)$ will be suitably extendable provided the functions $W_\pm(t) = w_1(t) \pm iw_0'(t)$ can be extended to be analytic functions $W_\pm(x,t)$ for $x > 0$ and $t > 0$. However neither u_0 nor u_1 can be prescribed arbitrarily when both w_0 and w_1 are to be specified.

Remark 2.26 Consider again the wave equation $\phi_{xx} = \phi_{tt}$ with $\phi(x,0) = f(x)$ and $\phi_t(x,0) = 0$ for $x \geq 0$. Extend ϕ_t by 0 and $f(x)$ by $\tilde{f}(x)$ for $x < 0$

where $\tilde{f}(-x) = g(x)$ with $g(0) = f(0)$ at least ($\tilde{f}(x) = f(x)$ for $x \geq 0$). Then $\phi(x,t) = \frac{1}{2}[\tilde{f}(x+t) + \tilde{f}(x-t)]$ and $\phi(0,t) = \frac{1}{2}[\tilde{f}(t) + \tilde{f}(-t)] = \frac{1}{2}[f(t) + g(t)] = h(t)$. If we consider the quarter plane problem of Example 2.24 with x and t interchanged it is seen that $h(t)$ could be prescribed along with $\phi(x,0) = f(x)$ and $\phi_t(x,0) = 0$. Hence there is an extension $\tilde{f}$ of f, namely $\tilde{f}(-x) = g(x) = 2h(x) - f(x)$ such that $B_g f(t) = h(t)$ where B_g denotes the transmutation operator determined by this extension. We note also that the extension $g = f$ corresponds to $h = f$ and $B_g = B_f$ is in this case a generalized translation operator of the type studied by Levitan [1; 3] (cf Section 2.3). Formally let us consider then $f \in \Psi$ with $\phi \in \Phi$ as in Remark 2.18 and look at $\phi(0,t) \in \Theta$ where $\Theta = \Psi$ here. Note first that $g = 0 \sim h = \frac{1}{2} f$ while $h = 0 \sim g = -f$ and if $f(0) = 0$ such g appear admissible. Thus $B_{-f}(f) = 0$, $B_f(f) = f$, and $B_o(f) = \frac{1}{2} f$. In particular if B_g is the zero operator then $B_g(g) = 0 = g$ and $g \to B_g$ is 1-1. Further

$$(2.57) \qquad B_g(f + h) = \frac{1}{2}[f + h + g] = B_g(f) + B_g(h) - B_o(g)$$

$$(2.58) \qquad B_g(\alpha f) = \frac{1}{2}[\alpha f + g] = \alpha B_g(f) - (\alpha - 1)B_o(g).$$

Therefore the operator defined by $A_g(f) = B_g(f) - B_o(g)$ is linear, $A_g \in L(\Psi)$, and $g \to A_g : \Psi \to L(\Psi)$ is 1-1 since $A_g = 0$ implies $A_g(g) = g - \frac{1}{2} g = \frac{1}{2} g = 0$. Note also that from $B_g(f - g) = B_o f$ follows $A_g(f - g) = B_o(f - g)$. From $A_g(f) = \frac{1}{2}[f + g] - \frac{1}{2} g = \frac{1}{2} f$ of course we see that $A_g = \frac{1}{2}$ is trivial but in particular $A_g \in L(\Psi)$ (continuity). Thus $g \to B_g : \Psi \to A(\Psi)$ where $A(\Psi)$ denotes continuous affine operators $\Psi \to \Psi$ and $B_g(f) = f$ implies $f = g$. It could be of interest to examine such mappings in more generality for other equations (cf Section 2.5).

2.3. Separation of variables and transforms. Let us first make some observations about transmutation and separation of variables which lead to a systematic formalism for handling many problems of interest in a unified manner. The presentation will be essentially formal at first but will be illustrated by numerous examples to justify it and a more rigorous treatment will follow. A preliminary version appears in Carroll [35] (cf. also Levitan [1; 5] and Lions [2; 3; 4]). Thus suppose we can transmute $P(D)$ into $Q(D)$ via a solution of

$$(3.1)\qquad P(D_x)\phi(x,y) = Q(D_y)\phi(x,y);\quad \phi(x,0) = f(x)$$

with appropriate conditions $\mathcal{C}$ (unspecified at the moment) to guarantee a unique solution $\phi(x,y)$, which determines then B by $(Bf)(y) = \phi(0,y)$. We assume $(Bf)(y) = \langle b(y,x), f(x)\rangle$ for suitable brackets on $(0,\infty)$ or $(-\infty,\infty)$ ($\langle\, ,\, \rangle_+$ will denote $(0,\infty)$). The conditions $\mathcal{C}$ (cf Remark 2.18) will involve some combination of growth requirements on ϕ or suitable derivatives plus stipulations that $D_y^i\phi(x,0) = 0$ for $i \in \{1, m-1\}$ where $m =$ order $Q(D)$. Suppose $QB = BP$ acting on functions $f \in \Psi$ for some reasonably large space Ψ and consider the eigenvalue equations

$$(3.2)\qquad P(D_x)h(x,\lambda) = -\lambda^2 h(x,\lambda)$$

$$(3.3)\qquad Q(D_y)\theta(y,\lambda) = -\lambda^2\theta(y,\lambda).$$

If $h(\cdot,\lambda) \in \Psi$ then $QBh = BPh = -\lambda^2 Bh$ so that $(Bh)(y,\lambda) = \langle b(y,x), h(x,\lambda)\rangle$ is an eigenvalue for $Q(D)$. Conversely consider $\phi(x,y) = w(x,y,\lambda) = \phi(x,0) = h(x,\lambda)\theta(0,\lambda)$ (cf. Levitan [1]). Given conditions $\mathcal{C}$ for this ϕ, which could usually be provided by conditions on θ, we obtain $w(0,y,\lambda) = \phi(0,y) = (Bh)(y,\lambda) = h(0,\lambda)\theta(y,\lambda)$. Consequently

$$(3.4)\qquad \theta(y,\lambda) = \frac{\theta(0,\lambda)}{h(0,\lambda)}\langle b(y,x), h(x,\lambda)\rangle.$$

If $\theta(0,\lambda) = 0$ or ∞ we get nothing whereas $h(0,\lambda) = 0$ corresponds to $h \in N(B)$ (null-space) which is $\{0\}$ if B is 1-1. In practice we will usually work with $h(0,\lambda) = \theta(0,\lambda) = 1$ and summarize this in

<u>Lemma 3.1</u> Under the conditions indicated the eigenfunctions of $Q(D)$ are characterized to be of the form $\theta(y,\lambda) = \langle b(y,x), h(x,\lambda)\rangle$ (from now on $h(0,\lambda) =$

$\theta(0,\lambda) = 1$ is assumed unless otherwise stated).

Now for $P(D) = \Sigma p_j(x)D^j$ the formal adjoint $P^*(D)$ is defined by $P^*(D)\psi = \Sigma(-1)^j D^j(p_j\psi)$ and we let

$$(3.5)\qquad P^*(D)\Omega(x,\lambda) = -\lambda^2\Omega(x,\lambda).$$

Consider then the formal expression

$$(3.6)\qquad \beta(y,x) = \langle\Omega(x,\lambda),\ \theta(y,\lambda)\rangle$$

where $P^*(D_x)\beta = Q(D_y)\beta$ and define $\mathcal{B}f(y) = \langle\beta(y,x),\ f(x)\rangle$. Since

$$\mathcal{B}Pf = \langle\beta(y,x),\ P(D_x)f(x)\rangle = \langle P^*(D_x)\beta(y,x),\ f(x)\rangle$$
$$= \langle Q(D_y)\beta(y,x),\ f(x)\rangle = Q\mathcal{B}f.$$

<u>Theorem 3.2</u> The operator $\mathcal{B}f(y) = \langle\beta(y,x),\ f(x)\rangle$ transmutes P into Q; i.e. $Q\mathcal{B} = \mathcal{B}P$.

Given B defined via (3.1) it will be of interest to know when $B = \mathcal{B}$ and we proceed as follows. Let $U(x,\xi)$ be the solution of

$$(3.7)\qquad P(D_x)U(x,\xi) = P(D_\xi)U(x,\xi)$$

satisfying $U(x,0) = f(x)$; conditions $D_\xi^k U(x,0) = 0$ may be imposed for uniqueness (or growth conditions). The function $U(x,\xi)$ will be a generalized displacement or translation operator in the sense of Levitan [1; 3] which could be written $U(x,\xi) = T_x^\xi f(x)$ (cf. Remark 1.2) and in general we will have $U(0,\xi) = f(\xi)$. Explicit constructions via Riemann functions are often possible and examples will be given below. Then we form the generalized convolution

$$(3.8)\qquad \phi(x,y) = \langle\beta(y,\xi),\ U(x,\xi)\rangle$$

and observe that formally

$$P(D_x)\phi(x,y) = \langle\beta(y,\xi),\ P(D_x)U(x,\xi)\rangle = \langle\beta(y,\xi),\ P(D_\xi)U(x,\xi)\rangle$$

$$= \langle P^*(D_\xi)\beta(y,\xi),\ U(x,\xi)\rangle = Q(D_y)\phi(x,y).$$

Suppose now that $\Omega(x,\lambda)$ is chosen so that $\beta(0,x) = \langle\Omega(x,\lambda),\ 1\rangle = \delta(x)$ (this is shown to be natural from the examples). Then $\phi(x,0) = \langle\delta(\xi),\ U(x,\xi)\rangle = U(x,0) = f(x)$ so, by definition of B,

$$\phi(0,y) = (Bf)(y) = \langle b(y,x),\ f(x)\rangle = \langle\beta(y,\xi),\ U(0,\xi)\rangle$$

$$= \langle\beta(y,\xi),\ f(\xi)\rangle$$

provided conditions $\mathcal{C}$ are satisfied (which we assume). Since this holds for $f \in \Psi$ with Ψ reasonably large we can conclude that $b(y,x) = \beta(y,x)$. Hence

Theorem 3.3 Under the conditions indicated if $\beta(0,x) = \langle\Omega(x,\lambda),\ 1\rangle = \delta(x)$ then the kernel of B is precisely $\beta(y,x) = \langle\Omega(x,\lambda),\ \theta(y,\lambda)\rangle$ and $B = \mathcal{B}$.

Remark 3.4 Generalized convolutions such as (3.8) should be useful in solving applied problems and we will comment further on this below (cf. Theorem 3.7 for example). This shows the importance of solving equations of the type (3.7) aside from the treatment of Hutson-Pym [1; 2], Leblanc [1], Povzner [1], etc. (cf. Section 2.5).

Let us define now the transforms (see Appendix 2 for general transforms).

(3.9) $\mathcal{P}f(\lambda) = \langle h(x,\lambda),\ f(x)\rangle$

(3.10) $\mathbb{P}F(x) = \langle\Omega(x,\lambda),\ F(\lambda)\rangle$

(3.11) $\mathcal{Q}f(\lambda) = \langle\theta(y,\lambda),\ f(y)\rangle = \hat{f}(\lambda).$

We suppose that $b(y,x) = \beta(y,x)$ so that taking $h(0,\lambda) = \theta(0,\lambda) = 1$ we have from (3.4) $\theta(y,\lambda) = \langle\beta(y,x),\ h(x,\lambda)\rangle = \mathcal{P}\beta(y,\cdot)$ while by definitions $\beta(y,x) = \mathbb{P}\theta(y,\cdot)$ (y is essentially a parameter here). Now there will usually be a variety of choices of Q for which the above procedures are valid and we ask that Q be chosen so that $\mathcal{Q}$ in (3.11) is an isomorphism $H \to \hat{H}$ for suitably large spaces H and $\hat{H}$. Then for $\hat{f} \in \hat{H}$, $P\mathbb{P}\hat{f} = \langle P\mathbb{P}\theta(y,\lambda),\ f(y)\rangle = \hat{f}$. Next we can write

$$(3.12)\qquad B^*f(x) = \langle \beta(y,x),\ f(y)\rangle = \langle\langle \Omega(x,\lambda),\ \theta(y,\lambda)\rangle,\ f(y)\rangle$$

$$= \langle \Omega(x,\lambda),\ \langle \theta(y,\lambda),\ f(y)\rangle\rangle = \mathbb{P}Qf(x)$$

and for $g = B^*f = \mathbb{P}Qf = \mathbb{P}\hat{f} \in \mathbb{P}\hat{H}$ we have $\mathbb{P}Pg = \langle \mathbb{P}P\beta(y,x),\ f(y)\rangle = \langle \beta(y,x)\ f(x)\rangle = g$. Hence $P = \mathbb{P}^{-1} : \mathbb{P}\hat{H} \to \hat{H}$ and since $P\mathbb{P}\hat{H} = \hat{H}$ there results also $\mathbb{P} = P^{-1} : \hat{H} \to \mathbb{P}\hat{H}$. We state this as

Theorem 3.5 If $\beta(y,x) = b(y,x)$ then under the conditions indicated $\mathbb{P} = P^{-1}$.

Remark 3.6 It follows also that $Q^{-1}PB^*f = f$ from (3.12) and Theorem 3.5 so that $(B^*)^{-1} = Q^{-1}P : \mathbb{P}\hat{H} \to H$.

We show next the relation of (3.8) to a generalization of a procedure of Levitan [1] for solving (3.1). First let us write (3.8) as

$$(3.13)\qquad \phi(x,y) = \langle\langle \Omega(\xi,\lambda),\ \theta(y,\lambda)\rangle,\ U(x,\xi)\rangle$$

$$= \langle \theta(y,\lambda),\ \langle \Omega(\xi,\lambda),\ U(x,\xi)\rangle\rangle.$$

Now for $\psi(x,\lambda) = \langle \Omega(\xi,\lambda),\ U(x,\xi)\rangle$ we have $P(D_x)\psi = \langle \Omega(\xi,\lambda),\ P(D_x)U(x,\xi)\rangle = \langle \Omega(\xi,\lambda),\ P(D_\xi)U(x,\xi)\rangle = \langle P^*(D_\xi)\Omega(\xi,\lambda),\ U(x,\xi)\rangle = -\lambda^2\psi$. If $Ph = -\lambda^2 h$ has unique solutions with $h(0,\lambda) = 1$, perhaps subject to some growth conditions, then $\psi(x,\lambda) = \psi(0,\lambda)h(x,\lambda)$ with $\psi(0,\lambda) = \langle \Omega(\xi,\lambda),\ U(0,\xi)\rangle = \langle \Omega(\xi,\lambda),\ f(\xi)\rangle = F(\lambda)$. Using this we can write from (3.13)

$$(3.14)\qquad \phi(x,y) = \langle \theta(y,\lambda),\ F(\lambda)h(x,\lambda)\rangle.$$

We can arrive at this formula in another way by generalizing a procedure of Levitan [1]. Thus as a solution to (3.1) try

$$(3.15)\qquad \phi(x,y) = \langle c(y,\lambda),\ h(x,\lambda)\rangle.$$

Given $\mathbb{P} = P^{-1}$ as in Theorem 3.5, from the relation

$$\theta(y,\lambda) = \mathcal{P}\mathbb{P}\theta(y,\cdot) = \langle h(x,\lambda), \langle \Omega(x,\mu), \theta(y,\mu)\rangle\rangle$$

$$= \langle\langle h(x,\lambda), \Omega(x,\mu)\rangle, \theta(y,\mu)\rangle$$

it is suggested that

(3.16) $\langle h(x,\lambda), \Omega(x,\mu)\rangle = \delta(\lambda - \mu)$

and this will be confirmed in numerous examples below. Using this we get formally from (3.15)

(3.17) $c(y,\mu) = \langle \phi(x,y), \Omega(x,\mu)\rangle$.

Then $Q(D_y)c = \langle Q(D_y)\phi(x,y), \Omega(x,\mu)\rangle = \langle P(D_x)\phi(x,y), \Omega(x,\mu)\rangle = \langle \phi(x,y), P^*(D_x)\Omega(x,\mu)\rangle = -\mu^2 c$. If $Q\theta = -\mu^2\theta$ has unique solutions with $\theta(0,\mu) = 1$, perhaps subject to some growth conditions, it follows that $c(y,\mu) = F(\mu)\theta(y,\mu)$ where $F(\mu) = c(0,\mu) = \langle f(x), \Omega(x,\mu)\rangle$ from (3.17). Putting this $c(y,\mu)$ in (3.15) we obtain (3.14) again. Thus

Theorem 3.7 Let $F(\lambda) = \langle \Omega(x,\lambda), f(x)\rangle$ and suppose $\beta(0,x) = \delta(x)$ so that $\phi(x,y)$ given by (3.8) represents the solution of (3.1). If $h(x,\lambda)$ is unique as indicated the ϕ can be written in the form (3.14). Similarly if (3.16) holds and $\theta(y,\lambda)$ is unique as indicated then (3.14) represents the solution of (3.1).

Remark 3.8 The question of inverting a transform such as $\mathcal{P}$ in (3.9) by a formula $f(x) = H(x,\lambda), F(\lambda)\rangle$ is studied in Mercer [1; 2; 3] (cf. also Titchmarsh [1]). The general eigenfunction expansion theory due to Weyl, Kodaira, Titchmarsh, et. al. will be exploited later (cf. Titchmarsh [2]).

Before developing more general results or refining the above let us indicate how these formal theorems work in applications; this will provide some useful guidelines.

Example 3.9 Take first $P(D) = D^2 = P^*(D)$ with $h(x,\lambda) = \exp(i\lambda x)$ and $\Omega(x,\lambda) = \frac{1}{2\pi}\exp(-i\lambda x)$. Let $Q(D)$ be arbitrary such that the requirements indicated above are satisfied and we will say in general that such Q are P-admissible. Take

$\theta(0,\lambda) = 1$ and from (3.6) we have $\beta(y,x) = \frac{1}{2\pi} <\theta(y,\lambda), \exp(-i\lambda x)>$ while it is known that $\beta(0,x) = \frac{1}{2\pi} <\exp(-i\lambda x), 1> = \delta(x)$. Recall here that $\frac{1}{2\pi}\int_{-\infty}^{\infty} \exp(-i\lambda x)d\lambda = \lim_{N\to\infty} (\text{Sin } Nx/\pi x) = \delta(x)$. Hence Theorem 3.3 applies and from (3.4) we have $\theta(y,\lambda) = <\beta(y,x), \exp(i\lambda x)>$. The formulas (3.9) - (3.10) with Theorem 3.5 then yield the Fourier inversion formulas as a by product. This cannot really be considered as a new derivation however since a traditional demonstration uses the formula $\delta(x - \xi) = \lim_{N\to\infty} (\text{Sin } N(x - \xi)/\pi(x - \xi))$ (cf. Sansone [1]) and this is equivalent to $\beta(0,x) = \delta(x)$.

Remark 3.10 Let us define

(3.18) $\quad \mathcal{B}f(\lambda) = <\Omega(x,\lambda), f(x)>$

(this transform was used in Theorem 3.7). Then under the conditions of Lemma 3.1 and Theorem 3.3 we can write

(3.19) $\quad (Bf)(y) = <\beta(y,x), f(x)> = <<\Omega(x,\lambda), \theta(y,\lambda)>, f(x)>$

$$= <<\beta(y,\xi), h(\xi,\lambda)>, <\Omega(x,\lambda), f(x)>> = <P\beta(y,\lambda), \mathcal{B}f(\lambda)>$$

and this is a kind of Parseval formula. Indeed for Example 3.9 we have $Pf = Ff$ and $\mathcal{B}f = F^{-1}f$ where F denotes the Fourier transform. Evidently $\mathcal{B}f = \frac{1}{2\pi} F[f(-x)] = F^{-1}f$ so we can identify with the Parseval formula written as $<FQ, F\psi> = 2\pi<Q,\psi(-x)>$ by putting $f(-x)$ in place of $f(x)$ in (3.19).

Example 3.11 Let $P(D) = D^2$ and $Q(D) = L_m(D)$ as in Example 2.6. Let $h(x,\lambda) = \exp(i\lambda x)$ and $\theta(y,\lambda) = \hat{R}^m(y,\lambda)$ so by (3.4)

(3.20) $\quad \theta(y,\lambda) = (B_m h)(y,\lambda) = F_x b_m(y,x) = \hat{R}^m(y,\lambda)$

and condition C can be taken as $D_y\theta(0,\lambda) = 0$ which holds; here $L_m B_m = B_m D^2$ as in Example 2.6 and $\beta(0,x) = \delta(x)$ by Example 3.9 when $\Omega(x,\lambda) = \frac{1}{2\pi}\exp(-i\lambda x)$. Thus $Pf = Ff$, $\mathcal{B}f = F^{-1}f$, and we take $\beta(y,x) = <\Omega(x,\lambda), \theta(y,\lambda)>_+$ so that if $F(\lambda) = f(\lambda)$ with f and F even

(3.21) $$\frac{1}{2}(B_m f)(y) = \langle \hat{R}^m(y,\lambda),\ F(\lambda)\rangle_+ = 2^m\Gamma(m+1)\int_0^\infty (\lambda y)^{-m} J_m(\lambda y)F(\lambda)d\lambda$$

$$= 2^m\Gamma(m+1)y^{-m-\frac{1}{2}}\,\mathbb{H}_m(\lambda^{-m-\frac{1}{2}}F(\lambda))$$

where $\mathbb{H}_m$ denotes the Hankel transform. The Hankel transform works on functions

$$g \in H_m = \{g \in C^\infty(0,\infty);\ \ \gamma^m_{n,k}(g) = \sup|x^n(x^{-1}D)^k(x^{-m-\frac{1}{2}}g(x))| < \infty\}$$

or its dual H'_m (see Zemanian [1] and cf. also Dube-Pandey [1; 2], Lee [1], A. Schwartz [1], Trione [1]). Hence for suitable f such that $\lambda^{-m-\frac{1}{2}}F(\lambda) \in H'_m$ we can compute the Hankel transform in (3.21) with inversion formula $(\lambda \geq 0)$

(3.22) $$2F(\lambda) = \frac{\lambda^{m+\frac{1}{2}}}{\Gamma(m+1)2^m}\int_0^\infty y^{m+\frac{1}{2}}(\lambda y)^{\frac{1}{2}}J_m(\lambda y)(B_m f)(y)dy$$

$$= \frac{1}{\Gamma^2(m+1)2^{2m}}\int_0^\infty (\lambda y)^{2m+1}\hat{R}^m(y,\lambda)(B_m f)(y)dy.$$

Let us write (recall $F(\lambda) = (\mathcal{P}f)(\lambda)$)

(3.23) $$(R^m F)(y) = \frac{1}{2}(B_m f)(y) = \langle \hat{R}^m(y,\lambda),\ F(\lambda)\rangle_+$$

$$= \langle \theta(y,\lambda),\ F(\lambda)\rangle_+ = (QF)(x).$$

Then an inverse S^m for R^m is defined by (3.22); setting $k_m = 1/2^{2m}\Gamma^2(m+1)$ let for $\lambda \geq 0$

(3.24) $$(S^m g)(\lambda) = k_m\langle (\lambda y)^{2m+1}\hat{R}^m(y,\lambda),\ g(y)\rangle_+$$

for suitable g and $F(\lambda) = S^m(R^m F)(\lambda)$ (cf. also Levitan-Sargsyan [2]). Since $F(\lambda) = \mathcal{P}f(\lambda) = F^{-1}f$ we have a formula $f = FF = \frac{1}{2}FS^m B_m f$ so that $B_m = B_m^{-1} = \frac{1}{2}FS^m$. There is a technical point involved here in the interplay between brackets on $(0,\infty)$ and $(-\infty,\infty)$ which requires $S^m g(\lambda)$ even in λ; we have chosen to work with even f and F and introduced a factor of $1/2$ in (3.21). Alternatively we could work solely on $(0,\infty)$ with $h(x,\lambda) = \Omega(x,\lambda) = \text{Cos}\ \lambda x$ (cf. also Example 3.14). Now the operator B_m should be a transmutation operator such that $D^2 B_m = B_m L_m$ and we can check this directly (cf Lions [2; 3; 4] and Example

3.14 below). Thus let $g(y)$ be given for $y \geq 0$ ($g'(0) = 0$ is not necessary) and from the definitions

$$2(\mathcal{B}_m g)(x) = (\mathcal{F}\mathcal{S}^m g)(x) = k_m \langle \exp i\lambda x, \langle (\lambda y)^{2m+1}\hat{R}^m(y,\lambda), g(y)\rangle_+ \rangle$$

we obtain immediately $D^2_m \mathcal{B}_m g = -\lambda^2 \mathcal{B}_m g$. On the other hand

$$(3.25) \qquad \langle (\lambda y)^{2m+1}\hat{R}^m(y,\lambda), L_m g(y)\rangle_+ = \langle L^*_m[(\lambda y)^{2m+1}\hat{R}^m(y,\lambda)], g(y)\rangle_+$$

$$= -\lambda^2 \langle (\lambda y)^{2m+1}\hat{R}^m(y,\lambda), g(y)\rangle_+$$

where $L^*_m \psi = D^2\psi - (2m+1)D(y^{-1}\psi)$. Here one has $L^*_m[(\lambda y)^{2m+1}\hat{R}^m(y,\lambda)] = (\lambda y)^{2m+1} L_m \hat{R}^m(y,\lambda)$ and a simple calculation shows that the terms at $y = 0$ arising from integration by parts will vanish (recall $2m+1 > 0$ or $\mathrm{Re}(2m+1) > 0$). Hence we can state (cf. also Example 3.14)

<u>Theorem 3.12</u> $B_m = \mathcal{Q}\mathcal{B} = R^m \mathcal{F}^{-1}$ satisfies $L_m B_m = B_m D^2$ while $\mathcal{B}_m = B_m^{-1} = \frac{1}{2}\mathcal{F}\mathcal{S}^m$ satisfies $D^2 \mathcal{B}_m = \mathcal{B}_m L_m$. The Parseval relation (3.19) written in the form $2\pi \langle \beta(y,x), f(-x)\rangle = \langle \mathcal{F}\beta(y,\cdot), \mathcal{F}f(\cdot)\rangle$ establishes the connection between (3.21) and (2.11) and the latter can be rewritten in terms of

$$(3.26) \qquad b_m(y,x) = \frac{\Gamma(m+1)}{\Gamma(\frac{1}{2})\Gamma(m+\frac{1}{2})} y^{-2m}(y^2 - x^2)_+^{m-\frac{1}{2}}.$$

The formula (3.26) is listed for comparison purposes later (cf (3.61)). For perspective we will organize here a table of the various transforms coming into play. Thus

<u>Remark 3.13</u> We take h and θ as in (3.2) - (3.3) with $h(0,\lambda) = \theta(0,\lambda) = 1$ so that if the roles of P and Q were reversed then θ would correspond to a new h. Take Ω as in (3.5) so that $\beta(0,x) = \delta(x)$, $\beta(y,x) = b(y,x)$, and Theorem 3.5 holds. Then take $\hat{\Omega}$ so that

$$(3.27) \qquad Q^*(D)\hat{\Omega} = -\lambda^2\hat{\Omega}$$

with $\langle \hat{\Omega}(x,\lambda), 1\rangle = \delta(x)$. In the case of Example 3.11 we have $\hat{\Omega}(y,\lambda) = k_m(\lambda y)^{2m+1}\hat{R}^m(y,\lambda)$. Relative to P we have transforms

(3.28) $\mathcal{P}f(\lambda) = \langle h(x,\lambda), f(x)\rangle$; $\mathbb{P}F(x) = \langle \Omega(x,\lambda), F(\lambda)\rangle$;

$\mathfrak{P}f(\lambda) = \langle \Omega(x,\lambda), f(x)\rangle$; $\mathbb{P}F(x) = \langle h(x,\lambda), F(\lambda)\rangle$

while the corresponding Q transforms are

(3.29) $\mathcal{Q}f(\lambda) = \langle \theta(y,\lambda), f(y)\rangle$; $\mathbb{Q}F(y) = \langle \hat{\Omega}(y,\lambda), F(\lambda)\rangle$;

$\mathcal{Q}f(\lambda) = \langle \hat{\Omega}(y,\lambda), f(y)\rangle$; $\mathbb{Q}F(y) = \langle \theta(y,\lambda), F(\lambda)\rangle$.

In Example 3.11 the transforms in (3.29) involve $\langle\ ,\ \rangle_+$ and some factors of 1/2 appear with $\mathcal{B}_m$ in connection with our using even functions but this should cause no confusion later. Now if everything fits together we should have

(3.30) $\mathbb{P} = \mathcal{P}^{-1}$; $\mathbb{Q} = \mathcal{Q}^{-1}$

whereas in Example 3.11 since the kernels are all functions of $x\lambda$ or $y\lambda$ (i.e. Fourier kernels) we have in addition

(3.31) $\mathfrak{P} = \mathbb{P}^{-1}$; $\mathcal{Q} = \mathbb{Q}^{-1}$.

Since $\frac{1}{2} B_m = \mathbb{Q}\mathfrak{P}$ by (3.23) (resp $B_m^* = \mathbb{P}\mathcal{Q}$ by (3.12)) we have $2B_m^{-1} = \mathfrak{P}^{-1}\mathbb{Q}^{-1} = \mathbb{P}\mathcal{Q}$ (resp. $(B_m^*)^{-1} = \mathcal{Q}^{-1}\mathbb{P}^{-1} = \mathbb{Q}\mathcal{P}$ as in Remark 3.6). In Example 3.11 $\mathfrak{P} = F^{-1}$ and $\mathbb{P} = F$ while $\mathcal{Q} = S^m$ so $B_m^{-1} = \frac{1}{2} F\mathcal{Q}$ as indicated. We refer also to Sections 2.5 and 2.6 for examples of interplay between transforms (also Appendix 2).

<u>Example 3.14</u> Let now $P(D) = L_m(D)$ with $Q(D)$ unspecified but such that B is well defined by (3.1) (such Q will be called P admissible). Let $h(x,\lambda) = \hat{R}^m(x,\lambda)$ and, setting $\theta(0,\lambda) = 1$, $\theta(y,\lambda) = \langle b(y,x), \hat{R}^m(x,\lambda)\rangle_+$. Thus

(3.32) $\theta(y,\lambda) = 2^m\Gamma(m+1)\lambda^{-m-\frac{1}{2}}\mathbb{H}_m[x^{-m-\frac{1}{2}}b(y,x)]$

and this is consistent with the formula for $\mathcal{B}_m$ in Example 3.11 which arises when $Q(D) = D^2$ with $\theta(y,\lambda) = \mathrm{Cos}\,\lambda y$. Indeed from $\mathcal{B}_m = \frac{1}{2}FS^m$ (with $S^m g(\lambda)$ treated as even in λ) with kernel $\beta_m(y,x)$ we have

$$(3.33)\quad \langle \beta_m(y,x),\ f(x)\rangle_+ = (\mathcal{B}_m f)(y) =$$

$$= k_m\varepsilon_m \int_0^\infty \int_0^\infty (\lambda x)^{m+1} \mathrm{Cos}\,\lambda y\ J_m(\lambda x) f(x)\,dx\,d\lambda$$

$$= k_m\varepsilon_m \int_0^\infty x^{m+\frac{1}{2}} f(x)\ \mathbb{H}_m[\lambda^{m+\frac{1}{2}} \mathrm{Cos}\,\lambda y]\,dx$$

where $\varepsilon_m = 2^m\Gamma(m+1)$ and $k_m = \varepsilon_m^{-2}$. Hence we have formally

$$(3.34)\quad \varepsilon_m\beta_m(y,x) = x^{m+\frac{1}{2}}\ \mathbb{H}_m[\lambda^{m+\frac{1}{2}} \mathrm{Cos}\,\lambda y]$$

which is equivalent to (3.32) for $\theta(y,\lambda) = \mathrm{Cos}\,\lambda y$ and $b = \beta_m$. Now a precise description of such distributions $\beta_m(y,x)$ is worth displaying, especially since in the case of general P admissible Q we take $\Omega(x,\lambda) = k_m(\lambda x)^{2m+1}\hat{R}^m(x,\lambda)$ with

$$(3.35)\quad \beta(y,x) = \langle \Omega(x,\lambda),\ \theta(y,\lambda)\rangle_+ = k_m\varepsilon_m x^{m+\frac{1}{2}}\ \mathbb{H}_m[\lambda^{m+\frac{1}{2}}\theta(y,\lambda)]$$

so that $\varepsilon_m\beta(0,x) = x^{m+\frac{1}{2}}\ \mathbb{H}_m[\lambda^{m+\frac{1}{2}}] = \varepsilon_m\delta(x)$ must make sense (cf. Trione [1] for a similar formula where a relation between $\delta(x)$ and $\delta(x^2)$ must be used - cf. also Bochner [2]). For this it clearly suffices to show $\beta_m(0,x) = \delta(x)$.

Thus let us summarize Lions' construction of kernels $\beta_m(y,x)$ of the type required in (3.34). He works directly with the transmutation problem $D^2\mathcal{B}_m = \mathcal{B}_m L_m$ for more general m and begins by setting

$$(3.36)\quad b_m = \frac{2\sqrt{\pi}}{\Gamma(m+1)\Gamma(-m-\frac{1}{2})} = \frac{(-1)^{m+1}2\Gamma(m+{}^3/_2)}{\sqrt{\pi}\ \Gamma(m+1)}$$

and then for $-1 < \mathrm{Re}\ m < -\frac{1}{2}$

$$(3.37)\quad \mathcal{B}_m f(y) = b_m y\int_0^y (y^2 - x^2)^{-m-{}^3/_2}\, x^{2m+1} f(x)\,dx$$

so that the kernel $\beta_m(y,x)$ has the form

$$(3.38)\quad \beta_m(y,x) = b_m y(y^2 - x^2)_+^{-m-{}^3/_2}\, x^{2m+1}$$

(note that $\mathcal{B}_m f(0) = f(0)$ or $\beta_m(0,x) = \delta(x)$). One has a formula then with $-1 < \mathrm{Re}\ m < -\frac{1}{2}$ for comparison with (3.34)

(3.39) $\mathbb{H}_m[\lambda^{m+\frac{1}{2}}\text{Cos } \lambda y] = 2^m\Gamma(m+1)x^{-m-\frac{1}{2}}\beta_m(y,x)$

$$= \frac{2^{m+1}\sqrt{\pi}}{\Gamma(-m-\frac{1}{2})}\, yx^{m+\frac{1}{2}}(y^2 - x^2)_+^{-m-3/2}$$

(cf. Erdelyi et. al. [1]). Starting here one wants to analytically continue $\beta_m(y,x)$ in m to establish formulas for all $\text{Re } m > -\frac{1}{2}$. To do this we write first

(3.40) $$\mathcal{B}_m f(y) = b_m \int_0^1 t^{2m+1}(1 - t^2)^{-m-3/2}\, f(ty)dt$$

and define recursively

$$T_1 f(t,y) = D_t[tf(ty)],\quad T_2 f(t,y) = D_t[t^3 T_1 f(t,y)],\ \ldots,\ T_n f(t,y)$$
$$= D_t[t^3 T_{n-1} f(t,y)],\ \ldots .$$

Then for $-1 < \text{Re } m < n - \frac{1}{2}$ the continuation of $\mathcal{B}_m$ is given by

(3.41) $$\mathcal{B}_m^n f(y) = \hat{c}_m \int_0^1 t^{2m-(2n-3)}(1 - t^2)^{-m+\frac{2n-3}{2}}\, T_n f(t,y)dt$$

where $\hat{c}_m = (-1)^n b_m/(2m+1)(2m-1)\ \ldots\ (2m-(2n-3))$. We note that for $n \geq 1$

(3.42) $$\mathcal{B}_m^n f(0) = (2n-1)(2n-3)\ \ldots\ 1\hat{c}_m f(0)\int_0^1 \frac{t^{2m+1}dt}{(1-t^2)^{m-\frac{2n-3}{2}}} = f(0)$$

since $(2n-1)(2n-3)\ \ldots\ 3\cdot 1 = 2^n\Gamma(n+\frac{1}{2})/\Gamma(\frac{1}{2})$, $(2m+1)\ \ldots\ (2m-(2n-3)) = 2^n\Gamma(m+3/2)/\Gamma(m-\frac{2n-3}{2})$, $\Gamma(-m+\frac{2n-1}{2})\Gamma(m-\frac{2n-3}{2}) = (-1)^{m-n+1}\pi$, $\Gamma(m+3/2)\Gamma(-m-\frac{1}{2}) = (-1)^{m+1}\pi$, and the integral in (3.42) equals $\Gamma(m+1)\Gamma(-m+\frac{2n-1}{2})/2\Gamma(n+\frac{1}{2})$. Let us give a different expression for the kernel $\beta_m^n(y,x)$ of $\mathcal{B}_m^n$ as follows. It is easy to see that

(3.43) $$T_n f(t,y) = \sum_{k=1}^{n+1} c_{nk} t^{2n+k-3} y^{k-1} f^{(k-1)}(ty)$$

where $c_{n1} = (2n-1)(2n-3)\ \ldots\ 3\cdot 1 = 2^n\Gamma(n+\frac{1}{2})/\Gamma(\frac{1}{2})$ so that $\mathcal{B}_m^n$ can be written for $m < n - \frac{1}{2}$ as

$$(3.44)\qquad \mathcal{B}_m^n f(y) = \tilde{c}_m \sum_{k=1}^{n+1} c_{nk} y^{k-1} \int_0^1 t^{2m+k}(1-t^2)^{-m+\frac{(2n-3)}{2}} f^{(k-1)}(ty)\,dt$$

$$= \tilde{c}_m \sum_{k=1}^{n+1} c_{nk} y^{-2n+1} \int_0^y x^{2m+k}(y^2-x^2)^{-m+\frac{(2n-3)}{2}} f^{(k-1)}(x)\,dx$$

where $\tilde{c}_m = (-1)^{m+n+1}\Gamma(m-\frac{2n-3}{m})/\sqrt{\pi}\,\Gamma(m+1)2^{n-1} = \hat{c}_m$.

Thus we have

$$(3.45)\qquad <\beta_m^n(y,x),\ f(x)> =$$

$$\tilde{c}_m y^{-2n+1} \sum_{k=1}^{n+1} c_{nk} < x^{2m+k}(y^2-x^2)_+^{-m+\frac{2n-3}{2}},\quad D^{k-1}f(x)>_+.$$

Summarizing we have $\mathcal{B}_m$ defined for all $m > -\frac{1}{2}$ and $D^2\mathcal{B}_m = \mathcal{B}_m L_m$. The kernel $\beta_m^n(y,x)$ is a distribution given by (3.41) or (3.45) with $\beta_m^n(0,x) = \delta(x)$. For $-1 < m < -\frac{1}{2}$ we have by (3.39)

$$(3.46)\qquad \lambda^{m+\frac{1}{2}}\text{Cos}\,\lambda y = \mathbb{H}_m[2^m\Gamma(m+1)x^{-m-\frac{1}{2}}\beta_m(y,x)]$$

$$= 2^m\Gamma(m+1)\lambda^{\frac{1}{2}}\mathcal{B}_m(x^{-m}J_m(\lambda x))(y).$$

Now $f(x) = x^{-m}J_m(\lambda x)$ is an analytic function of x and $\mathcal{B}_m(f)$ is equal to any of its determinations via formulas (3.41) or (3.45) for $n \geq 1$. Hence for example express $\mathcal{B}_m f(y)$ by $\mathcal{B}_m^1 f(y) = <\beta_m^1(y,x),\ f(x)>$ so that (3.46) holds with $\mathcal{B}_m^1$ replacing $\mathcal{B}_m$. But $\mathcal{B}_m^1 f$ makes sense for $-1 < m < \frac{1}{2}$ and hence so does (3.46); note here that $x^{-m}J_m(\lambda x) = \lambda^m g_m(x,\lambda)$ where

$$(3.47)\qquad g_m(x,\lambda) = 2^{-m}\sum_{n=0}^{\infty}\frac{(-1)^n(\frac{\lambda x}{2})^{2n}}{n!\,\Gamma(m+n+1)} = 2^{-m}\sum_{n=0}^{\infty}a_n(m)(\lambda x)^{2n}$$

so that (3.46) can be written

$$(3.48)\qquad \text{Cos}\,\lambda y = \Gamma(m+1)\mathcal{B}_m^1(\sum_{n=0}^{\infty}a_n(m)(\lambda x)^{2n}).$$

If $E = C^\infty[0,\infty)$ one knows $m \to \mathcal{B}_m$ is holomorphic with values in $L(E)$ (cf. Lions [2]) while for $\text{Re}\,z > 0$ both $\Gamma(z)$ and $1/\Gamma(z)$ are holomorphic. Hence (3.48) holds for $-1 < \text{Re}\,m < \frac{1}{2}$ and continuing we obtain (3.34) for $\text{Re}\,m > -\frac{1}{2}$ when $\beta_m(y,x)$ is identified with a suitable determination $\beta_m^n(y,x)$.

We remark here also that the function $U(x,\xi)$ indicated in (3.7) satisfying $L_m(D_x)U(x,\xi) = L_m(D_\xi)U(x,\xi)$ with $U(x,0) = f(x)$ (and $U(0,\xi) = f(\xi)$, $U_x(0,\xi) = U_\xi(x,0) = 0$) can be written explicitly in the form

$$(3.49) \qquad U(x,\xi) = \frac{\Gamma(m+1)}{\sqrt{\pi}\,\Gamma(m+\frac{1}{2})}\int_{-1}^{1}(1-t^2)^{m-\frac{1}{2}}f(w)\,dt$$

where $w^2 = x^2 + \xi^2 + 2x\xi t$ (cf. Copson-Erdelyi [1]).

Remark 3.15 Using a P admissible Q such as $Q(D) = D^2$ and knowing that $\beta(0,x) = \langle\Omega(x,\lambda), 1\rangle_+ = \delta(x)$ the inversion theory for P in Theorem 3.5 is equivalent to the Hankel inversion formulas and seems to represent a new derivation. However a certain amount of information about Hankel inversion is implicit in the fact that $\beta(0,x) = \delta(x)$.

Remark 3.16 We will want to deal with formulas of the form (3.16) for example and record here the following observations. Thus take $h(x,\lambda) = \hat{R}^m(x,\lambda)$ and $\Omega(x,\lambda) = k_m(\lambda x)^{2m+1}\hat{R}^m(x,\lambda)$ to obtain by the Hankel formulas

$$(3.50) \qquad \langle\Omega(x,\lambda), h(\xi,\lambda)\rangle_+ = \delta(x-\xi) = x^{m+1}\xi^{-m}\int_0^\infty \lambda J_m(\lambda x)J_m(\lambda\xi)\,d\lambda$$

$$= (x/\xi)^{m+\frac{1}{2}}\int_0^\infty (\lambda x)^{\frac{1}{2}}(\lambda\xi)^{\frac{1}{2}}J_m(\lambda x)J_m(\lambda\xi)\,d\lambda .$$

Similarly by symmetry

$$(3.51) \qquad \langle\Omega(x,\mu), h(x,\lambda)\rangle_+ = \delta(\lambda-\mu) = \mu^{m+1}\lambda^{-m}\int_0^\infty xJ_m(\lambda x)J_m(\mu x)\,dx .$$

We note here also that formally the truth of (3.50) implies $\langle\Omega(x,\lambda), 1\rangle_+ = \delta(x)$ since $h(0,\lambda) = 1$ but the converse does not seem to follow immediatly by analytical methods.

Example 3.17 The following differential equation arises in the study of sperical type functions on pseudo symmetric spaces (cf. Carroll [33; 34], Carroll-Showalter [8], Carroll-Silver [9; 10; 11])

$$(3.52) \qquad R_{tt} + [(2\beta+1)\tanh t + (2p+2\alpha+1)\coth t]R_t$$

$$+ [(p+\rho)^2 + \lambda^2 + q(q+2\beta)\mathrm{sech}^2 t]R = 0.$$

The solution satisfying $R(0) = 1$ and $R_t(0) = 0$ is

(3.53) $$R = R^{p,q}(\lambda,t) =$$

$$\cosh^{-\ell-p} t\ F\left(\frac{\ell+p+q}{2}, \frac{\ell+p-q}{2} - \beta, p+\alpha+1, \tanh^2 t\right)$$

where $\ell = i\lambda + \rho$ and $\rho = \alpha + \beta + 1$. For convenience we will take the special case where $\alpha = 0$, $\beta = -\frac{1}{2}$, $\ell = \frac{1}{2} + i\lambda$, and $q = 0$, which corresponds to working in a symmetric space based on $SL(2, \mathbb{R})$, and setting $p = m$ with $R^{m,0} = S^m$ this yields ($ch = \cosh$, $sh = \sinh$, etc)

(3.54) $$S^m(\lambda,t) = ch^{-\ell-m} t\ F\left(\frac{\ell+m}{2}, \frac{\ell+m+1}{2}, m+1, th^2 t\right)$$

$$= F\left(\frac{m+\frac{1}{2}+i\lambda}{2}, \frac{m+\frac{1}{2}-i\lambda}{2}, m+1, -sh^2 t\right)$$

$$= 2^m \Gamma(m+1) sh^{-m} t\ P^{-m}_{-\frac{1}{2}+i\lambda}(cht)$$

where P^{-m}_{μ} denotes the standard associated Legendre function. Setting

(3.55) $$Q_m(D) = D^2 + (2m+1)\coth t\ D + (m+\tfrac{1}{2})^2$$

the function $S^m(\lambda,t)$ satisfies

(3.56) $$Q_m(D)S^m = -\lambda^2 S^m.$$

We remark also that the formal adjoint of $Q_m(D)$ is given by

(3.57) $$Q^*_m(D)\psi = \psi_{tt} - (2m+1)D(\psi \coth t) + (m+\tfrac{1}{2})^2\psi.$$

It is worth noting here that in analogy with L_m one has for $\Omega^m(\lambda,t) = sh^{2m+1} t\ S^m(\lambda,t)$

(3.58) $$Q^*_m(D)\Omega^m(\lambda,t) = -\lambda^2\Omega^m(\lambda,t).$$

First take $P(D) = D^2$ with $h(x,\lambda) = \exp i\lambda x$ and $\Omega(x,\lambda) = \frac{1}{2\pi}\exp(-i\lambda x)$ while $Q(D) = Q_m(D)$ with $\theta(y,\lambda) = S^m(\lambda,y)$. Then $b(y,x) = \beta(y,x)$ and (3.6) yields

(3.59) $$\beta(y,x) = \langle\Omega(x,\lambda), \theta(y,\lambda)\rangle = \frac{1}{\pi}\int_0^\infty S^m(\lambda,y)\text{Cos}\ \lambda x d\lambda$$

(note $S^m(\lambda,y)$ is even in λ by the second expression in (3.54)). We are in a situation here analogous to Example 3.11 and the same kinds of remarks relative to $\langle\,,\,\rangle_+$ brackets and even functions should apply. Now it is known that if $f(x) = (\text{chy} - \text{chx})_+^{m-\frac{1}{2}}$ then (cf. Magnus-Oberhettinger-Soni [1])

(3.60) $$F_c f = \left(\frac{\pi}{2}\right)^{\frac{1}{2}}\Gamma(m+\tfrac{1}{2})\,\text{sh}^m y\; P^{-m}_{-\frac{1}{2}+i\lambda}(\text{chy})$$

where F_c denotes the Fourier cosine transform. Consequently

(3.61) $$\beta(y,x) = \frac{1}{\pi}\int_0^\infty S^m(\lambda y)\,\text{Cos}\,\lambda x d\lambda$$

$$= \frac{2^{m-\frac{1}{2}}\Gamma(m+1)}{\sqrt{\pi}\;\Gamma(m+\frac{1}{2})}\text{sh}^{-2m}y(\text{chy} - \text{chx})_+^{m-\frac{1}{2}}$$

and it is of interest to compare this formula with (3.26). We recall at this point also the generalized Mehler inversion formulas

(3.62) $$g(\text{cht}) = \int_0^\infty f(\lambda)P^{-m}_{-\frac{1}{2}+i\lambda}(\text{cht})d\lambda$$

(3.63) $$f(\lambda) = c(\lambda)\int_0^\infty g(\text{cht})P^{-m}_{-\frac{1}{2}+i\lambda}(\text{cht})\,\text{sht}dt$$

where $c(\lambda) = \pi^{-1}\lambda\,\text{sh}\pi\lambda\,\Gamma(m+\frac{1}{2}+i\lambda)\Gamma(m+\frac{1}{2}-i\lambda)$ (cf. Lowndes [1], Magnus-Oberhettinger-Soni [1], Oberhettinger-Higgins [1]).

Now let $P(D) = Q_m(D)$ with $h(x,\lambda) = S^m(x,\lambda)$ and $\Omega(x,\lambda) = \hat{c}(\lambda)\text{sh}^{2m+1} \times S^m(x,\lambda)$ and let $Q(D)$ be P admissible with $\theta(0,\lambda) = 1$. From (3.62) - (3.63) we obtain

(3.64) $$\int_0^\infty c(\lambda)P^{-m}_{-\frac{1}{2}+i\lambda}(\text{cht})P^{-m}_{-\frac{1}{2}+i\lambda}(\text{ch}\tau)d\lambda = \frac{\delta(t-\tau)}{\text{sht}}$$

so $\hat{c}(\lambda)$ can be chosen as

(3.65) $$\hat{c}(\lambda) = k_m c(\lambda) = \frac{\lambda\,\text{sh}\pi\lambda\,\Gamma(m+\frac{1}{2}+i\lambda)\Gamma(m+\frac{1}{2}-i\lambda)}{\pi 2^{2m}\Gamma^2(m+1)}$$

from which follows

(3.66) $$\langle\Omega(x,\lambda),\, h(\xi,\lambda)\rangle_+ = \delta(x-\xi)$$

$$2^{2m}\Gamma^2(m+1)\text{sh}^{m+1} \times \text{sh}^{-m}\xi\langle\hat{c}(\lambda)P^{-m}_{-\frac{1}{2}+i\lambda}(\text{chx}),\, P^{-m}_{-\frac{1}{2}+i\lambda}(\text{ch}\xi)\rangle_+.$$

Letting $\xi \to 0$ we have $\langle\Omega(x,\lambda), 1\rangle_+ = \beta(0,x) = \delta(x)$ formally which can be written

$$(3.67)\qquad \delta(x) = \frac{sh^{m+1}x}{2^m\Gamma(m+1)}\int_0^\infty c(\lambda)P^{-m}_{-\frac{1}{2}+i\lambda}(chx)d\lambda.$$

Similarly the formula $\langle\Omega(x,\mu), h(x,\lambda)\rangle_+ = \delta(\lambda-\mu)$ follows from (3.62) - (3.63) (cf (3.16)). In particular if we know somehow that the choice of $\hat{c}(\lambda)$ in (3.65) will yield $\beta(0,x) = \delta(x)$ then $\beta(y,x) = b(y,x)$ and the inversion theory for P in Theorem 3.5 yields the generalized Mehler inversion (3.62) - (3.63).

Remark 3.18 We mention here some work of Chebli [1] which connects several points of view (cf also Bochner [1]). Let L be an operator of the form

$$(3.68)\qquad Lu = -\frac{1}{A(x)}\frac{d}{dx}(A(x)u')$$

on a domain $D(L) \subset L^2(A(x)dx)$ over $(0,\infty)$ where $A \in C^\infty(0,\infty)$, $A(0) = 0$, $A(x) > 0$ for $x > 0$, and $(A'/A)(x) = \alpha/x + B(x)$ with $B \in C^0[0,\infty)$. Further let A be increasing with $A(x) \to \infty$ as $x \to \infty$ and A'/A decreasing; set $\rho = \frac{1}{2}\lim A'/A$ as $x \to \infty$. These hypotheses are satisfied by the radial part of the Laplace-Beltrami operator on noncompact Riemannian symmetric spaces of rank 1. Let $H^1 = \{u,u' \in L^2(A(x)dx)\}$, $H^2 = \{u \in H^1;\ Lu \in L^2(A(x)dx)\}$, and $D(L) = \{u \in H^2;\ A(x)u'(x) \to 0$ at $\infty\}$ or equivalently $D(L) = \{u \in H^1;\ v \to \int_0^\infty u'\bar{v}A(x)dx : H^1 \to \mathbb{C}$ is continuous in the norm of $L^2(A(x)dx)\}$ (note

$$\int Lu\bar{v}\, Adx = \int u'\bar{v}'\, Adx = Au'\bar{v}\Big|_0^\infty - \int \frac{1}{A}(Au')'\bar{v}dx).$$

Let $\mathcal{D}_0(\mathbb{R})$ be C^∞ even functions on $\mathbb{R}$ with compact support; $\mathcal{D}_0(\mathbb{R})$ is dense in $D(L)$ in graph norm and for $\phi \in \mathcal{D}_0(\mathbb{R})\int_0^\infty L\phi\bar{\phi} - Adx = \int_0^\infty |\phi'|^2 Adx \geq 0$ so $\sigma(L) \subset [0,\infty)$. Recall now Remark 1.2 defining generalized translation operators and take for example $L = -(D_x^2 + \frac{1}{x}D_x) = -\frac{1}{x}D_x(xD_x)$. Then the solution of $L\phi = \lambda\phi, \phi(0,\lambda) = 1$, $\phi'(0,\lambda) = 0$ written as $\phi(y,\lambda) = \sum_{k=0}^\infty \phi_k(y)\lambda^k$ is given by

$$(3.69)\qquad \phi(y,\lambda) = J_0(y\sqrt{\lambda}) = \sum_{k=0}^\infty (-1)^k\frac{y^k\lambda^k}{2^k(k!)^2}$$

and $T^y = \phi(y,L)$ so that one obtains after some calculation

(3.70) $$T_x^y f = \frac{1}{2\pi}\int_0^{2\pi} f(\sqrt{x^2 + y^2 + 2xy\,\mathrm{Cos}\,\theta})\,d\theta$$

(cf (3.49)). A second way to define T^y is via the Cauchy problem $L_x u = L_y u$, $u(x,0) = f(x)$, with $u(x,y) = T^y f(x) = T_x^y f$ (cf. (3.7)). Now in general if L has a spectral decomposition $L = \int \lambda dE_\lambda$ then in the spirit of $T^y = \phi(y,L)$ one can define

(3.71) $$T^y = \int \phi(y,\lambda)\,dE_\lambda$$

which one can make precise as follows for L of the type above (see Titchmarsh [2] for details and cf. Section 2.4). Let $f \in \mathcal{D}_o(\mathbb{R})$ and define a generalized Fourier transform $\hat{f}$ by

(3.72) $$\hat{f}(\lambda) = \int_0^\infty f(x)\phi(x,\lambda)A(x)\,dx.$$

Then there is a tempered positive measure σ on $\sigma(L) \subset [0,\infty)$ such that $\int_0^\infty |f|^2 A dx = \int_0^\infty |\hat{f}|^2 d\sigma$ and $f \to \hat{f}$ extends to an isometric isomorphism $L^2(Adx) \to L^2(\sigma)$ with inverse

(3.73) $$f(x) = \int_0^\infty \hat{f}(\lambda)\phi(x,\lambda)\,d\sigma(\lambda).$$

Further $D(L) = \{f; \int \lambda^2 |\hat{f}(\lambda)|^2 d\sigma(\lambda) < \infty\}$ and $\widehat{Lf} = \lambda\hat{f}$. One checks from the first definition of $T^y = \phi(y,L)$ that

$$T^y \phi(x,\lambda) = \phi(y,\lambda)\phi(x,\lambda)$$

so that in a spectral form we define a self adjoint T^y by

(3.74) $$\widehat{T^y f}(\lambda) = \phi(y,\lambda)\hat{f}(\lambda).$$

Using this background Chebli [1] obtains some results of interest in potential theory and harmonic analysis; we will not dwell on this but mention only a few properties here of general interest. Define T^y by (3.74); then T^y is a bounded self adjoint operator in $L^2(Adx)$.

<u>Theorem 3.19</u> T^y is sub-Markov in the sense that if $0 \leq f \leq 1$ a.e. then

$0 \le T^y f \le 1$ a.e. T^y is bounded in $L^p(Adx)$ for $0 \le p \le \infty$ (p integral) and $\| T^y f\|_p \le \| f\|_p$. If $f * g$ is defined for $f,g \in \mathcal{D}_0(\mathbb{R})$ by

$$(3.75) \qquad (f * g)(x) = \int_0^\infty T^y f(x) g(y) A(y) dy$$

then $*$ extends to $L^1(Adx)$ which becomes a commutative Banach algebra with characters $\phi(x,\lambda)$ for $\lambda \in \Sigma = \{\lambda = \lambda_1 + i\lambda_2;\ \lambda_2^2 \le 4\rho^2\lambda_1\}$.

The commutativity of $*$ is related to the property $T_x^\xi g(x) = T_\xi^x g(\xi)$ (cf. Section 2.5 for further discussion).

Remark 3.20 Let us expand the formulation of Remark 3.18 to relate it to some of the formulas developed earlier in Section 2.3 (cf. also Section 2.4 where such examples are indicated in a general manner). Thus take $L = -P$ in (3.68) so that the solution $\phi(x,\mu)$ of $L\phi = \mu\phi$, $\phi(0,\mu) = 1$, $\phi'(0,\mu) = 0$ is precisely $h(x,\lambda)$ for $\mu = \lambda^2$ (i.e. $h(x,\lambda) = \phi(x,\lambda^2)$). The equation $P(D_x)u = Q(D_y)u$ becomes $Lu = -Qu$ and we write

$$\hat{u}(\mu,y) = \int_0^\infty u(x,y)\phi(x,\mu)A(x)dx$$

(cf (3.72)). It follows that $\widehat{Lu} = \mu\hat{u}$ and $-\widehat{Qu} = -Q\hat{u}$ so

$$(3.76) \qquad Q\hat{u} = -\mu\hat{u};\quad \hat{u}(\mu,0) = \hat{f}(\mu)$$

since $u(x,0) = f(x)$. Thus $\hat{u}(\mu,y) = \hat{f}(\mu)\theta(y,\sqrt{\mu}) = \hat{f}(\mu)\psi(y,\mu)$ where $Q\theta = -\lambda^2\theta$, $\theta(0,\lambda) = 1$, and $\mu = \lambda^2$. Using the inversion formula (3.73) we obtain

$$(3.77) \qquad u(x,y) = \int_0^\infty \hat{f}(\mu)\psi(y,\mu)\phi(x,\mu)d\sigma(\mu).$$

Now L is self adjoint relative to $A(x)dx$ so $P^* = P$ and we can take $\Omega(x,\lambda) = w(\lambda)h(x,\lambda)$. Thus (3.77) has the form (3.14) with suitable definition of the brackets. We note that the convolution (3.75) is the same operation indicated by our generalized convolution of (3.8) since $U(x,\xi) = T_x^\xi f(x)$ from (3.7) and (3.75) can be written

$$(3.78) \qquad (f * g)(x) = \langle U(x,\xi), g(\xi)\rangle.$$

2.4. More on Kernels and transforms. There are now several aspects of the formal development in the last section which must be investigated in more detail. In particular it was assumed that all of the bracket constructions and operations upon them made sense (justified for special cases in the Examples). Also no directive for choosing $\Omega(x,\lambda)$ was given and we will examine that first. Suppose that $P(D) = D^2 + p(x)D + q(x)$. Then it is well known that $P(D)$ can be put into a self adjoint form by writing $P(D)y = \exp(-\int p)(y' \exp\int p)' + qy$ but we refrain from doing this since it is convenient to deal with solutions $h(x,\lambda)$ of $P(D)h = -\lambda^2 h$ satisfying $h(0,\lambda) = 1$. Now in the Examples 3.14 and 3.17 we used functions $\Omega(x,\lambda)$ satisfying $P^*(D)\Omega = -\lambda^2\Omega$ which had the form $\Omega = \alpha h$ where $P(D)h = -\lambda^2 h$ and it turns out that this is generally possible. Indeed let us try to find $\phi = \alpha\psi$ satisfying $P^*(D)\phi = 0$ given that $P(D)\psi = 0$. Computing $P^*(D)\phi$ we find that $\psi T' + 2\psi' T = 0$ where $T = \alpha' - p\alpha$. Hence $T = k\psi^{-2}$ and

$$(4.1) \qquad \phi = \alpha\psi = k\psi e^{\int p} \int \psi^{-2} e^{-\int p}.$$

But by a standard reduction of order technique it is known that $k\psi\int\psi^{-2}\exp(-\int p) = \gamma$ is a second solution of $P(D)\gamma = 0$ and hence there is a solution ϕ of $P^*(D)\phi = 0$ of the form $\phi = \gamma \exp\int p$ where $P(D)\gamma = 0$. This can also be confirmed directly and we record this surely well known fact as

<u>Lemma 4.1</u> Given $P(D) = D^2 + p(x)D + q(x)$ there is a solution $\phi = \gamma \exp\int p$ of $P^*(D)\phi = 0$ when $P(D)\gamma = 0$.

The functions Ω in Examples 3.14 and 3.17 are of this form. Next we observe that with γ and ϕ as above $(P(D)\gamma)\phi - \gamma P^*(D)\phi = \Xi = \gamma''\phi - \gamma\phi'' + p(\gamma'\phi + \gamma\phi') + p'\gamma\phi$ and $\gamma''\phi - \gamma\phi'' = -(2\gamma'\gamma p + \gamma^2 p' + \gamma^2 p^2)\exp\int p$ while $\gamma'\phi + \gamma\phi' = (2\gamma'\gamma + \gamma^2 p)\exp\int p$ so we have $\Xi = 0$. Similarly if $P(D)\gamma = -\lambda^2\gamma$ and $P(D)\tilde{\gamma} = -\mu^2\tilde{\gamma}$ with $\tilde{\phi} = \tilde{\gamma}\exp\int p$ $(\gamma = \gamma(x,\lambda),\ \tilde{\gamma} = \gamma(x,\mu))$ then $(P(D)\gamma)\tilde{\phi} - \gamma P^*(D)\tilde{\phi} = (\mu^2 - \lambda^2)\gamma\tilde{\gamma}\exp\int p = [(\gamma''\tilde{\gamma} - \gamma\tilde{\gamma}'') + p(\gamma'\tilde{\gamma} - \gamma\tilde{\gamma}')]\exp\int p$. Setting $W = \gamma'\tilde{\gamma} - \gamma\tilde{\gamma}'$ for the Wronskian there results

$$(4.2) \qquad (W' + pW)e^{\int p} = (P(D)\gamma)\tilde{\phi} - \gamma P^*(D)\tilde{\phi} = (\mu^2 - \lambda^2)\gamma\tilde{\gamma}e^{\int p}.$$

Since $(W' + pW)e^{\int p} = (We^{\int p})'$ we obtain

$$(4.3)\qquad \langle P(D)\gamma, \tilde{\phi}\rangle - \langle\gamma, P^*(D)\tilde{\phi}\rangle = (\mu^2 - \lambda^2)\int_a^b \gamma\tilde{\gamma}e^{\int p}dx = We^{\int p}\Big|_a^b$$

for brackets over an interval (a,b). The equation in the second line of (4.3) also follows directly from $P(D)\gamma = -\lambda^2\gamma$ and $P(D)\tilde{\gamma} = -\mu^2\tilde{\gamma}$. We summarize this in

Lemma 4.2 Let $P(D)\gamma = -\lambda^2\gamma$ and $P(D)\tilde{\gamma} = -\mu^2\tilde{\gamma}$ with $\tilde{\phi} = \tilde{\gamma}\exp\int p$. Then (4.3) holds where $W = \gamma'\tilde{\gamma} - \gamma\tilde{\gamma}'$. In particular if $We^{\int p} = 0$ at $x = a$ and $x = b$ then $\langle P(D)\gamma, \tilde{\phi}\rangle = \langle\gamma, P^*(D)\tilde{\phi}\rangle$ while γ and $\tilde{\gamma}$ are orthogonal relative to the weight function $\exp\int p$.

Example 4.3 Let us look at these formulas when $P(D) = L_m(D) = D^2 + \frac{2m+1}{x}D$. Let $\gamma = \hat{R}^m(x,\lambda)$ and $\tilde{\gamma} = \hat{R}^m(x,\mu)$ with $\tilde{\phi} = x^{2m+1}\hat{R}^m(x,\mu)$ since $\exp\int p = x^{2m+1}$. By (1.3.9) $\hat{R}^m_x(x,\lambda) = -x\lambda^2\hat{R}^{m+1}(x,\lambda)/2(m+1)$ so that

$$(4.4)\qquad W = \frac{x}{2(m+1)}[\mu^2\hat{R}^m(x,\lambda)\hat{R}^{m+1}(x,\mu) - \lambda^2\hat{R}^{m+1}(x,\lambda)\hat{R}^m(x,\mu)]$$

$$= \frac{2^{2m}\Gamma^2(m+1)}{x^{2m}\lambda^m\mu^m}[\mu J_m(\lambda x)J_{m+1}(\mu x) - \lambda J_{m+1}(\lambda x)J_m(\mu x)].$$

We recall some formulas often used in connection with Hankel inversion (cf. Sneddon [1]). First for $m > -\frac{1}{2}$

$$(4.5)\qquad \int_0^x \xi J_m(\lambda\xi)J_m(\mu\xi)d\xi = h_m(\lambda,\mu,x)$$

$$= \frac{x}{\mu^2 - \lambda^2}[\mu J_m(\lambda x)J_{m+1}(\mu x) - \lambda J_{m+1}(\lambda x)J_m(\mu x)].$$

Consequently, since $e^{\int p} = x^{2m+1}$, we obtain from (4.4)

$$(4.6)\qquad We^{\int p} = \frac{2^{2m}\Gamma^2(m+1)}{\lambda^m\mu^m}(\mu^2 - \lambda^2)\int_0^x \xi J_m(\lambda\xi)J_m(\mu\xi)d\xi.$$

Clearly $We^{\int p} = 0$ at $x = a = 0$ from (4.4) and we consider its behavior at $x = b = \infty$. It is known that

(4.7) $$\int_0^\infty \xi e^{-\varepsilon^2\xi^2} J_m(\lambda\xi) J_m(\mu\xi) d\xi = H_m(\lambda,\mu,\varepsilon)$$
$$= \frac{1}{2\varepsilon^2} I_m\left(\frac{\lambda\mu}{2\varepsilon^2}\right) \exp\left(-\frac{\lambda^2+\mu^2}{4\varepsilon^2}\right)$$

while, for small ε, $I_m\left(\frac{\lambda\mu}{2\varepsilon^2}\right) \sim \varepsilon \exp\left(\frac{\lambda\mu}{2\varepsilon^2}\right)/\sqrt{\pi\lambda\mu}$. Hence for small ε

(4.8) $$H_m(\lambda,\mu,\varepsilon) \sim \frac{1}{2\varepsilon\sqrt{\pi\lambda\mu}} \exp\left(-\frac{(\lambda-\mu)^2}{4\varepsilon^2}\right).$$

Now let us define $h_m(\lambda,\mu,\infty) = H_m(\lambda,\mu,0)$ so that for $\lambda \neq \mu$ $H_m(\lambda,\mu,0) = 0$ implies $We^{\int p} = 0$ at $x = \infty$. On the other hand as $\varepsilon \to 0$

(4.9) $$\int_{\mu-\delta}^{\mu+\delta} (\lambda\mu)^{\frac{1}{2}} H_m(\lambda,\mu,\varepsilon) d\lambda \sim \frac{1}{\sqrt{\pi}} \int_{-\delta/2\varepsilon}^{\delta/2\varepsilon} e^{-z^2} dz \to 1$$

so that $(\lambda\mu)^{\frac{1}{2}} H_m(\lambda,\mu,0) = \delta(\lambda-\mu)$. This seems to say that $W \exp \int p =$ $2^{2m}\Gamma^2(m+1)(\lambda\mu)^{-m}(\mu^2-\lambda^2)h_m \to 2^{2m}\Gamma^2(m+1)(\lambda\mu)^{-m-\frac{1}{2}}(\mu^2-\lambda^2)(\lambda\mu)^{\frac{1}{2}}H_m(\lambda,\mu,0)$ which equals zero. However from (4.3)

(4.10) $$\int_0^\infty \gamma(x,\lambda)\tilde{\gamma}(x,\mu)x^{2m+1}dx = 2^{2m}\Gamma^2(m+1)(\lambda\mu)^{-m-\frac{1}{2}}\delta(\lambda-\mu)$$

and this is precisely (3.51) (note $(\lambda\mu)^{-m-\frac{1}{2}}$ in (4.10) can be interperted as μ^{-2m-1} and $k_m = 2^{-2m}\Gamma^{-2}(m+1)$ with $h(x,\lambda) = \gamma(x,\lambda)$ and $\Omega(x,\mu) = k_m(\mu x)^{2m+1}\tilde{\gamma}(x,\mu)$).

We recall here in passing someother standard reductions of $ay'' + by' + cy = 0$ to self adjoint form. Thus if $\xi = \int a^{-\frac{1}{2}}dx$ then $(D_\xi^2 + \beta D_\xi + c)y = 0$ where $\beta = (b - \frac{1}{2}a')/a^{\frac{1}{2}}$. Then setting $y = Y\exp(-\frac{1}{2}\int\beta)$ we will obtain $D_\xi^2 Y + [c - \frac{1}{4}\beta^2 - \frac{1}{2}\beta']Y = 0$. The latter is a standard form often used in studying self adjoint eigenvalue problems. If we start with $y'' + py' + qy = P(D)y = 0$ then $y = Y\exp(-\frac{1}{2}\int p)$ yields $Y'' + [q - \frac{1}{4}p^2 - \frac{1}{2}p']Y = P_0(D)Y = 0$. If $p = (2m+1)/x$ and $q = \lambda^2$ we have $Y'' + [\lambda^2 - (m^2 - 1/4)/x^2]Y = 0$ with a solution $x^{\frac{1}{2}}J_m(\lambda x) = Y$; since $\exp(-\frac{1}{2}\int p) = x^{-m-\frac{1}{2}}$ this gives $y = x^{-m}J_m(\lambda x)$. We note also that under this reduction if $P(D)y = -\lambda^2 y$ and $P(D)\tilde{y} = -\mu^2\tilde{y}$ with $W = \tilde{y}y' - y'\tilde{y}$ then $\tilde{W} = Y\tilde{Y}' = Y'\tilde{Y} = W\exp(\int p)$ where $y = Y\exp(-\frac{1}{2}\int p)$ and $\tilde{y} = \tilde{Y}\exp(-\frac{1}{2}\int p)$. Another reduction starts from $-(py')' + \ell y = \lambda r y$ on $[a,b]$;

setting $z = \frac{1}{c}\int_a^x (r/p)^{\frac{1}{2}}d\xi$, $u = (rp)^{\frac{1}{4}}y$, and $\mu = c\lambda$ with $c = \frac{1}{\pi}\int_a^b (r/p)^{\frac{1}{2}}d\xi$ we get $-u'' + u = \mu u$ on $[0,\pi]$ where $q = \theta''/\theta - c^2\ell/r$ for $\theta = (rp)^{\frac{1}{4}}$ (cf. Levitan-Sargsyan [2]).

Remark 4.4 It seems appropriate at this point to recall also a few general facts about eigenfunction expansions for second order linear differential operators (in view of Lemmas 4.1 and 4.2). There are several accounts of the theory in book form to which we refer for motivation and details (cf. Akhiezer-Glazman [1], Coddington-Levinson [1], Levitan-Sargsyan [2], Naimark [1], Titchmarsh [2] and also Section 2.5). As the basic equation one usually takes

$$(4.11)\qquad Ly = -y'' + qy = \lambda y.$$

Let $\phi(x,\lambda)$ and $\theta(x,\lambda)$ satisfy (4.11) with $\phi(0,\lambda) = \text{Sin}\,\alpha$, $\phi'(0,\lambda) = -\text{Cos}\,\alpha$, $\theta(0,\lambda) = \text{Cos}\,\alpha$, and $\theta'(0,\lambda) = \text{Sin}\,\alpha$. We work first on the interval $[0,\infty)$ and let ℓ be defined by

$$[\theta(b,\lambda) + \ell\phi(b,\lambda)]\text{Cos}\,\beta + [\theta'(b,\lambda) + \ell\phi'(b,\lambda)]\text{Sin}\,\beta = 0.$$

As $\text{Ctn}\,\beta$ varies from $-\infty$ to ∞ ℓ describes a circle C_b in $\mathbb{C}$ and as $b \to \infty$ C_b either tends to a limit point or a limit circle. If $m(\lambda)$ is the limit point or any point on the limit circle then $\psi(x,\lambda) = \theta(x,\lambda) + m(\lambda)\phi(x,\lambda) \in L^2(0,\infty)$. One defines then

$$(4.12)\qquad \Phi(x,\lambda) = \psi(x,\lambda)\int_0^x \phi(y,\lambda)f(y)dy + \phi(x,\lambda)\int_x^\infty \psi(y,\lambda)f(y)dy$$

and from natural considerations

$$(4.13)\qquad f(x) = \lim_{R\to\infty}\Big[-\frac{1}{\pi i}\int_{-R+i\delta}^{R+i\delta}\Phi(x,\lambda)d\lambda\Big].$$

Define then a nondecreasing function $k(\lambda)$ by

$$(4.14)\qquad k(\lambda) = \lim_{\delta\to 0}\int_0^\lambda [-\text{Im}\,m(u + i\delta)]du$$

and passing R and δ to the limits we obtain

(4.15) $$f(x) = \frac{1}{\pi}\int_{-\infty}^{\infty} \phi(x,\lambda)dk(\lambda)\int_0^{\infty}\phi(y,\lambda)f(y)dy.$$

For the interval $(-\infty,\infty)$ let $\phi(x,\lambda)$ and $\theta(x,\lambda)$ satisfy (4.11) with $\phi(0,\lambda) = -1$, $\phi'(0,\lambda) = -1$, $\theta(0,\lambda) = 1$, and $\theta'(0,\lambda) = 0$. There exist functions $m_1(\lambda)$ and $m_2(\lambda)$ as before such that $\psi_1(x,\lambda) = \theta(x,\lambda) + m_1(\lambda)\phi(x,\lambda) \in L^2(-\infty,0)$ and $\psi_2(x,\lambda) = \theta(x,\lambda) + m_2(\lambda)\phi(x,\lambda) \in L^2(0,\infty)$. A function $\Phi(x,\lambda)$ is defined as

(4.16) $$\Phi(x,\lambda) = \frac{\psi_2(x,\lambda)}{m_1(\lambda) - m_2(\lambda)}\int_{-\infty}^{x}\psi_1(y,\lambda)f(y)dy + \frac{\psi_1(x,\lambda)}{m_1(\lambda) - m_2(\lambda)}\int_{x}^{\infty}\psi_2(y,\lambda)f(y)dy$$

and nondecreasing functions $\xi(\lambda)$ and $\zeta(\lambda)$ are defined by

(4.17) $$\xi(\lambda) = \lim_{\delta\to 0}\int_0^{\lambda} \mathrm{Im}\left[\frac{-1}{m_1(u+i\delta) - m_2(u+i\delta)}\right]du;$$

$$\zeta(\lambda) = \lim_{\delta\to 0}\int_0^{\lambda} \mathrm{Im}\left[\frac{-m_1(u+i\delta)m_2(u+i\delta)}{m_1(u+i\delta) - m_2(u+i\delta)}\right]du$$

while a function $\eta(\lambda)$ of bounded variation is defined by

(4.18) $$\eta(\lambda) = \lim_{\delta\to 0}\int_0^{\lambda} \mathrm{Im}\left[\frac{-m_1(u+i\delta)}{m_1(u+i\delta) - m_2(u+i\delta)}\right]du.$$

for the expansion theorem let

(4.19) $$E(\lambda) = \int_{-\infty}^{\infty}\theta(y,\lambda)f(y)dy; \quad F(\lambda) = \int_{-\infty}^{\infty}\phi(y,\lambda)f(y)dy$$

and then it follows that

(4.20) $$f(x) = \frac{1}{\pi}\int_{-\infty}^{\infty}E(\lambda)\theta(x,\lambda)d\xi(\lambda) + \frac{1}{\pi}\int_{-\infty}^{\infty}F(\lambda)\theta(x,\lambda)d\eta(\lambda) + \frac{1}{\pi}\int_{-\infty}^{\infty}E(\lambda)\phi(x,\lambda)d\eta(\lambda) + \frac{1}{\pi}\int_{-\infty}^{\infty}F(\lambda)\phi(x,\lambda)d\zeta(\lambda).$$

There are situations when the formulas take a simpler form. For example if $m_1(\lambda)$ tends to a real limit when $\mathrm{Im}\,\lambda \to 0$ so that $\psi_1(x,\lambda) \in L^2(-\infty,0)$ we have $d\eta(u) = m_1(u)d\xi(u)$ and $d\zeta(u) = m_1^2(u)d\xi(u)$ so (4.20) becomes

(4.21) $$f(x) = \frac{1}{\pi}\int_{-\infty}^{\infty}\psi_1(x,\lambda)d\xi(\lambda)\int_{-\infty}^{\infty}\psi_1(y,\lambda)f(y)dy.$$

On the other hand if $q(x)$ is even $m_1(\lambda) = -m_2(\lambda)$ and $\eta(\lambda) = 0$ so (4.20) becomes

$$(4.22) \qquad f(x) = \frac{1}{\pi}\int_{-\infty}^{\infty}\theta(x,\lambda)d\xi(\lambda)\int_{-\infty}^{\infty}\theta(y,\lambda)f(y)dy$$

$$+ \frac{1}{\pi}\int_{-\infty}^{\infty}\phi(x,\lambda)d\zeta(\lambda)\int_{-\infty}^{\infty}\phi(y,\lambda)f(y)dy.$$

We note also that (4.15) is often written in another form by taking

$$(4.23) \qquad \chi(x,\lambda) = \int_0^{\lambda}\phi(x,u)dk(u); \quad F(\lambda) = \int_0^{\infty}\chi(y,\lambda)f(y)dy.$$

Then one has

$$(4.24) \qquad f(x) = \frac{1}{\pi}\int_{-\infty}^{\infty}\phi(x,\lambda)dF(\lambda).$$

<u>Example 4.5</u> In the notation of Remark 4.4 the classical Fourier formulas can be written as follows. For $[0,\infty)$

$$(4.25) \qquad \theta(x,\lambda) = \text{Cos }\alpha\ \text{Cos } x\sqrt{\lambda} + \lambda^{-\frac{1}{2}}\text{Sin }\alpha\ \text{Sin } x\sqrt{\lambda};$$

$$\theta(x,\lambda) = \text{Sin }\alpha\ \text{Cos } x\sqrt{\lambda} - \lambda^{-\frac{1}{2}}\text{Cos }\alpha\ \text{Sin } x\sqrt{\lambda}.$$

The function $\psi(x,\lambda)$ must be a multiple of $e^{ix\sqrt{\lambda}}$ if $\text{Im }\lambda > 0$ to be in $L^2(0,\infty)$ and thus

$$(4.26) \qquad m(\lambda) = \frac{\text{Sin }\alpha - i\sqrt{\lambda}\,\text{Cos }\alpha}{\text{Cos }\alpha + i\sqrt{\lambda}\,\text{Sin }\alpha}.$$

Hence $\text{Im } m(\lambda) = -\sqrt{\lambda}/\text{Cos}^2\alpha + \lambda\ \text{Sin}^2\alpha$ for $\lambda > 0$ and is 0 for $\lambda < 0$. For $\alpha = \pi/2$

$$(4.27) \qquad \theta(x,\lambda) = \lambda^{-\frac{1}{2}}\text{Sin } x\sqrt{\lambda}; \quad \phi(x,\lambda) = \text{Cos } x\sqrt{\lambda}$$

with $m(\lambda) = -i\lambda^{-\frac{1}{2}}$, $-\text{Im } m(\lambda) = \lambda^{-\frac{1}{2}}$, and

$$(4.28) \qquad k(\lambda) = \int_0^{\lambda} u^{-\frac{1}{2}}du \qquad (\lambda > 0).$$

Hence from (4.23) - (4.24) for example

(4.29) $$\chi(x,\lambda) = \int_0^\lambda \operatorname{Cos} x\sqrt{u}\, u^{-\frac{1}{2}} du = \frac{2 \operatorname{Sin} x\sqrt{\lambda}}{x}$$

(4.30) $$F(\lambda) = \int_0^\infty \frac{2 \operatorname{Sin} x\sqrt{\lambda}}{x} f(x)dx \qquad (\lambda > 0)$$

(4.31) $$f(x) = \frac{1}{\pi} \int_0^\infty \operatorname{Cos} x\sqrt{\lambda}\, d[\int_0^\infty \frac{2 \operatorname{Sin} x\sqrt{\lambda}}{x} f(x)dx] .$$

Writing $s^2 = \lambda$ we obtain the standard Fourier cosine formula. Note also that (4.15) written directly is the same, namely

(4.32) $$f(x) = \frac{1}{\pi} \int_0^\infty \operatorname{Cos} x\sqrt{\lambda} \frac{d\lambda}{\sqrt{\lambda}} \int_0^\infty \operatorname{Cos} y\sqrt{\lambda}\, f(y)dy.$$

On $(-\infty,\infty)$ we have similarly

(4.33) $$\theta(x,\lambda) = \operatorname{Cos} x\sqrt{\lambda}; \quad \phi(x,\lambda) = -\lambda^{-\frac{1}{2}}\operatorname{Sin} x\sqrt{\lambda}$$

(4.34) $$m_1(\lambda) = i\lambda^{\frac{1}{2}}; \quad \psi(x,\lambda) = \exp(-ix\sqrt{\lambda})$$

(4.35) $$m_2(\lambda) = -i\lambda^{\frac{1}{2}}; \quad \psi_2(x,\lambda) = \exp(ix\sqrt{\lambda})$$

(4.36) $$\xi'(\lambda) = \frac{1}{2}\lambda^{-\frac{1}{2}}; \quad \zeta'(\lambda) = \frac{1}{2}\lambda^{\frac{1}{2}}; \quad \eta(\lambda) = 0$$

and (4.22) gives the standard Fourier inversion in the form

(4.37) $$f(x) = \frac{1}{\pi} \int_0^\infty \operatorname{Cos} x\sqrt{\lambda} \frac{d\lambda}{2\sqrt{\lambda}} \int_0^\infty \operatorname{Cos} y\sqrt{\lambda}\, f(y)dy$$
$$+ \frac{1}{\pi} \int_0^\infty \operatorname{Sin} x\sqrt{\lambda} \frac{d\lambda}{2\sqrt{\lambda}} \int_0^\infty \operatorname{Sin} y\sqrt{\lambda}\, f(y)dy.$$

<u>Example 4.6</u> Consider the Bessel equation

(4.38) $$y'' + (s^2 - \frac{\nu^2 - \frac{1}{4}}{x^2})y = 0$$

on $(0,\infty)$. Thus $q(x) = (\nu^2 - 1/4)/x^2$ and $\lambda = s^2$. This has solutions $x^{\frac{1}{2}}J_\nu(xs)$ and $x^{\frac{1}{2}}Y_\nu(xs)$, where $Y_\nu(z) = [J_\nu(z)\operatorname{Cos} \pi\nu - J_{-\nu}(z)]/\operatorname{Sin} \pi\nu$. Both endpoints are singular so one takes $a \in (0,\infty)$ as a point to specify data and then

(4.39) $$\phi(x,\lambda) = \frac{1}{2}\pi a^{\frac{1}{2}}x^{\frac{1}{2}}[J_\nu(xs)Y_\nu(as) - Y_\nu(xs)J_\nu(as)];$$

$$\theta(x,\lambda) = \frac{1}{2}\pi a^{\frac{1}{2}}x^{\frac{1}{2}}s\,[J_\nu(xs)Y_\nu'(as) - Y_\nu(xs)J_\nu'(as)] + \frac{1}{2a}\phi(x,\lambda).$$

At $x = 0$ we have the limit point case for $\nu \geq 1$ and the limit circle case for $0 \leq \nu < 1$ ($\nu = \frac{1}{2}$ is not singular). Now for $\nu > 1$ The solutions in $L^2(0,a)$ and $L^2(a,\infty)$ are respectively $x^{\frac{1}{2}}J_\nu(xs)$ and $x^{\frac{1}{2}}H_\nu^1(xs)$ where $H_\nu^1(z) = J_\nu(z) + iY_\nu(z)$ (we are in the limit point case at $x = \infty$ also). Hence $m_1(\lambda) = -sJ_\nu'(as)/J_\nu(as) - \frac{1}{2a}$ and

$$(4.40) \qquad \psi_1(x,\lambda) = x^{\frac{1}{2}}a^{-\frac{1}{2}}J_\nu(xw)J_\nu^{-1}(as).$$

Similarly $m_2(\lambda) = -sH_\nu^1(as)'/H_\nu^1(as) - \frac{1}{2a}$ and

$$(4.41) \qquad -\mathrm{Im}\,\frac{1}{m_1(\lambda) - m_2(\lambda)} = \begin{cases} \frac{1}{2}\pi a J_\nu^2(as), & s > 0 \\ 0, & s = it, \quad t > 0 \end{cases}.$$

We are in the case of (4.21) where $m_1(\lambda)$ is real for λ real and (4.21) yields the Hankel formula

$$(4.42) \qquad f(x) = \frac{1}{2}\int_0^\infty x^{\frac{1}{2}}J_\nu(xs)\,d\lambda \int_0^\infty y^{\frac{1}{2}}J_\nu(ys)f(y)\,dy$$

where $s^2 = \lambda$. For $0 < \nu < 1$ $m_2(\lambda)$ is the same as above and

$$(4.43) \qquad m_1(\lambda) = -s\,\frac{cs^{-\nu}J_\nu'(as) - s^\nu J_{-\nu}'(as)}{cs^{-\nu}J_\nu(as) - s^\nu J_{-\nu}(as)} - \frac{1}{2a}$$

where c is arbitrary. Hence

$$(4.44) \qquad \psi_1(x,\lambda) = \frac{x^{\frac{1}{2}}}{a^{\frac{1}{2}}}\left[\frac{cJ_\nu(xs) - s^{2\nu}J_{-\nu}(xs)}{cJ_\nu(as) - s^{2\nu}J_{-\nu}(as)}\right].$$

For $\lambda > 0$

$$(4.45) \qquad -\mathrm{Im}\,\frac{1}{m_1(\lambda) - m_2(\lambda)} = \frac{1}{2}\pi a\,\frac{[cJ_\nu(as) - s^{2\nu}J_{-\nu}(as)]^2}{c^2 - 2cs^{2\nu}\mathrm{Cos}\,\pi\nu + S^{4\nu}}$$

while for $\lambda < 0$ $[m_1(\lambda) - m_2(\lambda)]^{-1}$ is real and for $c < 0$ it is continuous. Hence from (4.21) again

$$(4.46) \qquad f(x) = \int_0^\infty \frac{x^{\frac{1}{2}}[cJ_\nu(xs) - s^{2\nu}J_{-\nu}(xs)]}{c^2 - 2cs^{2\nu} - \mathrm{Cos}\,\nu\pi + s^{4\nu}}\,s\,ds \cdot \int_0^\infty y^{\frac{1}{2}}[cJ_\nu(ys) - s^{2\nu}J_{-\nu}(ys)]f(y)\,dy.$$

Letting $c \to -\infty$ we get (4.42). Further Bessel function formulas can be found in Titchmarsh [2] from which these were taken.

Let now $P(D)y = y'' + py' + qy = -\lambda^2 y$ so that $Y = y \exp(\frac{1}{2}\int p)$ satisfies $P_o(D)Y = Y'' - QY = Y'' - [\frac{1}{4}p^2 + \frac{1}{2}p' - q]Y = -\lambda^2 Y$. Set $\eta = y \exp \int p$ so that, as in Lemma 4.1, $P^*(D)\eta = -\lambda^2\eta$. Then the relation

$$(4.47) \qquad \langle \eta, y\rangle = \langle ye^{\int p}, y\rangle = \langle Y, Y\rangle$$

provides a natural connection of the P-P* theory with the self adjoint theory for P_o. In the case $P(D) = L_m(D)$ we would have $y \sim x^{-m}J_m(\lambda x)$, $Y \sim x^{\frac{1}{2}}J_m(\lambda x)$, and $\eta \sim x^{2m+1}J_m(\lambda x)$ for example.

More precisely for $m > 1$ consider ψ_1 given by (4.40) so for $\lambda = s^2$ we have

$$(4.48) \qquad h(x,s) = \hat{R}^m(x,s) = \frac{2^m\Gamma(m+1)a^{\frac{1}{2}}J_m(as)}{s^m} e^{-\frac{1}{2}\int p}\psi_1(x,\lambda)$$

$$(4.49) \qquad \Omega(x,s) = k_m(sx)^{2m+1}\hat{R}^m(x,s)$$

$$= \frac{s^{m+1}a^{\frac{1}{2}}J_m(as)}{2^m\Gamma(m+1)} e^{\frac{1}{2}\int p}\psi_1(x,\lambda).$$

Now $\frac{1}{\pi}\,d\xi(\lambda) = as\ J_m^2(as)ds$ by (4.41) and denoting $\frac{1}{\pi}\,d\xi$ generically by $d\rho$ we can write

$$(4.50) \qquad \langle h(x,s),\ \Omega(\xi,s)\rangle = \langle \psi_1(x,\lambda),\ \psi_1(\xi,\lambda)\rangle_{d\rho}$$

which by (4.21) can be identified with $\delta(x - \xi)$ in accordance with Section 3. For $[0,\infty)$ with 0 regular take now $\alpha = \pi/2$ in the definition of $\phi(x,\lambda)$ and for $\lambda = s^2$ set $h(x,s) = \phi(x,\lambda)\exp(-\frac{1}{2}\int_0^x p(\xi)d\xi)$ so that $h(0,s) = 1$; further take $d\rho = \frac{1}{\pi}\,dk$. There are also two situations where (4.21) may apply. For $(-\infty,\infty)$ take $a = 0$ with $h(x,s) = \psi_1(x,\lambda)\exp(-\frac{1}{2}\int_0^x p(\xi)d\xi)$ and $d\rho = \frac{1}{\pi}\,d\xi$ so again $h(0,s) = 1$. For $(0,\infty)$ with 0 singular (as in the Bessel case above) the behavior of ψ_1 and $\exp(-\frac{1}{2}\int p)$ as $x \to 0$ must be ascertained; if $\psi_1(x,\lambda)\exp(-\frac{1}{2}\int p) \to \ell(a,s) \neq 0$ or ∞ then we set $h(x,s) = \psi_1(x,\lambda)\exp(-\frac{1}{2}\int p)/\ell(a,s)$ (as in (4.48)) and take $d\rho = \frac{1}{\pi}\,d\xi$. Then in (4.49)

$$\Omega(x,s) = \psi_1(x,\lambda)\exp(\frac{1}{2}\int p)\ell(a,s)d\rho/ds$$

to produce (4.50).

Similarly whenever $d\rho/d\lambda$ is a function with $d\rho/ds = 2s\, d\rho/d\lambda$ we take $\Omega(x,s)$ to be of this general form in all the cases indicated (i.e. $\Omega = \psi_1\exp(\frac{1}{2}\int p)\ell\; d\rho/ds$, $\Omega = \psi_1\exp(\frac{1}{2}\int p)d\rho/ds$, or $\Omega = \phi\, \exp(\frac{1}{2}\int p)d\rho/ds$). There results in an obvious manner

Theorem 4.7 The constructions of Remark 4.4 give rise to a variety of situations where $P(D)h = -s^2h$ and $P^*(D)\Omega = -s^2\Omega$ with

$$(4.51)\qquad \langle h(x,s),\, \Omega(x,t)\rangle = \delta(s - t);\quad \langle h(x,s),\, \Omega(\xi,s)\rangle = \delta(x - \xi).$$

Reading off $\delta(x) = \langle 1,\, \Omega(x,s)\rangle$ from (4.51) as before we have an abundent source of models for the developments in Section 2.3 (cf also Section 2.5).

Remark 4.8 It is worth collecting here some information about general transformations following Titchmarsh [1] and Mercer [1; 2; 3] (cf. also Braaksma-Schurtman [1], Dijksma-deSnoo [1], Eringen [1; 2], Nasim [1], Walton [1]). First let f and g be connected by the so called Fourier kernels

$$(4.52)\qquad f(x) = \int_0^\infty k(xu)g(u)du;\quad g(u) = \int_0^\infty h(yu)f(y)dy$$

and let Mellin transform be defined by

$$(4.53)\qquad M(f)(s) = \hat{f}(s) = \int_0^\infty f(x)x^{s-1}dx$$

with inversion formula

$$(4.54)\qquad f(x) = \frac{1}{2\pi i}\int_{c-i\infty}^{c+i\infty}\hat{f}(s)x^{-s}ds.$$

Then the condition for (4.52) is

$$(4.55)\qquad \hat{k}(s)\hat{h}(1 - s) = 1.$$

This situation is discussed extensively in Titchmarsh [1] where numerous examples are given. There is no concern there however with the possible origin of such kernels k and h.

Another approach followed by Mercer [1; 2; 3] examines transforms of the type

$$(4.56)\qquad \hat{f}(\lambda) = \int_0^\infty K(\lambda,x)f(x)dx; \quad f(x) = \int_0^\infty \hat{f}(\lambda)H(\lambda,x)d\lambda$$

partly with a view of the possible origin of such kernels in differential problems and we propose to discuss this. Thus first assume $K(\lambda,x)$ satisfies

$$(4.57)\qquad \frac{1}{p}[pK']' + [\lambda^2 u - v]K = 0$$

with (4.56) in effect; then the form of $H(\lambda,x)$ can be determined in certain cases. Indeed let

$$(4.58)\qquad \frac{1}{p}(pw')' + [-\mu^2 u - v]w = 0$$

while for $W(w,K) = wK' - w'K$ there holds

$$(4.59)\qquad \Phi(\lambda,\mu) = -pW(w,K)\Big|_0^\infty = L(\lambda)M(\mu).$$

In this event multiplying (4.57) by w and (4.58) by K respectively and subtracting we obtain upon integration

$$(4.60)\qquad (\lambda^2 + \mu^2)\int_0^\infty p(x)u(x)w(\mu,x)K(\lambda,x)dx = L(\lambda)M(\mu).$$

Thus given the inversion of (4.56) there results formally

$$(4.61)\qquad \int_0^\infty \frac{L(\lambda)H(\lambda,x)d\lambda}{\lambda^2+\mu^2} = \frac{p(x)u(x)w(\mu,x)}{M(\mu)}.$$

In view of the integral in (4.61) possibly diverging Mercer uses the identity

$$(4.62)\qquad 1 = (\lambda^2 + \mu^2)\int_0^\infty \exp[-(\lambda^2 + \mu^2)\rho]d\rho$$

to write (4.61) as

$$(4.63)\qquad \int_0^\infty e^{-\mu^2\rho}\int_0^\infty e^{-\lambda^2\rho}L(\lambda)H(\lambda,x)d\lambda d\rho = \frac{p(x)u(x)w(\mu,x)}{M(\mu)}.$$

Setting $H_o(\lambda,x) = \frac{1}{2}\lambda^{-1}L(\lambda)H(\lambda,x)$ and $w_o(\mu,x) = p(x)u(x)w(\mu,x)/M(\mu)$ with

$t = \lambda^2$ and $s = \mu^2$ one can write (4.63) as

$$\text{(4.64)} \qquad LL[H_o(t^{\frac{1}{2}},x)] = w_o(s^{\frac{1}{2}},x)$$

where LL denotes an iterated Laplace transform $t \to p \to s$ (or a Stieltjes transform). A standard sort of argument (cf. Sneddon [1], Widder [1]) yields (cf. also Theorem 4.14)

$$\text{(4.65)} \qquad H_o(t^{\frac{1}{2}},x) = \frac{i}{2\pi}\,[w_o(t^{\frac{1}{2}}e^{i\pi/2},x) - w_o(t^{\frac{1}{2}}e^{-i\pi/2},x)]$$

$$\text{(4.66)} \qquad H(\lambda,x) = \frac{i\lambda p(x)u(x)}{\pi L(\lambda)}\left[\frac{w(\lambda e^{i\pi/2},x)}{M(\lambda e^{i\pi/2})} - \frac{w(\lambda e^{-i\pi/2},x)}{M(\lambda e^{-i\pi 2})}\right].$$

<u>Theorem 4.9</u> Let (4.56) hold with $K(\lambda,x)$ satisfying (4.57) and assume (4.59) where $w(\mu,x)$ satisfies (4.58). Then (4.66) determines $H(\lambda,x)$.

<u>Example 4.10</u> Take $p = u = 1$, $v = 0$, $K(\lambda,x) = \text{Sin}\,\lambda x$, and $w(\mu,x) = \exp(-\mu x)$. Then $L(\lambda) = \lambda$, $M(\mu) = 1$, and $H(\lambda,x) = \frac{2}{\pi}\,\text{Sin}\,\lambda x$ from (4.66). If $p = u = 1$, $v = 0$, $K(\lambda,x) = \text{Cos}\,\lambda x$, and $w(\mu,x) = \exp(-\mu x)$ we have $L(\lambda) = 1$ with $M(\mu) = \mu$, and (4.66) yields $H(\lambda,x) = \frac{2}{\pi}\,\text{Cos}\,\lambda x$.

<u>Example 4.11</u> Take $p = x$, $u = 1$, $v = m^2/x^2$, $K(\lambda,x) = J_m(\lambda x)$, and $w(\mu,x) = K_m(\mu x)$. One obtains $L(\lambda) = \lambda^m$ and $M(\mu) = \mu^{-m}$ and $\text{Re}\, m > -1$ is required if the integral in (4.60) is to converge. Here $K_m(z) = \int_0^\infty \cosh mt \exp(-z \cosh t)dt$ for example or $K_m(z) = \frac{\pi}{2}\,\text{csc}m\pi[I_{-m}(z) - I_m(z)]$ where $I_m(z) = \exp(-\frac{im\pi}{2})J_m(iz)$. Then from (4.66)

$$\text{(4.67)} \qquad H(\lambda,x) = \frac{ix\lambda^{1-m}}{\pi}\,[(i\lambda)^m K_m(i\lambda x) - (-i\lambda)^m K_m(-i\lambda x)].$$

Using the identities

$$\text{(4.68)} \qquad K_m(z) = \frac{\pi i}{2}\,e^{im\pi/2}H_m^1(ze^{i\pi/2}) = -\frac{\pi i}{2}\,e^{-im\pi/2}H_m^2(ze^{-i\pi/2})$$

and $J_m(z) = \frac{1}{2}\,[H_m^1(z) + H_m^2(z)]$ one can write (4.67) as $H(\lambda,x) = x\lambda J_m(\lambda x)$.

In anotherpaper Mercer assumes (4.56) plus the condition (for $\lambda > 0$ and $\text{Re}\,\mu > 0$)

(4.69) $$\int_0^\infty K(\lambda,x)w(\mu,x)dx = \frac{\lambda}{\lambda^2+\mu^2}$$

for some function $w(\mu,x)$; nothing is assumed now about differential equations. Then defining

(4.70) $$H_s(t,x) = \frac{i}{\pi}\,[w(e^{\frac{is\pi}{2}}t,x) - w(e^{\frac{-is\pi}{2}}t,x)]$$

it is shown that for suitable $\hat{f}(t^2\hat{f}(t) \in L^1(0,\gamma)$ and $t^{-2}\hat{f}(t) \in L^1(\gamma,\infty)$ for each $\gamma \in (0,\infty))$

(4.71) $$\frac{1}{2}\,[\hat{f}(\lambda - 0) + \hat{f}(\lambda + 0)] =$$

$$\lim_{s\to 1^-} \int_0^\infty \hat{f}(t)\int_0^\infty K(\lambda,x)H_s(t,x)dxdt.$$

Theorem 4.12 Under the conditions indicated formally $H(t,x) = \lim_{s\to 1^-} H_s(t,x)$.

Example 4.13 For $K(\lambda,x) = \mathrm{Sin}\,\lambda x$ and $w(\mu,x) = \exp(-\mu x)$ we get the Fourier sine formulas and one notes that (4.60) with $p = u = 1$, $L(\lambda) = \lambda$, and $M(\mu) = 1$ is the same as (4.69).

Another direction pursued by Mercer involves the extension of Theorem 4.9 to kernels arising from higher order differential equations. The proof is different and somewhat nicer so we include it; the procedure used for Theorem 4.9 is somewhat more refined however in treating integrals such as (4.61). Thus let P_n denote any differential operator (linear) of even order n with formal adjoint P_n^*. Suppose

(4.72) $$[P_n + \lambda^n u]K = 0; \quad [P_n^* - \mu^n u]w = 0.$$

Let $P(K,w)$ be the bilinear concomitant of K and w defined by

(4.73) $$wP_nK - KP_n^*w = \frac{d}{dx}\,P(K,w)$$

(cf. Ince [1]) and assume

(4.74) $$-P(K,w)\Big|_0^\infty = L(\lambda)M(\mu).$$

Theorem 4.14 Suppose (4.56) holds and that the bilinear concomitant is separable in the sense expressed by (4.74). Assume in addition that for $x > 0$ the function $\lambda^{n-1}w(\lambda e^{-i\pi/n},x)/M(\lambda e^{-i\pi/n})$ has no poles for $0 \leq \arg\lambda \leq 2\pi/n$ while

$$(4.75)\qquad \lim_{R\to\infty}\int_\Gamma \frac{\lambda^{n-1}w(\lambda e^{-i\pi/n},x)\,d\lambda}{(\lambda^n+\mu^n)M(\lambda e^{-i\pi/n})} = 0$$

uniformly in $x > 0$ and $\mu > 0$ where Γ is the circular arc defined by $|\lambda| = R$ and $0 \leq \arg\lambda \leq 2\pi/n$. Then

$$(4.76)\qquad H(\lambda,x) = \frac{-n\lambda^{n-1}u(x)}{\pi L(\lambda)}\,\mathrm{Im}\left[\frac{w(\lambda e^{i\pi/n},x)}{M(\lambda e^{i\pi/n})}\right].$$

Proof: From (4.72) - (4.73) one has

$$(4.77)\qquad -P(K,w)\Big|_0^\infty = (\lambda^n + \mu^n)\int_0^\infty u(x)w(\mu,x)K(\lambda,x)\,dx$$

and using (4.74) with (4.56) there results

$$(4.78)\qquad \int_0^\infty \frac{L(\lambda)H(\lambda,x)}{\lambda^n+\mu^n}\,d\lambda = \frac{u(x)w(\mu,x)}{M(\mu)}.$$

Consider now the region $|\lambda| < R$, $0 < \arg\lambda < 2\pi/n$ bounded by a contour C. Then

$$(4.79)\qquad \frac{nu(x)}{2\pi i}\int_C \frac{\lambda^{n-1}w(\lambda e^{-i\pi/n},x)\,d\lambda}{(\lambda^n+\mu^n)M(\lambda e^{-i\pi/n})} = \frac{u(x)w(\mu,x)}{M(\mu)}$$

under the hypotheses stated. Writing $\int_C = \int_0^R + \int_\Gamma + \int_{R\exp\frac{2\pi i}{n}}^0$ put $\lambda =$ $t\exp(2\pi i/n)$ in the last integral and then set $\lambda = t$. Since $\int_\Gamma \to 0$ the left side of (4.79) becomes

$$(4.80)\qquad \frac{nu(x)}{2\pi i}\int_0^\infty \frac{\lambda^{n-1}}{\lambda^n+\mu^n}\left[\frac{w(\lambda e^{-i\pi/n},x)}{M(\lambda e^{-i\pi/n})} - \frac{w(\lambda e^{i\pi/n},x)}{M(\lambda e^{i\pi/n})}\right]d\lambda.$$

But by (4.79) and (4.78) we can then conclude that (4.76) holds. QED

Examples 4.15 Take $P_2 = D_x^2$, $u = 1$, $K(\lambda,x) = \mathrm{Cos}\,\lambda x$, and $w(\mu,x) = \exp(-\mu x)$. Then $L(\lambda) = 1$ and $M(\mu) = \mu$ so the integrand in (4.75) is $i\exp(i\lambda x)/(\lambda^2 + \mu^2)$ and (4.75) holds, while $i\exp(i\lambda x)$ has no poles for $0 \leq \arg\lambda \leq \pi$. Hence $H(\lambda,x) = -\frac{2\lambda}{\pi}\,\mathrm{Im}(\exp(-i\lambda x)/i\lambda) = \frac{2}{\pi}\,\mathrm{Cos}\,\lambda x$.

Remark 4.16 The requirement that the bilinear concomitant be separable as indicated is somewhat ad hoc and seems worthy of further investigation. From our point of view the interest of a formula such as (4.76) is that for $u(x) = 1$ $H(\lambda,x)$ satisfies $P_n^*(D)H = -\lambda^n H$ with $P_n(D)K = -\lambda^n K$ while the inversion (4.56) holds. This feature establishes contact with the formalism developed in Section 3.

2.5. Generalized translations. We begin by describing some work of Hutson-Pym [1; 2] which generalizes the methods of Levitan [1; 3] and Povzner [1] (cf. also Ehrenpreis [3]) where generalized translation operators are obtained by solving equations

$$[-D_x^2 + q(x)]u = [-D_y^2 + q(y)]u \tag{5.1}$$

on intervals $[0,\pi]$ for example or $[0,\infty)$ with initial conditions $u(x,0) = f(x)$, $D_y u(x,0) = hf(x)$, and suitable boundary conditions $D_y u - hu = 0$ at $x = 0$ and $D_y u + Hu = 0$ at $x = \pi$ say. Such problems can be solved by eigenfunction methods or by techniques of integral equations (cf also Lions [2; 3; 4]) and the Hutson-Pym approach deals with more general and abstract versions of the matter which reduce to and include the classical situations as special cases (some classical formulations are indicated later). Let h and v be two functions in say L^∞_{loc} and consider the integral equation

$$u(x,y) = v(x,y) + \frac{1}{2}\int_0^y \int_{x-y+t}^{x+y-t} h(s,t)u(s,t)\,ds\,dt \tag{5.2}$$

for $(x,y) \in \mathbb{R}^2$. If $\Delta(x,y)$ is the triangle with vertices $(x-y,\ 0)$, (x,y), and $(x+y,\ 0)$ the integral in (5.2) is over $\Delta(x,y)$ and one writes

$$Hw(x,y) = \frac{1}{2}\int_{\Delta(x,y)} hw(s,t)\,ds\,dt. \tag{5.3}$$

Formally a solution of (5.2) is $u = (I - H)^{-1}v = \sum_0^\infty H^n v$ with $H^0 = I$ when this makes sense. Let $p_k(f) = \sup|f(x)|$ for $x \in K$, K compact, and give L^∞_{loc}

the topology determined by such seminorms; for L^1_{loc} use seminorms $q_k(f) = \int_K |f(x)|dx$. Let E^n be C^∞ functions in $\mathbb{R}$ or $\mathbb{R}^2$ with the Schwartz topology of uniform convergence of $D^\alpha f$ on compact sets for $|\alpha| \leq n$. Some routine calculation yields immediately for $h,w \in L^\infty_{loc}(\mathbb{R}^2)$ and $n \geq 1$

$$(5.4) \qquad |H^n w| \leq p_\Delta(w)[p_\Delta(h)]^n y^{2n}/(2n)!;$$

$$|H^n w| \leq q_\Delta(w)[p_\Delta(h)]^n y^{2n-2}/2(2n-2)!.$$

Now for $a \geq 0$ let $\square(a)$ be the square with vertices $(a,0)$, $(-a,0)$, $(0,a)$, and $(0,-a)$ so that $\square(a) = \Delta(0,a) \cup \Delta(0,-a)$ and $(x,y) \in \square(a)$ implies $\Delta(x,y) \subset \square(a)$. From (5.4) follows $p_k(H^n v) \leq p_\square(v)[p_\square(h)]^n a^{2n}/(2n)!$ for $K \subset \square(a) = \square$ and hence $\Sigma H^n v$ converges uniformly on K. There results the convergence in L^∞_{loc} and thus $(I - H)^{-1} : L^\infty_{loc} \to L^\infty_{loc}$ is well defined and continuous since

$$p_k((I - H)^{-1}v) \leq p_\square(v)\sum_0^\infty [p_\square(h)]^n a^{2n}/(2n)!.$$

Similarly one proves easily that for $h \in L^\infty_{loc}$ (resp. $h \in E^n$) $H : L^\infty_{loc} \to E^0$ (resp $H : E^n \to E^{n+1}$) and $(I - H)^{-1} : E^0 \to E^0$ (resp. $E^{n+1} \to E^{n+1}$) are continuous.

The integral equation (5.2) written as $u = v + Hu$ is related to a differential equation similar to (5.1) since formally

$$(5.5) \qquad D_x^2 u - D_y^2 u + hu = D_x^2 v - D_y^2 v$$

and if v is a suitable combination $g(x + y) + k(x - y)$ with $h = -q(x) + q(y)$ we have (5.1).

Now take linear $E_i : L^\infty_{loc}(\mathbb{R}) \to L^\infty_{loc}(\mathbb{R})$ (resp. $E^n(\mathbb{R}) \to E^n(\mathbb{R})$), $i = 1,2$, and set

$$(5.6) \qquad Ef(x,y) = E_1 f(x + y) + E_2 f(x - y)$$

mapping $L^\infty_{loc}(\mathbb{R}) \to L^\infty_{loc}(\mathbb{R}^2)$ (resp. $E^n(\mathbb{R}) \to E^n(\mathbb{R}^2)$). One defines then a generalized translation operator $T : L^{\infty}_{loc}(\mathbb{R}) \to L^{\infty}_{loc}(\mathbb{R}^2)$ (resp $\mathcal{E}^n(\mathbb{R}) \to \mathcal{E}^n(\mathbb{R}^2)$) as the unique solution of

(5.7) $\quad Tf = Ef + HTf$

for $f \in D(T)$. Since $T = (I - H)^{-1}E$ this makes sense. Next one wants a suitable notion of boundary operators as follows. For continuous u evidently $Hu(x,0) = D_yHu(x,0) = 0$ and for $Tf \in C^1$ one writes

(5.8) $\quad Af(x) = Tf(x,0) = Ef(x,0); \quad Bf(x) = D_yTf(x,0) = D_yEf(x,0)$

so that $2E_1f = Af + \int_0^x Bf$ and $2E_2f = Af - \int_0^x Bf$. One proves then easily

<u>Theorem 5.1</u> Let $n \geq 2$ and $h \in \mathcal{E}^{n-1}(\mathbb{R}^2)$ with $A : \mathcal{E}^n \to \mathcal{E}^n$ and $B : \mathcal{E}^n \to \mathcal{E}^{n-1}$ continuous linear maps. Then there is a unique continuous generalized translation operator $T : \mathcal{E}^n(\mathbb{R}) \to \mathcal{E}^n(\mathbb{R}^2)$ satisfying for $f \in \mathcal{E}^n$

(5.9) $\quad D_x^2Tf - D_y^2Tf + hTf = 0; \quad Tf(x,0) = Af(x); \quad D_yTf(x,0) = Bf(x).$

We recall now that convolution of distributions is defined by the rule $\langle S * T,\phi\rangle = \langle S \otimes T, \phi(\xi+\eta)\rangle$ for say $S \in \mathcal{E}'$ and $T \in \mathcal{D}'$. Here $\phi(\xi + \eta) = T^{\eta}_{\xi}\phi(\xi)$ is a standard translation. The map T of Theorem 5.1 is also continuous from $\mathcal{E}(\mathbb{R}) \to \mathcal{E}(\mathbb{R}^2)$ so its adjoint $T^* : \mathcal{E}'(\mathbb{R}^2) \to \mathcal{E}'(\mathbb{R}^1)$ is continuous in the weak or strong topologies. Therefore a generalized convolution is naturally defined by the rule

(5.10) $\quad \mu * \nu = T^*(\mu \otimes \nu)$

for $\mu,\nu \in \mathcal{E}'(\mathbb{R})$ and is a map $\mathcal{E}' \times \mathcal{E}' \to \mathcal{E}'$ continuous in the strong topologies and in the weak topologies on bounded sets. If $\mu = f$ and $\nu = g$ we have

(5.11) $\quad \langle f * g,\phi\rangle = \int f(x)g(y)(T\phi)(x,y)dxdy.$

Note that in a case such as (3.75) with T^x_y real and self adjoint satisfying $T^{\xi}_x g(x) = T^x_{\xi}g(\xi)$ one can write for example

(5.12) $\quad \int(T\phi)(x,y)g(y)dy = \int T^y_x\phi(x)g(y)dy = \int T^x_y\phi(y)g(y)dy = \int\phi(y)T^x_y g(y)dy$

so that (5.11) becomes

$$(5.13) \qquad \langle f * g, \phi\rangle = \int\int f(x) T_y^x g(y)\phi(y)\,dy\,dx$$

$$= \int \phi(y) \int T_y^x g(y) f(x)\,dx\,dy$$

allowing one to identify $f * g$ with the integral $\int T_y^x g(y) f(x)\,dx$ as in (3.75) (note here that

$$\int T_y^x g(y) f(x)\,dx = \int T_x^y g(x) f(x)\,dx$$

$$= \int g(x) T_x^y f(x)\,dx = \int g(x) T_y^x f(y)\,dy).$$

Now Hutson-Pym (loc. cit.) study the convolution of (5.10) to see when it is commutive or associative for example and establish a number of results some of which are indicated below.

First let $M(K)$ be the Banach space of Radon measures with support in K so that $\mathcal{K}' = (\mathcal{E}^0)' = \cup M(K)$ as a set and the strong topology on $\mathcal{K}'$ is the finest locally convex topology inducing the norm topology on each $M(K)$. For $T : \mathcal{E}^0(\mathbb{R}) \to \mathcal{E}^0(\mathbb{R}^2)$ continuous arising as in Theorem 5.1 one can easily show that for $\mu,\nu \in \mathcal{K}'$ with supports in K there is a constant c_K such that $\|\mu * \nu\| \le c_K \|\nu\|\,\|\mu\|$. Let now $L^1(K) \subset M(K)$ be the absolutely continuous measures with respect to Lebesgue measure $d\lambda$ with $L^1_{\mathcal{K}} = \cup L^1(K)$ as a set. For $f \in L^1_{loc}$ one writes $f\lambda$ or $fd\lambda$ for the associated (possibly unbounded) measure. Using properties indicated above one proves that if $T : \mathcal{E}^0(\mathbb{R}) \to \mathcal{E}^0(\mathbb{R}^2)$ is continuous with $E_1, E_2 : \mathcal{E}^0 \cap L^1 \to \mathcal{E}^0$ continuous when $\mathcal{E}^0 \cap L^1$ has the L^1 topology and $\mathcal{E}^0$ the topology of L^1_{loc} then for $\mu \in \mathcal{K}'$ and $\nu \in L^1_{\mathcal{K}}$ both $\mu * \nu$ and $\nu * \mu$ belong to $L^1_{\mathcal{K}}$. The desired properties of E_1, E_2 can be assured by similar properties of A, B for $A,B : \mathcal{E}^0 \to \mathcal{E}^0$. For commutativity and associativity one has first

<u>Proposition 5.2</u> Convolution in $\mathcal{E}'$, $\mathcal{K}'$, or $L^1_{\mathcal{K}}$ is commutative if and only if $Tf(x,y) = Tf(y,x)$ for all f in $\mathcal{E}$ or $\mathcal{E}^0$. It is associative if and only if $(T \otimes I)Tf = (I \otimes T)Tf$ for all f in $\mathcal{E}$ or $\mathcal{E}^0$.

Proof: For commutativity consider $\delta_x(f) = f(x)$ so that $\delta_x * \delta_y(f) = \delta_x \otimes \delta_y(Tf) = Tf(x,y)$ while $\delta_y * \delta_x(f) = Tf(y,x)$ so symmetry of Tf is necessary in E' or K' and the L^1_K case is routine. The converse is obvious. For associativity one thinks of $(T \otimes I)(g)(x,y,z) = (Tg(\cdot,z))(x,y)$ and looks at for example

$$(\delta_x * \delta_y) * \delta_z(f) = ((\delta_x * \delta_y) \otimes \delta_z)(Tf) = \delta_x \otimes \delta_y \otimes \delta_z (T \otimes I)Tf.$$

More relevant since it deals with the structure of T is the following

Theorem 5.3 If convolution is commutative there is a function $h \in L^\infty_{loc}$ determining T such that $h(x,y) + h(y,x) = 0$.

Proof: Suppose T is defined by $h \in L^\infty_{loc}$. Let $S(x,y,\delta)$ be the square with vertices (x,y), $(x+\delta, y+\delta)$, $(x,y+2\delta)$, and $(x-\delta, y+\delta)$. Then $\chi_S = \chi(\Delta(x,y)) - \chi(\Delta(x+\delta, y+\delta)) + \chi(\Delta(x,y+2\delta)) - \chi(\Delta(x-\delta, y+\delta))$. For g a function of two variables write $[g](x,y,\delta) = g(x,y) - g(x+\delta, y+\delta) + g(x,y+2\delta) - g(x-\delta, y+\delta)$. Thus if $g(x,y) = \int hTf$ over $\Delta(x,y)$ one has $[g](x,y,\delta) = \int hTf$ over $S(x,y,\delta)$. Since $Ef(x,y) = E_1 f(x+y) + E_2 f(x-y)$ clearly $[Ef](x,y,\delta) = 0$. Let $\tilde{S}(x,y,\delta)$ $(= S(y+\delta, x-\delta, \delta))$ be the reflection of $S(x,y,\delta)$ in the line $y = x$ and $\tilde{g}(x,y) = g(y,x)$. Thus if convolution is commutative $(Tf)^\sim = Tf$. Now since $[Ef](x,y,\delta) = 0$ we have

$$(5.14) \qquad [Tf](x,y,\delta) = \frac{1}{2}\int_S hTf; \quad [Tf]^\sim(x,y,\delta)$$
$$= -[Tf](y+\delta, x-\delta,\delta) = -\frac{1}{2}\int_{\tilde{S}} hTf = -\frac{1}{2}\int_S \tilde{h}(Tf)^\sim.$$

Hence if $Tf = (Tf)^\sim$ one obtains the equation $\int_S (h + \tilde{h})Tf = 0$ and hence $(h + \tilde{h})Tf = 0$ a.e. Now $Tf(x,y)$ may be zero but if $h_1 \neq h_2$ only when $Tf = 0$ then $\int h_1 Tf = \int h_2 Tf$ so modify h if necessary such that $h(x,y) = -h(y,x)$ $(h(x,x) = 0)$. QED

Assume now convolution is both associative and commutative with $h \in L^\infty_{loc}$. Then for f such that $(T \otimes I)Tf \in C^2$ one has a.e. in $\mathbb{R}^3$

$$(5.15) \qquad (h(x,y) + h(y,z) + h(z,x))(T \otimes I)Tf(x,y,z) = 0.$$

To see this let first $h \in E'$ and $f \in E^2$ and set $u(x,y,z) = (T \otimes I)Tf(x,y,z)$ so that using Proposition 5.2 one has immediately $u(x,y,z) = u(y,z,x) = u(z,x,y)$. From Theorem 5.1 there results $(D_x^2 - D_y^2)u(x,y,z) + h(x,y)u(x,y,z) = 0$. Since $u(x,y,z) = u(y,z,x)$ one has similarly $(D_y^2 - D_z^2)u(x,y,z) + h(y,z)u(x,y,z) = 0$ and $(D_z^2 - D_x^2)u(x,y,z) + h(z,x)u(x,y,z) = 0$ by the same token. Adding gives (5.15). For general h and f as indicated (5.15) follows by approximation. This result can now be further refined as follows. Write $F = \{(x,y,z);\ (T \otimes I)Tf(x,y,z) = 0$ for all f such that $(T \otimes I)Tf \in C^2\}$. Then F is closed and $U = \mathbb{R}^3/F$ is open. Let $U_z = \{(x,y);\ (x,y,z) \in U\}$ and $V \subset U_z$ be compact. Take I a compact interval with $z \in I$ such that $V \times I \subset U$. Let $\mu \in L^1_K(\mathbb{R})$ have support in I with $\mu(I) = 1$ and write $\pi(x) = \int h(x,s)d\mu(s)$. From Theorem 5.3 $h(x,y) + h(y,x) = 0$ a.e. and hence for $(x,y,z) \in V \times I$ $h(x,y) = -h(z,x) - h(y,z) = h(x,z) - h(y,z)$ a.e. Hence

$$h(x,y) = \int h(x,z)d\mu(z) - \int h(y,z)d\mu(z) = \pi(x) - \pi(y)$$

for almost all $(x,y) \in V$. This leads to

<u>Theorem 5.4</u> Let $h \in L^\infty_{loc}$ with convolution both commutative and associative. If $\cup U_z = \mathbb{R}^2$ $(z \in \mathbb{R})$ then there is a function $\pi \in L^\infty_{loc}$ such that $h(x,y) = \pi(x) - \pi(y)$ a.e.

In general $\cup U_z$ is at least dense in $\mathbb{R}^2$ and $h(x,y) = \pi(x) - \pi(y)$ locally a.e. in a dense open subset of $\mathbb{R}^2$ with π depending on the nbh chosen. Recall now from Theorem 5.1 that $Af(x) = Tf(x,0)$ and $Bf(x) = D_yTf(x,0)$. Let $\delta f = f(0)$ and $\delta_1(f) = (Df)(0) = -\delta'f$. Then $A = (I \otimes \delta)T$ and $B = (I \otimes \delta_1)T$. If convolution is commutative $Tf(x,y) = Tf(y,x)$ from Proposition 5.2 so $(D \otimes I)Tf(0,0) = (I \otimes D)Tf(0,0)$ which can be written as $\delta_1 \otimes \delta(Tf) = (\delta \otimes \delta_1)(Tf)$ (i.e. in E' $\delta_1 * \delta = \delta * \delta_1$).

<u>Theorem 5.5</u> If convolution is commutative then for $Tf \in C^1$ $\delta_1 Af = \delta Bf$. If convolution is both commutative and associative then for $Tf \in C^1$ $ABf = BAf$.

Proof: For the first statement one has

$$(5.16)\qquad \delta_1 Af = \delta_1(I\otimes\delta)Tf = \delta_1\otimes\delta(Tf)$$

$$= \delta\otimes\delta_1(Tf) = \delta(I\otimes\delta_1)Tf = \delta Bf.$$

For the second assertion one has then for $x\in\mathbb{R}$

$$(5.17)\qquad ABf(x) = A((I\otimes D)Tf(\cdot,0))(x) = T((I\otimes D)Tf(\cdot,0))(x,0)$$

$$= (T\otimes D)Tf(x,0,0) = (I\otimes I\otimes D)(T\otimes I)Tf(x,0,0)$$

$$(5.18)\qquad BAf(x) = B(Tf(\cdot,0))(x) = (I\otimes D)T(Tf(\cdot,0))(x,0)$$

$$= ((I\otimes D)T\otimes I)Tf(x,0,0) = (I\otimes D\otimes I)(T\otimes I)Tf(x,0,0)$$

$$= (I\otimes D\otimes I)(I\otimes T)Tf(x,0,0) = (I\otimes(D\otimes I)T)Tf(x,0,0)$$

$$= (I\otimes(I\otimes D)T)Tf(x,0,0) = (I\otimes I\otimes D)(I\otimes T)Tf(x,0,0)$$

$$= (I\otimes I\otimes D)(T\otimes I)Tf(x,0,0).$$ QED

Now for $h(x,y) = \pi(x) - \pi(y)$ we set $Lf = D^2f + \pi f$ and (5.9) becomes

$$(5.19)\qquad (L\otimes I)u = (I\otimes L)u.$$

For convenience in formulation assume here $h\in E$ so that $\pi\in E$ and $T : E(\mathbb{R})\to E(\mathbb{R}^2)$. For any continuous $S : E(\mathbb{R})\to E(\mathbb{R})$ the adjoint S' is called a centralizer for a convolution $*$ if $S'(\mu*\nu) = (S'\mu)*\nu$ for $\mu,\nu\in E'$. Thus for $f\in E$

$$S'(\mu*\nu)(f) = \mu*\nu(Sf) = \mu\otimes\nu(TSf)$$

while $(S'\mu)*\nu(f) = S'\mu\otimes\nu(Tf) = ((S'\otimes I)(\mu\otimes\nu)(Tf) = \mu\otimes\nu((S\otimes I)Tf)$. Consequently S' is a centralizer if and only if $(S\otimes I)Tf = TS$; if $*$ is commutative then also $TS + (I\otimes S)T$.

<u>Theorem 5.6</u> Let $S : E\to E$ commute with L, A, and B. Then S' is a central-

izer. If S' is a centralizer then S commutes with A and B. In particular L' is a centralizer if and only if L commutes with A and B.

Proof: If S commutes with L we have for $f \in E$

$$(5.20) \qquad (L \otimes I)(S \otimes I)Tf = (LS \otimes I)Tf = (S \otimes I)(L \otimes I)Tf =$$

$$(S \otimes I)(I \otimes L)Tf = (S \otimes L)Tf = (I \otimes L)(S \otimes I)Tf$$

where $(L \otimes I)Tf = (I \otimes L)Tf$ since Tf is a solution of (5.19). If S commutes with A $(S \otimes I)Tf$ satisfies $(S \otimes I)Tf(x,0) = SAf(x) = ASf(x) = TSf(x,0)$. If S commutes with B one has

$$(5.21) \qquad D_y(S \otimes I)Tf(x,0) = (I \otimes D)(S \otimes I)Tf(x,0) = (S \otimes I)(I \otimes D)Tf(x,0)$$

$$= SBf(x) = BSf(x) = D_y TSf(x,0).$$

Thus $(S \otimes I)Tf$ and TSf satisfy (5.9) and by uniqueness are equal. The second assertion follows by rearranging the above calculations. QED

Next one examines sufficient conditions for commutativity and associativity and we will simply state some results of Hutson-Pym [2] without proof.

Theorem 5.7 Suppose $h(x,y) + h(y,x) = 0$ a.e., $h(-x,y) = h(x,y)$ a.e., $Ef(x,y) = Ef(y,x)$ a.e. and $Ef(-x,y) = Ef(x,y)$ a.e. Then convolution is commutative.

Let now $h \in E$, $h(x,y) = \pi(x) - \pi(y)$ as above with $\pi \in E$, and $L = D^2 + \pi$.

Theorem 5.8 Let π be analytic with A and B mapping analytic functions to analytic functions. Let L commute with A and B with $\delta B = \delta_1 A$. Then convolution is commutative.

Theorem 5.9 Let $\pi \in E$ and suppose L, A, B all commute with convolution commutative. Then convolution is associative.

The smoothness requirements on π can be relaxed but we omit this. A simple concrete example of some of this is

Example 5.10 Let $Af(x) = \alpha[f(x) + f(-x)]$ and $Bf(x) = \beta[f(x) - f(-x)]$. Then A and B map analytic functions to analytic functions, $(Af)'(0) = 0 = Bf(0)$, and $AB = BA = 0$. L commutes with A and B when π is even. Hence the corresponding convolution will be commutative and associative.

Some aspects of the above theory are given a more abstract form in Hutson-Pym [1] and we will report on this here. One considers translations T associated with L and L is often a compact operator here; however very often n-th order differential operators D_n have compact resolvants (cf. Goldberg [1]) and without loss of generality one can assume $L = D_n^{-1}$ is compact. The corresponding results will then be indicated below in Example 5.17. Let E be a normed space with dual E' and bidual E''. One will have three continuous linear maps $A : E \to E''$, $B : E' \to E'$, and $C : E'' \to E''$ where $A = C$ restricted to E, B is the composition of the canonical map $E' \to E'''$ and $A^* : E''' \to E'$, and $C = B^*$. Thus sometimes B will be written as A^* and C as A when no confusion will arise. L will be a continuous linear map $E \to E''$ whose extension to E'' is again called L while $L^* : E' \to E'$ according to the conventions indicated. Let $B(E',E')$ denote continuous bilinear forms with its standard norm and one writes $E'\hat{\otimes}E'$ for the closure of $E'\otimes E'$ in $B(E',E')'$ so that $(E'\hat{\otimes}E')' = B(E',E')$ (thus $\hat{\otimes}$ means $\hat{\otimes}_\pi$). For $A,B \in L(E';E')$ define $A^*\otimes B^* : B(E',E') \to B(E',E')$ by

$$\langle (A^*\otimes B^*)f, (x,y)\rangle = \langle f, (Ax,By)\rangle = f(Ax,By).$$

A translation operator will be a norm continuous map $T : E \to B(E',E')$ so that $T^* : B(E',E')' \to E'$ and since $E'\hat{\otimes}E' \subset B(E',E')'$ we consider $T^* : E'\hat{\otimes}E' \to E'$ with $T^{**} : E'' \to (E'\hat{\otimes}E')' = B(E',E')$. One says T is L admissable (or T admits L) if $(L\otimes I)T = TL = (I\otimes L)T$. This can mean $(L^{**}\otimes I)T = T^{**}L = (I\otimes L^{**})T$ or $(L^{**}\otimes I)T^{**} = T^{**}L^{**} = (I\otimes L^{**})T^{**}$ but these are essentially the same and the meaning will be clear from context. A B-condition is a pair (a,A), $a \in E'$, $A \in L(E;E'')$, such that $Tf(a,\cdot) = Af$ for $f \in E$. Thus T satisfies

(a,A) if this is true or equivalently $Tf(a,\cdot) = Af$ for $f \in E''$. L will now be assumed compact unless otherwise stated so that L^* is compact, and the extension of L to E'' is compact. Let $\sigma_0(L^*)$ be $\sigma(L^*) - \{0\}$ and for $\lambda \in \sigma_0(L^*)$ there is a projection P^*_λ defined by

$$P^*_\lambda = \frac{1}{2\pi i}\int_\Gamma R(\zeta,L^*)d\zeta$$

where Γ surrounds λ with no other points of $\sigma_0(L^*) \cup \{0\}$ inside or on Γ; $R(\zeta,L^*) = (L^* - \zeta I)^{-1}$. If $V_\lambda = P^*_\lambda E'$ then

$$V_\lambda = \{x \in E';\ (L^* - \lambda I)^{\nu_\lambda} x = 0\}$$

for sufficiently large ν_λ and $\dim V_\lambda < \infty$. If $\lambda \neq \mu$ then $P^*_\lambda P^*_\mu = 0$ and if $A : E' \to E'$ commutes with L^* then it commutes with P^*_λ. Now the spectrum of the extension of L to E'' is the same as $\sigma(L^*)$ and for $\lambda \in \sigma_0(L^*)$ there is a projection $Q_\lambda : E'' \to E''$ defined as above such that $Q_\lambda = (P^*_\lambda)^*$. One will write P_λ for Q_λ restricted to E or in E'' and set $U_\lambda = P_\lambda E'' = \{f \in E'';\ (L - \lambda I)^{\nu_\lambda} f = 0\}$ (same ν_λ as above).

Now if T is an L-admissable translation operator set for $a \in E'$ ${}_aTf = Tf(a,\cdot)$ and $T_af = Tf(\cdot,a)$ $(f \in E)$. Taking adjoints the same requirements hold for $f \in E''$. The definition says that T satisfies the B-condition $(a,{}_aT)$ and the collection of such B-conditions determines T. Thus generically there will be many translations T associated with a given L and one wants further (B) conditions which determine T uniquely; for second order differential operators L these took the form of boundary conditions. First we collect some elementary facts and observe that if T satisfies (a,A) then A commutes with L. Indeed for $f \in E''$ and $x \in E'$

$$(5.22)\qquad ALf(x) = T(Lf)(a,x) = (I \otimes L)Tf(a,x)$$

$$= Tf(a,L^*x) = Af(L^*x) = LAf(x).$$

Thus ${}_xT$ and T_x commute with L. Next one has

Proposition 5.11 For $\lambda \in \sigma_o(L^*)$, $TP_\lambda = (I \otimes P_\lambda)T = (P_\lambda \otimes I)T = (P_\lambda \otimes P_\lambda)T$. For $\lambda \in \sigma_o(L^*)$, $TU_\lambda \subset U_\lambda \otimes U_\lambda$.

Proof: For $f \in E''$, $x,y \in E'$, one has

$$(5.23) \qquad TP_\lambda f(x,y) = ({}_xT)P_\lambda f(y) = P_\lambda({}_xT)f(y) =$$

$$({}_xT)f(P_\lambda^* y) = Tf(x,P_\lambda^* y) = (I \otimes P_\lambda)Tf(x,y).$$

In a similar manner $TP_\lambda = (P_\lambda \otimes I)T$ and since $P_\lambda^2 = P_\lambda$ one obtains $TP_\lambda = (P_\lambda \otimes I)(I \otimes P_\lambda)T = (P_\lambda \otimes P_\lambda)T$. For $f,g \in E''$ now an element $f \otimes g \in B(E',E')$ is determined by $f \otimes g(x,y) = f(x)g(y)$ and if $U_1,U_2 \subset E''$ $U_1 \otimes U_2$ is the subspace of $B(E',E')$ spanned by $\{f \otimes g;\ f \in U_1,\ g \in U_2\}$. The last assertion of the proposition will follow by showing $P_\lambda \otimes P_\lambda B(E',E') \subset U_\lambda \otimes U_\lambda$. Let $f_1,\dots,f_n$ be a vector space basis of U_λ and let $h \in P_\lambda \otimes P_\lambda B(E',E')$. Then for $y \in E'$

$$h(\cdot,y) = P_\lambda \otimes P_\lambda)h(\cdot,y) = P_\lambda(h(\cdot,P_\lambda^* y)) \in U_\lambda.$$

Hence there are scalars $g_1(y),\dots,g_n(y)$ such that $h(\cdot,y) = \Sigma g_k(y)f_k$. The f_k are linearly independent and in U_λ so for each k there are $x_k \in V_\lambda$ $(P_\lambda^* x_k = x_k)$ such that $f_i(x_k) = 0$ for $i \neq k$ and $f_k(x_k) = 1$. Then $g_k = h(x_k,\cdot)$ so $g_k \in E''$ and

$$P_\lambda g_k = P_\lambda(h(P_\lambda^* x_k,\cdot)) = P_\lambda \otimes P_\lambda(h(x_k,\cdot))$$

$$= h(x_k,\cdot) = g_k.$$

Thus $g_k \in U_\lambda$ and since $h = \Sigma f_i \otimes g_i$ the result follows. QED

One denotes by ΣU_λ the smallest vector subspace of E'' containing every U_λ. Then by simple calculation as above one can show that if T_1 and T_2 are two L-admissible translations and either (a) for each $\lambda \in \sigma_o(L^*)$ $f \in U_\lambda$ and $x,y \in V_\lambda$ implies $T_1f(x,y) = T_2f(x,y)$ or (b) for each $\lambda \in \sigma_o(L^*)$ $f \in E$ and $x,y \in V_\lambda$ implies $T_1f(x,y) = T_2f(x,y)$ then $T_1f = T_2f$ for $f \in \Sigma U_\lambda$. Thus if ΣU_λ is weakly dense (i.e. weak* dense) in E'' then under these hypotheses $T_1 = T_2$.

Let now $\mathrm{Comm}\, L^*$ be the set of operators on E' which commute with L^* and $\mathrm{Comm}^2 L^*$ the set of operators commuting with $\mathrm{Comm}\, L^*$. $\mathrm{Comm}^2 L^*$ is a commutative ring with $\mathrm{Comm}^2 L^* \subset \mathrm{Comm}\, L^*$ and $P^*_\lambda \in \mathrm{Comm}\, L^*$ for $\lambda \in \sigma_0(L^*)$. The following basically algebraic facts are easily proved and we omit details. First the norm closure V of ΣV_λ is a module over $\mathrm{Comm}^2 L^*$ with each V_λ a submodule. Simply note that if $x \in V_\lambda$ and $A \in \mathrm{Comm}^2 L^*$ then $Ax = AP^*_\lambda x = P^*_\lambda Ax \in V_\lambda$. Now one says that $\{x_i;\ i \in I\} \subset V$ is a set of generators for the module V if the closed linear span of $\{Ax_i;\ A \in \mathrm{Comm}^2 L^*;\ i \in I\}$ is V. The index γ of V is the minimal cardinal of a set of generators. Let γ_λ be the index of V_λ ($\gamma_\lambda < \infty$). Then clearly if $\{x_i\}$ generates V $\{P^*_\lambda x_i\}$ generates V_λ so $\gamma_\lambda \leq \gamma$. Conversely if a set $\{x_i\}$ is such that $\{P^*_\lambda x_i\}$ generates V_λ then $\{x_i\}$ generates V. In fact one has $\gamma = \sup \gamma_\lambda$ with a little argument. Finally if $A \in \mathrm{Comm}^2 L^*$ and $f \in \Sigma U_\lambda$ then $A^* f \in \Sigma U_\lambda$ and $(A^* \otimes I)Tf = (I \otimes A^*)Tf = TA^* f$. To see this note that A^* commutes with P_λ and let $x \in E'$. Since T_x commutes with L, T^*_x commutes with L^* and A; hence T_x commutes with A^*. For $y \in E'$

$$(5.24) \qquad (A^* \otimes I)Tf(y,x) = Tf(Ay,x) = T_x f(Ay)$$

$$= A^* T_x f(y) = T_x A^* f(y) = TA^* f(y,x).$$

The following theorems now require some proving and we will indicate the idea without spelling out all the details.

<u>Theorem 5.12</u> Suppose ΣU_λ is weakly dense in E''. Then an L-admissible translation T can be determined by γ B-conditions.

Proof: Let $\gamma < \infty$ and suppose T_1 and T_2 are L-admissible translations satisfying B-conditions (a_i, A_i), $1 \leq i \leq \gamma$, where $\{a_i\}$ generates V. Then $T_1 = T_2$. To show this one need only show that if $f \in U_\lambda$ and $x,y \in V_\lambda$ then $T_1 f(x,y) = T_2 f(x,y)$ (by previous remarks). But for $x \in V_\lambda$ there are $B_i \in \mathrm{Comm}^2 L^*$ such that $x = \sum_1^\gamma B_i a_i$. Then

$$(5.25) \qquad T_1f(x,y) = T_1f(\textstyle\sum B_ia_i,y) = \sum T_1f(B_ia_i,y)$$

$$= \textstyle\sum (B_i^* \otimes I)T_1f(a_i,y) = \sum T_1B_i^*f(a_i,y).$$

Similarly $y = \sum C_ja_j$, $C_j \in \text{Comm}^2 L^*$, and one has $T_1f(x,y) = \Sigma\Sigma T_1C_j^*B_i^*f(a_i,a_j)$. Since T_1 satisfies the B-conditions (a_i,A_i) we have

$$(5.26) \qquad T_1f(x,y) = \sum_1^\gamma \sum_1^\gamma (A_iC_j^*B_i^*f)(a_j)$$

and the right side of (5.26) is independent of T_1 so $T_1 = T_2$ acting on f. QED

<u>Theorem 5.13</u> Let T be an L-admissible translation satisfying (a_i,A_i) for $1 \leq i \leq s$ where $\{a_i\}$ does not generate V. Then there is an L-admissible translation $T_1 \neq T$ satisfying these B-conditions. Thus if $\gamma < \infty$ T cannot be determined by $\gamma - 1$ B-conditions.

Proof: This is proved using some lemmas which we summarize here. First one shows that if W is a finite dimensional vector space, $L_0 : W \to W$ a linear map, $\{x_i\}$ a set of generators for W over $\text{Comm}^2 L_0$ $(i = 1,\ldots,s)$, and $\{x_2,\ldots,x_s\}$ not a set of generators of W then there is an $A \in \text{Comm}^2 L_0$ such that $Ax_1 \neq 0$ but $Ax_i = 0$ for $2 \leq i \leq s$. The proof is routine algebra. Secondly let $\{a_i\} \subset E'$ $(1 \leq i \leq s)$ be such that for some $\lambda \in \sigma_0(L^*)$ $\{P_\lambda^*a_i\}$ generates V_λ but $\{P_\lambda^*a_2,\ldots,P_\lambda^*a_s\}$ does not. Then there is an $A \neq 0$ and an L-admissible translation T satisfying the B-conditions (a_1,A), $(a_2,0),\ldots,(a_s,0)$. This requires some work but there are basically no new constructions in the proof. Finally note the obvious fact that if T_1, T_2 are L-admissible satisfying (a_i,A_i) and (a_i,B_i) then $T_1 + T_2$ is L-admissible and satisfies $(a_i,A_i + B_i)$. QED

Now (cf (5.10)) one defines a multiplication on E' as a continuous linear map $m : E' \hat{\otimes} E' \to E'$ so if T is a translation operator we define

$$(5.27) \qquad x \cdot y = T^*(x \otimes y).$$

It follows that T is L-admissible if and only if for $x,y \in E'$

$$(5.28) \qquad L^*(x \cdot y) = (L^*x) \cdot y = x \cdot (L^*y).$$

Indeed for $f \in E''$ one has $f(L^*(x \cdot y)) = Lf(x \cdot y) = TLf(x,y)$ while

$$f((L^*x) \cdot y) = Tf(L^*x,y) = (L \otimes I)Tf(x,y).$$

Thus $L^*(x \cdot y) = (L^*x) \cdot y$ if and only if $TL = (L \otimes I)T$ which is the admissibility condition. The other equation is similar. This shows that $x \to x \cdot y$ and $y \to x \cdot y$ commute with L^* and hence for $A \in \mathrm{Comm}^2 L^*$ $(Ax) \cdot y = A(x \cdot y) = x \cdot (Ay)$. One says now that T is commutative etc if the product (5.27) is.

Proposition 5.14 Let T be L-admissible and satisfy (a_i, A_i), $1 \leq i \leq s$, where $\{a_i\}$ generates V. Suppose ΣU_λ is weakly dense in E''. Then T is commutative if and only if $A_i^* a_j = A_j^* a_i$ for each (i,j).

Proof: Define T' by $T'f(x,y) = Tf(y,x)$ for $f \in E$, $x,y \in E'$. Then T' is L-admissible and $f(y \cdot x) = Tf(y,x) = T'f(x,y)$ while $f(x \cdot y) = Tf(x,y)$. Thus T is commutative if and only if $T = T'$ and this holds if and only if $Tf(x,y) = T'f(x,y)$ for $f \in U_\lambda$ and $x,y \in E'$ by previous remarks. It is enough to have $Tf(a_i,a_j) = T'f(a_i,a_j)$ for $f \in E''$ and here

$$\text{(5.29)} \qquad f(A_i^* a_j) = A_i^{**} f(a_j) = Tf(a_i,a_j) = T'f(a_j,a_i);$$

$$f(A_j^* a_i) = A_j^{**} f(a_i) = Tf(a_j,a_i).$$

The result follows. QED

One shows easily that T is associative if and only if $(T \otimes I)T = (I \otimes T)T$ where $T \otimes I : B(E',E') \to B(E',E',E')$ is defined by $(T \otimes I)f(x,y,z) = f(T^*(x,y),z) = f(x \cdot y,z)$ (similarly for $(I \otimes T)$). Then one has

Theorem 5.15 Under the hypotheses of Proposition 5.14 T is commutative and associative if and only if both $A_i^* a_j = A_j^* a_i$ and $A_i^* A_j^* = A_j^* A_i^*$ for all i,j.

Proof: First define $S_1, S_2 : E \to B(E',E')$ by $S_1 f(x,y) = (T \otimes I)Tf(x,y,z)$ and $S_2 f(x,y) = (I \otimes T)Tf(x,y,z)$ for $z \in E'$ fixed. Then S_i $(i = 1,2)$ is an L-admissible translation. This is clear for S_1 while for S_2 one can write

(5.30) $(L \otimes I)S_2 f(x,y) = S_2 f(L^*x,y) = (I \otimes T)Tf(L^*x,y,z) =$

$Tf(L^*x, y \cdot z) = f((L^*x) \cdot (y \cdot z)) = f(L^*(x \cdot (y \cdot z))) =$

$Lf(x \cdot (y \cdot z)) = (I \otimes T)TLf(x,y,z) = S_2 Lf(x,y).$

Similarly $(I \otimes L)S_2 = S_2 L$. Next observe that $f(a_i \cdot y) = Tf(a_i,y) = A_i f(y) = f(A_i^* y)$ so that $a_i \cdot y = A_i^* y$. If T is commutative we have then $a_i \cdot (y \cdot a_j) = a_i \cdot (a_j \cdot y) = A_i^* A_j^* y$ while $(a_i \cdot y) \cdot a_j = a_j \cdot (a_i \cdot y) = A_j^* A_i^* y$. Now given T commutative the following chain of equivalences shows that a necessary and sufficient condition for associativity is that $A_i^* A_j^* = A_j^* A_i^*$. Thus for appropriate f,x,y,z

(5.31) $(T \otimes I)T = (I \otimes T)T \iff (T \otimes I)Tf(x,y,z) =$

$(I \otimes T)Tf(x,y,z) \iff (T \otimes I)Tf(a_i,y,z) =$

$(I \otimes T)Tf(a_i,y,z) \iff (T \otimes I)Tf(a_i,y,a_j) =$

$(I \otimes T)Tf(a_i,y,a_j) \iff Tf(a_i \cdot y, a_j) =$

$Tf(a_i, y \cdot a_j) \iff f((a_i \cdot y) \cdot a_j) =$

$f(a_i \cdot (y \cdot a_j) \iff f(A_j^* A_i^* y) = f(A_i^* A_j^* y) \iff$

$A_j^* A_i^* y = A_i^* A_j^* y \iff A_j^* A_i^* = A_i^* A_j^*.$ QED

<u>Theorem 5.16</u> Every continuous nonzero homomorphism ϕ of E' is an eigenvector of L.

Proof: We can find $y \in E'$ such that $\phi(y) \neq 0$. Then for $x \in E'$

(5.32) $(L\phi)(x) \cdot \phi(y) = \phi(L^*x) \cdot \phi(y) = \phi((L^*x) \cdot y)$

$= \phi(x \cdot (L^*y)) = \phi(x)\phi(L^*y) = \phi(x)L\phi(y).$

Therefore $L\phi = \dfrac{L\phi(y)}{\phi(y)} \phi.$ QED

We note also that if ϕ is an eigenvector as above then $T\phi(x,y) = \phi(x \cdot y) = \phi(x)\phi(y) = (\phi \otimes \phi)(x,y)$ so $T\phi = \phi \otimes \phi$. If T is a commutative associative L-admissible translation, E' is semisimple, and ΣU_λ is weakly dense in E'' one can embed E' into a function algebra on the ideal space I of E' as in the classical Gelfand theory and some results in this direction are proved in Hutson-Pym [1].

Example 5.17 To relate the preceeding to differential problems let Δ be an operator in E with domain D and suppose Δ has a compact resolvant for some λ which without loss of generality can be taken to be $\lambda = 0$. Set $L = \Delta^{-1}$ and say $A : E \to E$ commutes with Δ if $AD \subset D$ and $A\Delta f = \Delta Af$ for $f \in D$. A translation T is Δ-admissible if for $f \in D$ and $x \in E'$, $Tf(x,\cdot)$ and $Tf(\cdot,x)$ belong to D with $\Delta(Tf(x,\cdot)) = (T\Delta f)(x,\cdot)$ and $\Delta(Tf(\cdot,x)) = (T\Delta f)(\cdot,x)$. Then A commutes with Δ if and only if A commutes with L and for E reflexive T is Δ-admissible if and only if it is L-admissible. Recall that $U_\lambda = \{f; (L - \lambda I)^{\nu_\lambda} f = 0\}$ for large enough ν_λ and set $W_\lambda = \{f; (\Delta - \lambda I)^{\nu_\lambda} f = 0\}$. Then it is easy to show that for $\lambda \neq 0$ $U_\lambda = W_{\lambda^{-1}}$. The preceeding results applied to this situation yield

Theorem 5.18 Let E be a reflexive Banach space and Δ an operator on E with compact resolvant. Suppose ΣW_μ is dense in E. Then a translation operator T associated with Δ cannot be determined by fewer than $\gamma = \max \dim W_\mu$ B-conditions (a,A) $(a \in E'$ and A commuting with $\Delta)$. There are sets of γ B-conditions which do determine T and T is then both commutative and associative if and only if $A_i^* a_j = A_j^* a_i$ while $A_i A_j = A_j A_i$. The continuous nonzero homomorphisms of E' are eigenvectors of Δ.

There is a great deal of material available on the normed rings associated with generalized translation operators and second order differential equations (see e.g. Leblanc [1], Levitan [1; 3], Marčenko [1], Naimark [2], Povzner [1] and references there). Related spectral considerations arise in the inverse Sturm-Liouville theory developed by Gelfand-Levitan [4] and in inverse scattering theory (cf. Agranovič-Marčenko [1], Faddeev [1], Lang [1], Lax-Phillips [2], Levin [1], Marčenko [2; 3]). We will recall some of the basic constructions here following Marčenko [3]. Let $L_i(D_x)u = D_x^2 u - q_i(x)u$, $i = 1,2$, and consider

$$(5.33) \qquad L_1(D_x)u = L_2(D_y)u; \quad u = u(x,y);$$

$$u(x,0) = \phi(x); \quad D_y u(x,0) = \psi(x)$$

$(-\infty < x < \infty;\ y \geq 0)$. One compares here with (5.1) for $L_1 = L_2$ and more generally with (3.1) for $P(D_x) = L_1(D_x)$ and $Q(D_y) = L_2(D_y)$. The original work of Levitan [1; 3] (cf also Lions [2; 3; 4], Marčenko [1], Povzner [1]) started from this point of view for suitable choices of ϕ, ψ. One works with a Riemann function $R(x,y,x_0,y_0)$ in a triangular region $\Delta(x_0,y_0)$ as before to obtain a formula

$$(5.34) \qquad u(x_0,y_0) = \frac{1}{2}\left[\phi(x_0 + y_0) + \phi(x_0 - y_0)\right] + \frac{1}{2}\int_{x_0 - y_0}^{x_0 + y_0} [\psi(x)R(x,0,x_0,y_0) - \phi(x)D_y R(x,0,x_0,y_0)]dx.$$

Let now $e_0(\lambda,y)$ satisfy $L(D_y)e_0 = -\lambda^2 e_0$ with $e_0(\lambda,0) = 1$ and $e_0'(\lambda,0) = i\lambda$. Then $u(x,y) = e^{i\lambda x}e_0(\lambda,y)$ satisfies $D_x^2 u = L(D_y)u$ with $u(x,0) = e^{i\lambda x}$ and $D_y u(x,0) = e^{i\lambda x})'$ $(L(D_y) = D_y^2 - q(y))$. Put this in (5.34) with $x_0 = 0$ and set $K(y_0,x) = -\frac{1}{2}[D_x R(x,0,0,y_0) + D_y R(x,0,0,y_0)]$ to obtain $(\psi = \phi')$

$$(5.35) \qquad e_0(\lambda,y_0) = e^{i\lambda y_0} + \int_{-y_0}^{y_0} K(y_0,x)e^{i\lambda x}dx.$$

The operator $I + K$ defined now by

$$(5.36) \qquad (I + K)f(x) = f(x) + \int_{-x}^{x} K(x,t)f(t)dt$$

will be called a transmutation operator in the spirit of Section 2.3. It maps the solution $e^{i\lambda x}$ of $D^2 e^{i\lambda x} = -\lambda^2 e^{i\lambda x}$ into the solution $e_0(\lambda,x)$ of $L(D_x)e_0 = -\lambda^2 e_0$ satisfying the same initial conditions.

Applying this fact to $e^{\pm i\lambda x}$ and selected initial conditions we obtain

Theorem 5.19 Let $w(\lambda,x,h)$ (resp. $w(\lambda,x,\infty)$) satisfy $L(D_x)w = -\lambda^2 w$ with $w(\lambda,0,h) = 1$, $w'(\lambda,0,h) = h$ (resp. $w(\lambda,0,\infty) = 0$, $w'(\lambda,0,\infty) = 1$) and set

(5.37) $$K(x,t,\infty) = K(x,t) = K(x,-t);\quad K(x,t,h) =$$

$$h + K(x,t) + K(x,-t) + h\int_t^x [K(x,\xi) - K(x,-\xi)]d\xi.$$

Then one has the formula

(5.38) $$w(\lambda,x,h) = \text{Cos}\,\lambda x + \int_0^x K(x,t,h)\,\text{Cos}\,\lambda t\,dt$$

(5.39) $$w(\lambda,x,\infty) = \frac{\text{Sin}\,\lambda x}{\lambda} + \int_0^x K(x,t,\infty)\,\frac{\text{Sin}\,\lambda t}{\lambda}\,dt.$$

The corresponding operators (cf. (5.36)) are denoted by $I + K_h$ and $I + K_\infty$ and as Volterra operators they will have inverses $I + L_h$ and $I + L_\infty$ where for example

(5.40) $$\frac{\text{Sin}\,\lambda x}{\lambda} = w(\lambda,x,\infty) + \int_0^x L(x,t,\infty)w(\lambda,t,\infty)dt.$$

If one sets now $K(t,\xi) = 0$ for $|\xi| > |t|$ it can be proved from the formula (variation of parameters)

(5.41) $$e_o(\lambda,x) = e^{i\lambda x} + \int_0^x \frac{\text{Sin}\,\lambda(x-t)}{\lambda}\,q(t)e_o(\lambda,t)dt$$

and (5.35) that $K(x,t)$ satisfies

(5.42) $$K(x,t) = \frac{1}{2}\int_0^{\frac{1}{2}(x+t)} q(u)du + \frac{1}{2}\int_0^x q(u)\int_{t-(x-u)}^{t+(x-u)} K(u,\xi)d\xi du.$$

Theorem 5.20 The integral equation (5.42) has the unique solution $K(x,t)$ which is continuous. If $g \in C^n$ then $K \in C^{n+1}$ in both variables and satisfies $K(x,-x) = 0$ with

(5.43) $$L(D_x)K = D_t^2 K;\quad K(x,x) = \frac{1}{2}\int_0^x q(\xi)d\xi.$$

Similar theorems apply to the other transmutation kernels. For example $L(x,t,h)$ satisfies

(5.44) $D_x^2 L = L(D_t)L; \quad L(x,x,h) = -h - \frac{1}{2}\int_0^x q(t)dt$

with $hL(x,0,h) = D_t L(x,0,h)$. Consider next $w(\lambda,x,h)$, $K(x,t,h)$, and $L(x,t,h)$ for example and denote them by $w(\lambda,x)$, $K(x,t)$, and $L(x,t)$ for simplicity. Write $C(\lambda,f) = \int_0^\infty f(x)\text{Cos}\,\lambda x dx$ for the cosine transform and

(5.45) $w(\lambda,f) = \int_0^\infty f(x)w(\lambda,x)dx$

(5.46) $\hat{f}(x) = f(x) + \int_x^\infty f(\xi)K(\xi,x)d\xi;$

$$\check{g}(x) = g(x) + \int_x^\infty g(\xi)L(\xi,x)d\xi.$$

Let K^2 (resp. $K^2(\sigma)$) be the space of L^2 functions on $[0,\infty)$ with compact support (resp. with support in $[0,\sigma]$). The space $CK^2(\sigma)$ of cosine transforms of K^2 consists of even entire functions $g(\lambda)$ with $g \in L^2$ for real λ and such that $|g(\lambda)| \leq c \exp \sigma|\text{Im}\lambda|$. Let $Z(\sigma)$ be the even entire functions $g \in L^1$ for λ real satisfying the same estimate. Let $Z = \cup Z(\sigma)$ with the standard topology and $CK^2 = \cup CK^2(\sigma)$; then $Z \subset CK^2$. Let Z' be the dual of Z and define C in Z' by duality for $f \in Z$ and $T \in Z'$ as

$$\langle f, C(\lambda,T)\rangle = \langle T, \int_0^\infty f(\lambda)\text{Cos}\,\lambda x d\lambda\rangle.$$

Now from (5.38) and the analogue of (5.40) for L_h plus the definition of $C(\lambda,f)$ and (5.45) there follows from (5.46)

(5.47) $w(\lambda,f) = C(\lambda,\hat{f}); \quad C(\lambda,g) = w(\lambda,\check{g}).$

One can prove now

Theorem 5.21 There exists $R \in Z'$ such that

(5.48) $\int_0^\infty f(x)g(x)dx = \langle R, w(\lambda,f)w(\lambda,g)\rangle$

for $f,g \in L^2$ of compact support; R is connected with the kernel $L(x,t)$ by the formula $R = \frac{2}{\pi}(1 + C(L))$ where $C(L) = C(\lambda, L(x,0))$.

This is a form of Parseval theorem where R is called the generalized spectral function. The proof is based on constructing a sequence $R_n(\lambda)$ such that as $n \to \infty$, $\int_0^\infty R_n(\lambda)w(\lambda,x)w(\lambda,y)d\lambda \to \delta(x - y)$ (cf. Sections 2.3 and 2.4) and we omit details here. As a consequence of Theorem 5.21 one can prove easily that

$$(5.49) \qquad f(x) = \langle w(\lambda,f)w(\lambda,x), R\rangle.$$

In the case when q and h are real (called the symmetric case) one has (cf Section 2.4)

<u>Theorem 5.22</u> In the symmetric case there exists a nondecreasing function $\rho(\mu)$ such that for $f \in L^2$

$$(5.50) \qquad f(x) = \int_{-\infty}^{\infty} w(\sqrt{\mu},f)w(\sqrt{\mu},x)d\rho(\mu)$$

where the Parseval formula

$$\int_0^\infty f(x)\overline{g(x)}dx = \int_{-\infty}^{\infty} w(\sqrt{\mu},f)\overline{w(\sqrt{\mu},g)}d\rho(\mu)$$

also holds.

<u>Remark 5.23</u> Let us underline the fact that we are not dealing with generalized translations here. These would arise if in (5.33) one took $L_1 = L_2$ and say $\phi = f$ with $\psi = 0$ (growth conditions on q are also imposed). The solution $u(x,y)$ can be obtained then via a Riemann function again and written as

$$(5.51) \qquad u(x,y) = T_x^y f = \frac{1}{2}[f(x + y) + f(x - y)] + \int_0^\infty f(t)R(t,x,y)dt$$

(cf Levitan [1; 3], Povzner [1]). The generalized convolution

$$(5.52) \qquad f \circ g = \int_0^\infty T_x^y(f)g(y)dy$$

can then be expressed in the form

$$(5.53) \qquad f \circ g = \frac{1}{2} f * g + \int_0^\infty \int_0^\infty R(t,x,y)f(x)g(y)dxdy.$$

More generally $f \circ g = h$ is that function which satisfies

(5.54) $$\int_0^\infty \int_0^\infty T_x^y(\phi)f(x)g(y)dxdy = \int_0^\infty \phi(t)h(t)dt$$

for an arbitrary C^∞ even function ϕ. Symmetry properties of R lead to commutativity and associativity - more or less immediate. If $L = L^1$ + an identity with norm $\|\lambda e + f\| = |\lambda| + \|f\|$ then L is a commutative normed ring under o.

Remark 5.24 In the spirit of Remark 5.23 and the spectral considerations preceeding this let us cite some results of Leblanc [1]. This work illustrates and refines some of the early Marčenko-Povzner ideas. Let $Lu = L(D_x)u = D_x^2 u - q(x)u$ and consider $Lu + \lambda^2 u = 0$ with $u(0) = a$ and $u'(0) = b$. A general solution is written as $w(\lambda^2,x) = a\alpha(\lambda^2,x) + b\beta(\lambda^2,x)$ where α (resp β) is the solution corresponding to $a = 1$, $b = 0$ (resp. $a = 0$, $b = 1$). Thus in the preceeding notation $\alpha(\lambda^2,x) = w(\lambda,x,0)$ while $\beta(\lambda^2,x) = w(\lambda,x,\infty)$. Povzner [1] studied the associated Banach algebras with characters $w(\lambda^2,x)$ and showed that when $\int_0^\infty (1 + x^2)|q(x)|dx < \infty$ the solutions of $Lu = 0$ can be written as

$$w(0,x) = A[x + \phi_1(x)] + B[1 + \phi_2(x)]$$

where $\phi_i(x)$ is bounded and tends to zero with $\phi_i'(x)$ when $x \to \infty$. Leblanc puts the associated Banach algebras into four classes (a,A) according as $a = 0$ or $A = 0$ and refers to Marčenko [1] for the case $a \neq 0$, $A = 0$. Leblanc studies the cases $a = 0$, $A \neq 0$ and $a \neq 0$, $A \neq 0$, leaving the case $a = 0$, $A = 0$ open since it seems less interesting. First consider $a = 0$, $A \neq 0$ and make the hypotheses (weaker than Povzner [1])

(5.55) $$\int_0^\infty x|q(x)|dx < \infty; \quad \lim_{x\to\infty} \frac{\partial\beta}{\partial x}(0,x) \neq 0.$$

It is mentioned that the condition on q is less natural than Povzner's condition above but it gives a better estimate at ∞ which is useful. We will state some of the results without giving proofs. First (cf (5.39) so $B(t,x) \sim K(x,t,\infty)$)

Proposition 5.25 Given the hypotheses above $\beta(\lambda^2,x)$ may be written

(5.56) $$\beta(\lambda^2,x) = \frac{\operatorname{Sin}\lambda x}{\lambda} + \int_0^x B(x,y)\,\frac{\operatorname{Sin}\lambda y}{\lambda}\,dy$$

where $B(x,y)$ is defined for $0 \le y \le x$ $(= 0$ for $y > x)$ and satisfies $\int_0^x |B(x,y)|dy \le K$ with $\int_y^\infty x|B(x,y)|dx \le 2Ky$. The map U defined by

(5.57) $$Uf(x) = f(x) + \int_0^\infty B(t,x)f(t)dt$$

sends $L^1 = L^1(xdx)$ into itself with $\| U(f) \| \le (K+1)\| f \|$.

Set $L^2 = L^2(dx)$ now and recall a standard theorem due to Weyl (cf Titchmarsh [2]).

<u>Theorem 5.26</u> There are a finite number of real negative values of λ^2 such that $x^{-1}\beta(\lambda^2,x)$ is bounded. Denote these by λ_n^2 and set for $f \in L^2$

(5.58) $$\overset{\frown}{f}(\lambda^2) = \int_0^\infty f(x)\beta(\lambda^2,x)dx;$$

$$M(\lambda) = 1 + \int_0^\infty e^{-i\lambda t}q(t)\beta(\lambda^2,t)dt.$$

Then in L^2 one has

(5.59) $$f(x) = \sum_1^N \frac{\overset{\frown}{f}(\lambda_n^2)\beta(\lambda_n^2,x)}{\| \beta(\lambda_n^2,x)\|_2^2} + \frac{2}{\pi}\int_0^\infty \overset{\frown}{f}(\lambda^2)\beta(\lambda^2,x)\,\frac{\lambda^2 d\lambda}{|M(\lambda)|^2}$$

together with the corresponding Parseval-Plancherel formula.

Define now a product

(5.60) $$f \circ g(x) = \iint_{u+v>x>|u-v|} f(u)g(v)dudv$$

and set for $\psi \in L^1$

(5.61) $$\tilde{\psi}(\lambda^2) = \int_0^\infty \psi(x)\,\frac{\operatorname{Sin}\lambda x}{\lambda}\,dx.$$

Then there is $\phi,\psi \in L^1$ such that

(5.62) $$|M(\lambda)|^2 = 1 + \tilde{\psi}(\lambda^2); \quad |M(\lambda)|^{-2} = 1 + \phi(\lambda^2)$$

provided $M(0) \neq 0$ (this condition is equivalent to $D_t\beta(0,t) \neq 0$). This requires some calculation which we omit. The product $\circ$ of (5.60) provides L^1 with a Banach algebra structure whose characters are $\sin \lambda x/\lambda$. Now invert (5.56) to obtain

$$(5.63) \qquad \frac{\sin \lambda x}{\lambda} = \beta(\lambda^2,x) + \int_0^x H(x,y)\beta(\lambda^2,y)dy$$

so that $H(x,y) \sim L(x,y,\infty)$ from (5.40). Set

$$(5.64) \qquad \mathcal{H}(x,y) = H(x,y) - \sum_0^N \frac{\mathrm{sh}|\lambda_n|x}{|\lambda_n|} \frac{\beta(\lambda_n^2,y)}{\|\beta(\lambda_n^2,y)\|_2^2}$$

and define for $f \in L^1$

$$(5.65) \qquad Wf(x) = f(x) + \int_0^\infty \mathcal{H}(t,x)f(t)dt.$$

<u>Proposition 5.27</u> W is a map $L^1 \to L^1$ such that if $M(0) \neq 0$ $\|Wf\| \leq c\|f\|$.

The proof involves using some facts which relate various transforms and we display some of this. Thus we have transforms $\frown$ and $\sim$ given by (5.58) and (5.61). Note that $\int_0^\infty \beta(\lambda^2,x)\beta(\lambda_n^2,x)dx = 0$ for $\lambda^2 \geq 0$ so H can be replaced by $\mathcal{H}$ in (5.63) and thus

$$(5.66) \qquad \tilde{f}(\lambda^2) = \overset{\frown}{Wf}(\lambda^2).$$

Since $\overset{\frown}{Wf}(\lambda_n^2) = 0$ (5.59) becomes

$$(5.67) \qquad Wf(x) = \frac{2}{\pi}\int_0^\infty \tilde{f}(\lambda^2)\beta(\lambda^2,x)\frac{\lambda^2 d\lambda}{|M(\lambda)|^2}.$$

We note here the correspondences $C(\lambda,f) \sim \tilde{f}$, $w(\lambda,f) \sim f$, $\check{g} \sim Wg$ (cf (5.45) - (5.46), (5.61), (5.65)). The relation $C(\lambda,g) = w(\lambda,\check{g})$ of (5.47) has an analogue here in

$$(5.68) \qquad \tilde{g} = \overset{\frown}{Wg}.$$

<u>Theorem 5.28</u> Let E be the space generated by the $\beta(\lambda_n^2,x)$ and $L^1 = E \oplus I$. Then the image of L^1 under W is equal to I. The product $f \cdot g = W[U(f) \circ U(g)]$ gives I a structure of commutative Banach algebra which is isomorphic to L^1 with the product $\circ$. The Gelfand transform algebras of I and L^1 are the same

relative to characters $\beta(\lambda^2,x)$ and $\text{Sin}\ \lambda x/\lambda$.

The proof involves examining the relation (5.68) plus

$$(5.69)\qquad \overbrace{f\cdot g} = \widetilde{U(f)\circ U(g)} = \widetilde{U(f)}\widetilde{U(g)}$$

$$(5.70)\qquad \overbrace{f} = \widetilde{U(f)}.$$

Similarly using general characters $w(\lambda^2,x)$ one can prove

Theorem 5.29 Let $\int_0^\infty (1+x^2)|q(x)|dx < \infty$ and consider the case $a \neq 0$, $A \neq 0$. One can provide $L^1((1+x)dx)$ with a commutative Banach algebra structure whose characters are $w(\lambda^2,x) = a\alpha(\lambda^2,x) + b\beta(\lambda^2,x)$ which is isomorphic to $L^1((1+x)dx) \oplus \mathbb{C}^N$ where $L^1((1+x)dx)$ has the product $f\circ g + \frac{1}{2} f * g$ (prolong f and g to be even for $*$) with characters $\text{Cos}\ \lambda x + \text{Sin}\ \lambda x/\lambda$.

2.6. Miscellaneous topics. We feel compelled to make a few remarks here about the application of some of the transmutation-transform theory of Section 2.5 to the inverse Sturm-Lionville and inverse scattering problems following Marčenko [3]. In addition to references already cited we mention a survey article by Durovin-Matveev-Novikov [1] which should indicate some interesting new connections (cf. also Maslov-Manin [1]). We use the notation of Marčenko in Section 2.5. The inverse Sturm-Lionville problem involves recovering the problem $L(D_x)u = -\lambda^2 u$, $u'(0) = hu(0)$, from the spectral function R of Theorem 5.21. Thus basically one wants $q(x)$ given R. We continue to suppress the h in writing $w(\lambda,x)$ for $w(\lambda,x,h)$ etc. and write formally

$$(6.1)\qquad \Phi(x) = \langle R, \frac{1-\text{Cos}\ \lambda x}{\lambda^2}\rangle$$

$$(6.2)\qquad f(x,y) = \frac{1}{2}[\Phi''(x+y) + \Phi''(|x-y|)].$$

It is shown that in fact $\Phi \in C^3$ with $\Phi'(0) = 1$ and $\Phi''(0) = -h$ so (6.2) makes sense. Further if $f(\lambda) \in CK^2(\sigma)$ and $\langle f(\lambda)y(\lambda), R\rangle = 0$ for all $y(\lambda) \in CK^2(\sigma)$

then $f(\lambda) = 0$. These two properties essentially characterize spectral functions. One now derives an integral equation for $K(x,t) = K(x,t,h)$ as follows. The function $w(\lambda,x)$ is a multiplier in Z and Z' so consider $(R - \frac{2}{\pi})w(\lambda,x)$ (recall $R - \frac{2}{\pi} = \frac{2}{\pi}\ C(\lambda,\ L(x,0)))$. From (5.38) one has then

$$(6.3)\qquad (R - \tfrac{2}{\pi})w(\lambda,x) = (R - \tfrac{2}{\pi})\text{Cos}\ \lambda x + \\ + (R - \tfrac{2}{\pi}) \int_0^x K(x,t)\text{Cos}\ \lambda t dt.$$

Note now that $\text{Cos}\ \lambda a C(\lambda,f) \sim \frac{1}{2}\ [f(x + y) + f(|x - y|)]$ while if $\phi(\lambda) = \int_0^\infty g(\xi)\text{Cos}\ \lambda\xi d\xi$ then $\phi(\lambda)C(\lambda,f) \sim \frac{1}{2} \int_0^\infty g(\xi)[f(x + \xi) + f(|x - \xi|)]d\xi$. It follows from (6.3) that

$$(6.4)\qquad (R - \tfrac{2}{\pi})w(\lambda,x) \sim \tfrac{1}{\pi}\ [L(x + y,\ 0) + L(|x - y|,\ 0)] \\ + \tfrac{1}{\pi} \int_0^x K(x,t)[L(t + y,\ 0) + L(|t - y|,\ 0)]dt.$$

Express now $\text{Cos}\ \lambda x$ through $w(\lambda,x)$ by a formula analogous to (5.40) so that if $F(\lambda) = C(\lambda,f)$

$$(6.5)\qquad F(\lambda) = \int_0^\infty [f(x) + \int_x^\infty L(y,x)f(y)dy]w(\lambda,x)dx.$$

Hence by Theorem 5.21 we have

$$(6.6)\qquad \langle F(\lambda),\ Rw(\lambda,x)\rangle = \langle F(\lambda)w(\lambda,x),\ R\rangle \\ = f(x) + \int_x^\infty f(y)L(y,x)dy.$$

Moreover $\langle F(\lambda),\ \frac{2}{\pi}\ w(\lambda,x)\rangle = \langle F(\lambda)w(\lambda,x),\ \frac{2}{\pi}\rangle = \frac{2}{\pi} \int_0^\infty F(\lambda)[\text{Cos}\ \lambda x + \int_0^x K(x,y)\text{Cos}\ \lambda y dy]d\lambda$ $= f(x) + \int_0^x K(x,y)f(y)dy$. Consequently

$$(6.7)\qquad \langle F(\lambda),\ (R - \tfrac{2}{\pi})w(\lambda,x)\rangle \\ = \int_x^\infty f(y)L(y,x)dy - \int_0^x K(x,y)f(y)dy = \int_0^\infty f(y)[L(y,x) - K(x,y)]dy.$$

Thus $(R - \frac{2}{\pi})w(\lambda,x) \sim \frac{2}{\pi}\ [L(y,x) - K(x,y)]$ and using (6.4) this gives

(6.8) $$L(y,x) - K(x,y) = \frac{1}{2}[L(x+y,0) + L(|x-y|,0)] + \frac{1}{2}\int_0^x K(x,t)[L(t+y,0) + L(|t-y|,0)]dt.$$

Thus for $0 \leq y \leq x$ one has the important integral equation satisfied by K

(6.9) $$f(x,y) + K(x,y) + \int_0^x K(x,t)f(t,y)dt = 0$$

where $f(x,y) = \frac{1}{2}[L(x+y,0) + L(|x-y|,0)]$ which is easily seen to be the same as the expression for $f(x,y)$ in (6.2).

Theorem 6.1 Given $R \in Z'$ satisfying the two properties indicated and Φ, f defined by (6.1) - (6.2) the equation (6.9) has a unique solution K. K is continuous and has in both variables as many continuous derivatives as Φ''.

Thus from (6.9) $K(x,y)$ is expressed by means of the spectral function R and K determines the Sturm-Lionville problem; the uniqueness of K implies there is a unique Sturm-Lionville problem associated with R. The uniqueness proof is based on showing that the homogeneous equation for any a

(6.10) $$g(y) + \int_0^a g(t)f(t,y)dt = 0$$

has only the zero solution and this follows from the first property of R indicated earlier. If $\Phi \in C^4$ one obtains from (6.9)

(6.11) $$L(D_x)K(x,y) = D_y^2 K(x,y), \quad 0 \leq y \leq x < \infty$$

(6.12) $$q(x) = 2D_x K(x,x); \quad D_y K(x,0) = 0.$$

That is q is defined by (6.12) and then appears in $L(D_x)$. Summarizing one has

Theorem 6.2 In order that $R \in Z'$ be the spectral function of a Sturm-Lionville problem $L(D_x)u = -\lambda^2 u$, $u'(0) = hu(0)$ with continuous q it is necessary and sufficient that (1) For any $\sigma > 0$ there is no nonzero $f(\lambda) \in CK^2(\sigma)$ such that $\langle f(\lambda)y(\lambda), R\rangle = 0$ for all $y(\lambda) \in CK^2(\sigma)$ (2) The function Φ defined by (6.1) belongs to C^3 with $\Phi'(0) = 1$. The function q (defined by (6.12)) will then have as many continuous derivatives as Φ''' and $h = K(0,0)$.

The proof requires some argument but we omit details since the important constructions have already been displayed. We will give next a brief introduction to some techniques needed for inverse scattering theory based on the preceeding constructions. This will serve to tie together some of this material and perhaps motivate the reader to pursue the theory further; it also continues to provide examples of the transform connections indicated in Section 2.3. The matter is beautifully developed in Marčenko [3] at some length and we refer to this for inspiration. Consider again $L(D_x)u = -\lambda^2 u$ with $u(0) = 0$ while $\int_0^\infty x|q(x)|dx < \infty$. As $x \to \infty$ the equation approaches $D_x^2 v = -\lambda^2 v$ and one idea is to extract information about q by comparing solutions at ∞ with v.

<u>Proposition 6.3</u> For λ in the closed upper half plane there is a solution $e(\lambda,x)$ of $L(D_x)e = -\lambda^2 e$, $e(\lambda,0) = 0$, given in the form

$$(6.13) \qquad e(\lambda,x) = e^{i\lambda x} + \int_x^\infty K(x,t)e^{i\lambda t}dt$$

with $K(x,x) = \frac{1}{2}\int_x^\infty q(t)dt$.

Extimates on $K(x,t)$ are also obtained but we omit this. Here $K(x,t) = 0$ for $t < x$ and K satisfies an equation analogous to (5.42). One writes $(I + K)f = f(x) + \int_x^\infty K(x,t)f(t)dt$ with inverse $I + L$ where

$$(6.14) \qquad (I + L)f = f(x) + \int_x^\infty L(x,t)f(t)dt$$

$$(6.15) \qquad L(x,y) + K(x,y) + \int_x^y L(x,t)K(t,y)dt = 0.$$

Again K is characterized by $L(D_x)K(x,t) = D_t^2 K(x,t)$, $D_x K(x,x) = -\frac{1}{2} q(x)$, and $\lim D_x K(x,t) = \lim D_t K(x,t) = 0$ as $x + t \to \infty$. Estimates on $e(\lambda,x)$ are also obtained which we omit in remarking that estimates on the various functions are however very important for the theory. Another fairly routine construction now provides for every λ a solution $w(\lambda,x,\infty)$ of $L(D_x)w = -\lambda^2 w$ satisfying as $x \to 0$

$$(6.16) \qquad w(\lambda,x,\infty) = x(1 + O(1)); \quad w'(\lambda,x,\infty) = 1 + O(1).$$

The function w is analytic for $\text{Im}\,\lambda \geq 0$ and estimates comparing it there to $\text{Sin}\,\lambda x/\lambda$ are given.

<u>Proposition 6.4</u> Let $S(\lambda) = e(-\lambda,0)/e(\lambda,0) = \overline{S(-\lambda)} = S^{-1}(-\lambda)$. For real $\lambda \neq 0$ one has

$$(6.17)\qquad -\frac{2i\lambda w(\lambda,x,\infty)}{e(\lambda,0)} = e(-\lambda,x) - S(\lambda)e(\lambda,x).$$

The function $e(\lambda,0)$ can have for $\text{Im}\,\lambda > 0$ only a finite number of zeros all of which are simple and lie on the imaginary axis. The function $\lambda e(\lambda,0)^{-1}$ is bounded in some nbh of 0. This takes some proving which we omit. Similarly one can show that $1 - S(\lambda)$ is the Fourier transform of $F_S(x) = F_S^1(x) + F_S^2(x)$ where $F_S^1 \in L^1(-\infty,\infty)$, $F_S^2 \in L^2(-\infty,\infty)$, and $\sup|F_S^2| < \infty$. Now let $i\lambda_k$, $k = 1,2,\ldots,n$, be the zeros of $e(\lambda,0)$ ordered by $0 < \lambda_1 < \lambda_2 < \ldots < \lambda_n$ and $m_k^{-1} = \| e(i\lambda_k,x)\|$ in L^2. In fact

$$m_k^{-2} = -e'(i\lambda_k,0)e(i\lambda_k,0)/2i\lambda_k.$$

The function $u(\lambda,x) = e(-\lambda,x) - S(\lambda)e(\lambda,x)$, $\lambda \in (0,\infty)$, is a bounded solution of $L(D_x)u = -\lambda^2 u$, $u(0) = 0$, with $u(i\lambda_k,x) = m_k e(i\lambda_k,x)$. In fact such functions form a complete set of eigenfunctions; setting $u(\lambda,f) = \int_0^\infty f(x)u(\lambda,x)dx$ one has a Parseval relation

$$(6.18)\qquad (f,g) = \sum_1^n u(i\lambda_k,f)\overline{u(i\lambda_k,g)} + \frac{1}{2\pi}\int_0^\infty u(\lambda,f)\overline{u(\lambda,g)}d\lambda.$$

One arrives at this by observing first that for $f^*(x) = (I + K^*)f$ it follows that $\tilde{f}^*(\lambda) = \int_0^\infty f(x)e(-\lambda,x)dx$ where $\tilde{f}^*(\lambda)$ denotes the ordinary Fourier transform of f^*. Indeed from (6.13)

$$(6.19)\qquad \int_0^\infty f(x)e(-\lambda,x)dx = \int_0^\infty f(x)[e^{-i\lambda x} + \int_x^\infty K(x,t)e^{-i\lambda t}dt]dx$$

$$= \int_0^\infty [f(x) + \int_0^x f(\xi)K(\xi,x)d\xi]e^{-i\lambda x}dx.$$

Hence $u(\lambda,x) = \tilde{f}^*(\lambda) - S(\lambda)\tilde{f}^*(-\lambda)$ with $u(i\lambda_k,x) = m_k\int_0^\infty f^*(x)\exp(-\lambda_k x)dx$. Write T for the right side of (6.18) and set

(6.20) $$F(x) = \sum_1^n m_k^2 e^{-\lambda_k x} + \frac{1}{2\pi}\int_{-\infty}^{\infty} (1 - S(\lambda))e^{i\lambda x} d\lambda$$

(6.21) $$F(f) = \int_0^{\infty} F(x + y)f(y)dy.$$

Then F is a bounded self adjoint operator in $L^2(0,\infty)$ and writing out T one obtains $T = ((I + K)(I + F)(I + K^*)f,g)$. Hence the Parseval relation (6.18) is equivalent to

(6.22) $$(I + K)(I + F)(I + K^*) = I.$$

Writing this out and using the fact that $K(x,y) = 0$ for $x > y$ it follows that (6.22) is equivalent to

(6.23) $$F(x + y) + K(x,y) + \int_x^{\infty} K(x,t)F(y + t)dt = 0.$$

This integral equation for K is basic in the theory and one can prove

Theorem 6.5 The kernel $K(x,y)$ satisfies (6.23) for $0 < x < y < \infty$ where F is given by (6.20) and from (6.23) follows the Parseval relation (6.22) or (6.18).

The proof uses (6.17) and other facts mentioned above. Now a typical problem in quantum scattering theory involves recovering a potential $q(x)$ from knowledge of $S(\lambda)$, λ_k, and m_k. Theorem 6.5 allows one to do this by finding $K(x,y)$ and setting $q(x) = -\frac{1}{2} D_x K(x,x)$. The question reduces to specifying the properties required of $S(\lambda)$, λ_k, and m_k so that they represent scattering data for an equation $L(D_x)u = -\lambda^2 u$ with $u(\lambda,0) = 0$. The crucial fact is

Property 6.6 The function $S(\lambda)$ is defined on the real axis with $S(\lambda) = \overline{S(-\lambda)} = S^{-1}(-\lambda)$ while $1 - S(\lambda) \to 0$ as $|\lambda| \to \infty$ and is the Fourier transform of a function

(6.24) $$F_S(x) = \frac{1}{2\pi}\int_{-\infty}^{\infty} (1 - S(\lambda))e^{i\lambda x} d\lambda$$

which is the sum of $F_S^1(x)$ and $F_S^2(x)$ with $F_S^1 \in L^1(-\infty,\infty)$ and $F_S^2 \in L^2(-\infty,\infty)$. For $x > 0$ one has $\int_0^{\infty} x|F_S^1(x)|dx < \infty$.

Another property which should be mentioned is that the number n of eigenvalues indicated above satisfies

(6.25) $$n = \frac{\log S(0^+) - \log S(\infty^+)}{2\pi i} - \frac{1 - S(0)}{4}.$$

Theorem 6.7 If Property 6.6 holds then (6.23) has for $\varepsilon > 0$ a unique solution $K(x,y) \in L^1(\varepsilon,\infty)$. The function $e(\lambda,x)$ defined by (6.13) satisfies $L(D_x)e = -\lambda^2 e$ with $q(x) = -\frac{1}{2} D_x K(x,x)$ and $\int_\varepsilon^\infty x|q(x)|dx < \infty$. Here $F(x)$ is given by (6.20) and, writing $F_a^+(f) = \int_0^\infty F(t + y + 2a)f(t)dt$, if $I + F_a^+$ has an inverse for $a = 0$ one can take $\varepsilon = 0$.

Another approximation to a "best" theorem (cf. Theorem 6.9) is obtained by defining

(6.26) $$F_S^+ f = \int_0^\infty F_S(t + y)f(t)dt;$$

$$F_S^- f = \int_{-\infty}^0 F_S(t + y)f(t)dt.$$

Theorem 6.8 Assume Property 6.6 and (A) the equation $f + F_S^+ f = 0$ has n linearly independent solutions in $L^2(0,\infty)$ (B) the equation $g - F_S^- g = 0$ has only the zero solution in $L^2(-\infty,0)$. Then the spectral data $S(\lambda)$, λ_k, m_k $(k = 1,\ldots,n)$ arises from a problem $L(D_x)u = -\lambda^2 u$, $u(0) = 0$, and q defined as above satisfies $\int_0^\infty x|q(x)|dx < \infty$.

We omit the proofs of these theorems but include them to indicate the circle of ideas. Finally putting all this together one has

Theorem 6.9 In order that the spectral data $S(\lambda)$, λ_k, m_k arise from a problem $L(D_x)u = -\lambda^2 u$, $u(0) = 0$, with $\int_0^\infty x|q(x)|dx < \infty$ it is necessary and sufficient that Property 6.6 hold with (6.25).

For much more material in this direction and applications to modern work in KdV equations see Marčenko [3].

We mention now a construction of the transmutation operator for Sturm-Lionville type equations of second order with operator coefficients following Androščuk [1] (cf. also Gorbačuk-Gorbačuk [2;3], Levitan [4;6], Rofe-Beketov [1], Solovev [1]). Consider $u(0) = h \in D(A)$, $u'(0) = 0$, and

$$(6.27) \qquad Lu = -D_x^2 u + q(x)u - Au; \quad Lu = \lambda u$$

where $u(x) \in H$, a separable Hilbert space, and A is a self adjoint operator semibounded below which can be taken as $A > 0$ with $A^{-1} \in L(H)$ without loss of generality. Assume also that $x \to A^{\frac{1}{2}}q(x)A^{-\frac{1}{2}} \in C^0(L_s(H))$ ($q(x) \in L(H)$ with $q(x)$ self adjoint and $x \to q(x) \in C^0(L_s(H))$). Then one wants to extend formulas such as (5.38) and an analogue of (5.40) in some manner.

Theorem 6.10 Under the hypotheses indicated the equation (6.27) can be solved in the form

$$(6.28) \qquad u(x,\lambda) = h\,\mathrm{Cos}\sqrt{\lambda}x + \int_0^x K(x,t)Ah\,\mathrm{Cos}\sqrt{\lambda}t\,dt$$

where $(x,t) \to K(x,t) \in C^0(L_s(H))$ with $K(x,t) = 0$ for $t > x$.

The proof involves working with the integral equation (by successive approximations)

$$(6.29) \qquad w(x,\lambda) = \mathrm{Cos}\sqrt{A+\lambda I}\;x + \int_0^x \frac{\mathrm{Sin}\sqrt{A+\lambda I}\;(x-t)}{\sqrt{A+\lambda I}}\,q(t)w(t,\lambda)dt.$$

Using the formula in operators

$$(6.30) \qquad \frac{\mathrm{Sin}\sqrt{A+\lambda I}\;(x-t)}{\sqrt{A+\lambda I}} = \frac{1}{2}\int_{-x+t}^{x-t} e^{i\sqrt{A}s} J_0(\sqrt{\lambda}\sqrt{(x-t)^2 - s^2})ds$$

one obtains a formula

$$(6.31) \qquad w(x,\lambda) = \mathrm{Cos}\sqrt{A+\lambda x}$$
$$+ \int_0^x\int_0^{x-t} R(x,t,\tau)\mathrm{Cos}\sqrt{\lambda}\;\tau\;\mathrm{Cos}\sqrt{A+\lambda I}\;t\;d\tau dt$$

where $R(x,t,\tau) \in L_s(H)$ and depends continuously there on (x,t,τ). If $h \in D(A)$ then $w(x,\lambda)h$ satisfies (6.27). Now $\mathrm{Cos}\sqrt{A+\lambda I}\;xh$ is the solution of (6.27) for $q = 0$ and $\mathrm{Cos}\sqrt{A+\lambda I}\;xh = h\,\mathrm{Cos}\sqrt{\lambda}\;x + g(x)$ with

$$(6.32)\qquad g(x) = \int_0^x \frac{\mathrm{Sin}\sqrt{A+\lambda I}\,(x-t)}{\sqrt{A+\lambda I}}\, Ah\, \mathrm{Cos}\sqrt{\lambda}\, t dt.$$

Since it is known that

$$(6.33)\qquad J_0(\sqrt{\lambda}\sqrt{(x-t)^2 - s^2}) = \frac{2}{\pi}\int_0^{\sqrt{(x-t)^2-s^2}} \frac{\mathrm{Cos}\sqrt{\lambda}\,\tau d\tau}{\sqrt{(x-t)^2 - s^2 - \tau^2}}$$

one can express $g(x)$ and hence $\mathrm{Cos}\sqrt{A+\lambda I}\, xh$ in terms of $h\,\mathrm{Cos}\sqrt{\lambda}\, x$ (using (6.32), (6.33), (6.30))). Putting all this in (6.31) gives (6.28).

<u>Theorem 6.11</u> Let $q(x)$, $A^{\frac{1}{2}}q(x)A^{-\frac{1}{2}}$, $\exp\sqrt{A}\, t$, and $\exp -\sqrt{A}\, t$ be strongly continuous operator functions with $q(\cdot) \in C^2$. For $h \in D(A)$ an inverse transmutation is expressed by (6.34) below where $H(x,t)$ is a weakly continuous operator function $H_o \to H_-$ ($H(x,t) = 0$ for $t > x$).

$$(6.34)\qquad h\,\mathrm{Cos}\sqrt{\lambda}\, x = u(x,\lambda) + \int_0^x H(x,t)Au(t,\lambda)dt.$$

Proof: Let $E(\Delta)$ be the spectral resolution associated with A and for such Δ on $E(\Delta)H$ define the scalar product

$$(f,g)_+ = (A^2\exp 2\sqrt{A}\, x_oA^2f,\ A^2\exp 2\sqrt{A}\, x_oA^2g)$$

($x_o > 0$ some fixed number); let H_+ be the resulting space. Set $H_o = H$ and let H_- be the corresponding negative norm space (i.e. let $f \in H_o$ so $(f,u)_o = \langle \ell_f,u\rangle$ for $u \in H_+$, $\ell_f \in H_+'$ (antidual) and $\|\ell_f\| = \sup|(f,u)_o|/\|u\|_+ \leq \|f\|_o$ - H_- is the completion of H_o in the norm $\|f\|_- = \|\ell_f\|$). Now the vector function

$$(6.35)\qquad g(x,y) = \mathrm{Cos}\sqrt{\lambda}\, xw(y,\lambda)h - \frac{1}{2}[w(x-y,\lambda)h + w(x+y,\lambda)h]$$

is a strong solution of $g_{xx} - g_{yy} - A_g + q(y)g = Af(x,y)$ where $g(0,y) = g_x(0,y) = 0$ and, taking $g(-y) = g(y)$,

$$(6.36)\qquad f(x,y) = (I - A^{-1}q(y))\left[\frac{w(x-y,\lambda)h + w(x+y,\lambda)h}{2}\right].$$

Let $(\exp\sqrt{A}\,\tau)^+$ denote the transpose of $(\exp\sqrt{A}\,\tau) : H_+ \to H_o$ so $(\exp\sqrt{A}\,\tau)^+ :$ $H_o \to H_-$ and consider $g_o(x,y) \in H_-$ defined by

$$(6.37) \qquad g_o(x,y) = \frac{1}{2\pi} \int_{Q<(0,y,x)} \frac{(\exp\sqrt{A}\,\tau)^+ Af(\xi,\eta)\,dQ}{\sqrt{(x-\xi)^2 - (y-\eta)^2 - \tau^2}} .$$

Here $(\tau,\eta,\xi) = Q < (0,y,x)$ is the interior of the cone $(x-\xi)^2 - (y-\eta)^2 - \tau^2 = 0$ with vertex $(0,y,x) < (t_o,y_o,x_o)$, $0 < \xi < x < x_o$, and $P_o = (t_o,y_o,x_o)$ fixed. The function g_o is a weak solution of the g equation above for $q = 0$, i.e.

$$(6.38) \qquad D_x^2(g_o(x,y), \phi) - D_y^2(g_o(x,y), \phi) -$$

$$- (g_o(x,y), A\phi) = (f(x,y), A\phi)$$

for $\phi \in H_+$. For $q(y) \not\equiv 0$ the function $g(x,y)$ above is a solution of the integral equation

$$(6.39) \qquad g(x,y) = \frac{-1}{2\pi} \int_{Q<(0,y,x)} \frac{(\exp\sqrt{A}\,\tau)^+ q(\eta) g(\xi,\eta)\,dQ}{\sqrt{(x-\xi)^2 - (y-\eta)^2 - \tau^2}} + g_o(x,y).$$

The series solution $g(x,y)$ of (6.39) converges over H_+ then and one can write

$$(6.40) \qquad g(x,y) = \int_0^x \int_{\xi+y-x}^{-\xi+y-x} \tilde{H}(x,y,\xi,\eta) Af(\xi,\eta)\,d\eta\,d\xi$$

where $\tilde{H}$ maps H_o into H_-. Using the form of $f(x,y)$ above and (6.35) one obtains

$$(6.41) \qquad \mathrm{Cos}\sqrt{\lambda}\,x w(y,\lambda)h = \frac{1}{2}\,[w(x-y, \lambda)h + w(x+y, \lambda)h]$$

$$+ \int_0^x \tilde{H}_1(x,y,t) Aw(t,\lambda)h\,dt.$$

Setting $y = 0$ (6.34) results where $H(x,t) = \tilde{H}_1(x,0,t)$. QED

<u>Remark 6.12</u> It seems appropriate here to make some comments about separation of variables and tensor products in the spirit of Berezanskij [1], Carroll [27], Cordes [2], B. Friedman [1], Ichinose [1; 2; 3; 4], Maurin-Maurin [3], Mohammed [1], Reed-Simon [2; 3], Schechter [2]. First let us review some of Ichinose's constructions; material on topological tensor products, crossnorms, etc can be found in

LaMadrid [1], Grothendieck [1], Shatten [1], Treves [1]. Let E and F be Banach spaces and α a "reasonable" crossnorm on $E \otimes F$ (i.e. the dual norm α' is a crossnorm on $E' \otimes F'$); we recall that a crossnorm is to satisfy $\alpha(e \otimes f) = \|e\| \|f\|$. A reasonable crossnorm α is said to be faithful if the natural continuous map $E \hat{\otimes}_\alpha F \to E \hat{\otimes}_\varepsilon F$ is 1-1; a crossnorm α is uniform if for any $(A,B) \in L(E) \times L(F)$ $\sup\|(A \otimes B)u\|_\alpha \leq \|A\| \|B\|$ where the sup is over $\|u\|_\alpha \leq 1$. Let $A : D(A) \subset E \to E$ and $B : D(B) \subset F \to F$ be closed densely defined linear operators (the theory can be developed more generally but we are only interested in such operators). Let $P(\xi,\eta) = \Sigma\, c_{jk}\xi^j\eta^k$ and consider

$$(6.42) \qquad P(A \otimes I,\ I \otimes B) = \Sigma\, c_{jk}A^j \otimes B^k$$

with domain $D(A^m \otimes B^n) = D(A^m) \otimes D(A^n)$ where m and n are the degrees of P in ξ and η respectively. Let $A \hat{\otimes}_\alpha B$ be the closure of $A \otimes B$ in $E \hat{\otimes}_\alpha F$ which makes sense if α is reasonable and faithful (which we shall assume in what follows - also for simplicity assume α is uniform). It can be shown then that $A \hat{\otimes}_\alpha B = (A \hat{\otimes}_\alpha I)(I \hat{\otimes}_\alpha B) = (I \hat{\otimes}_\alpha B)(A \hat{\otimes}_\alpha I)$; similarly $A^j \hat{\otimes}_\alpha I = (A \hat{\otimes}_\alpha I)^j$ and if one assumes $\rho(A)$ and $\rho(B)$ nonempty while it does not occur that one of the extended spectra $\sigma_e(A)$ and $\sigma_e(B)$ contains 0 with the other containing ∞ then

$$(6.43) \qquad P(A \hat{\otimes}_\alpha I,\ I \hat{\otimes}_\alpha B) = \Sigma\, c_{jk}A^j \hat{\otimes}_\alpha B^k.$$

Here $\sigma_e(T) = \sigma(T)$ for our purposes and we assume in general that $\rho(A)$ and $\rho(B)$ are nonempty. Let now $\mathcal{P}(A,B)$ be the class of polynomials such that: For any open nbh W in $\mathbb{C}$ of $P(\sigma(A), \sigma(B))$ there are nonempty open U (resp V) with $\complement U \subset \rho(A)$ (resp. $\complement V \subset \rho(B)$) with "nice" boundaries ∂U (resp. ∂V) such that (a) $P(U,V) \subset W$ (b) $\xi R(\xi,A)$ (resp $\eta R(\eta,B)$) is uniformly bounded in U (resp V) (c) $P(\xi,\eta)(|\xi| + |\eta|)^{-1}$ is bounded away from 0 on $\bar{U} \times \bar{V}$ for large $|\xi| + |\eta|$. Now an extension $\tilde{T}$ of an operator T is called maximal if $\overline{G(\tilde{T})} = \overline{G(T)}$ and $\tilde{T}$ has no proper extension with this property; if T is closable then $\bar{T} = \tilde{T}$. The following spectral mapping theorem then holds

Theorem 6.12 Let $P \in \mathcal{P}(A,B)$ and α be any reasonable uniform crossnorm. Then

(6.44) $$P(\sigma(A), \sigma(B)) = \sigma(P(A \otimes I, I \otimes B))$$
$$= \sigma(\tilde{P}(A \otimes I, I \otimes B)).$$

This means in particular that (6.44) holds if $\sigma(A) \neq \phi$ and $\sigma(B) \neq \phi$ while $\sigma(P(A \otimes I, I \otimes B)) = \phi$ if and only if at least one of $\sigma(A)$ or $\sigma(B)$ are empty. For α faithful with $\rho(A)$ and $\rho(B)$ nonempty $P(A \otimes I, I \otimes B)$ is closable so $\tilde{P} = \bar{P}$ and for $P(\sigma(A), \sigma(B)) \neq \mathbb{C}$

(6.45) $$\bar{P}(A \otimes I, I \otimes B) = (\Sigma\, c_{jk} A^j \hat{\otimes}_\alpha B^k)^-.$$

Thus when it does not occur that one of $\sigma(A)$ and $\sigma(B)$ contains 0 while the other contains ∞ there results $P(\sigma(A), \sigma(B)) = \sigma(P(A \hat{\otimes}_\alpha I, I \hat{\otimes}_\alpha B)) = \sigma(\bar{P}(A \hat{\otimes}_\alpha I, I \hat{\otimes}_\alpha B))$.

A closed densely defined operator T is said to be of type $(\theta_T, M(\theta))$, $0 \leq \theta_T < \pi$, if $\rho(T)$ contains the complement of the sector $S(\theta_T) = \{z;\ |\arg z| \leq \theta_T\}$ and $\| zR(z,T) \| \leq M(\theta)$, $\theta = \arg z$, outside $S(\theta_T')$ for $\theta_T < \theta_T' < \pi$. If α is a crossnorm as above and A_j is of type $(\theta_j, M_j(\theta))$, $0 \leq \theta_j < \pi$, $\theta_j + \theta_k < \pi$ $(j \neq k)$ then, taking $j = 1,2$ for simplicity and setting $\mathbb{A}_1 = A_1 \hat{\otimes}_\alpha I$, $\mathbb{A}_2 = I \hat{\otimes}_\alpha A_2$,

(6.46) $$\Sigma\, \sigma(A_j) = \Sigma\, \sigma(\mathbb{A}_j) = \sigma(\Sigma\, \mathbb{A}_j)$$
$$= \sigma((\Sigma\, \mathbb{A}_j)^-).$$

There are a number of results concerning when $\Sigma\, \mathbb{A}_j = (\Sigma\, \mathbb{A}_j)^-$ (cf. also Carroll [27]) and we cite (cf Mikhailov [1])

Theorem 6.13 Let A and B be m-accretive operators in Hilbert spaces E and F with α the standard Hilbert crossnorm and let either (a) A or B is self adjoint or (b) A and B are m-sectorial with semiangles θ_A, θ_B such that $\operatorname{Tan} \theta_A \operatorname{Tan} \theta_B \leq 1$. Then $(A \otimes I + I \otimes B)^- = A \hat{\otimes}_\alpha I + I \hat{\otimes}_\alpha B$ and $\sigma(A) + \sigma(B) =$

$\sigma(A \hat{\otimes}_\alpha I + I \hat{\otimes}_\alpha B)$.

Here an operator A is accretive if $-A$ is dissipative and $\rho(A)$ should contain the left half plane with $\| A + \lambda I)^{-1}\| \leq 1/\text{Re}\ \lambda$ for $\text{Re}\ \lambda > 0$. An m-sectorial operator A with vertex 0 and semiangle θ_A, $0 \leq \theta_A < \pi/2$ is an m-accretive operator with $|\text{Im}(Ax,x)| \leq \text{Tan}\ \theta_A \text{Re}(Ax,x)$. Now going back to E, F and G Banach let $\mathcal{E}$ be a bounded Boolean algebra of projections in G for example and α (resp. β) a uniform reasonable norm in $E \otimes G$ (resp. $F \otimes G$) which we assume faithful whenever necessary for a specific conclusion. The family $\mathcal{E}$ is said to satisfy the Gowurin property $\mathcal{G}$ is there is a constant K such that for any (finite) set $E_i \in \mathcal{E}$ of disjoint projections and any set $A_i \in L(E,F)$ one has for $u \in E \otimes F$

$$(6.47) \qquad \| \sum (A_i \otimes E_i)u\|_\beta \leq K \sup_i \| A_i\| \, \| u\|_\alpha .$$

This property is examined in Ichinose [3] and we omit the details here. In general if $B : D(B) \subset G \to G$ is a scalar type spectral operator (cf Dunford-Schwartz [1]) or a self adjoint operator with a countably additive resolution of the identity $E(d\lambda)$ then the property $\mathcal{G}$ guarantees the existence of integrals

$$(6.48) \qquad \int_{\sigma(B)} [F(\lambda) \otimes E(d\lambda)]v$$

in $F \hat{\otimes}_\beta G$ for $v \in E \otimes G$ when $F(\lambda) : \sigma(B) \to L(E,F)$ is continuous and uniformly bounded. Consider then polynomials such as (6.42) and write $P(\xi,\eta) = c_m(\eta)\xi^m + \ldots + c_o(\eta)$ where $c_m(\eta) \not\equiv 0$. Then

$$(6.49) \qquad P(A,\eta) = \sum_{j=0}^{m} c_j(\eta)A^j$$

is a closed densely defined operator in E with domain $D(A^{m(\eta)})$ where $m(\eta)$ is the largest integer such that $c_{m(\eta)}(\eta) \not\equiv 0$. Assume condition $(P) : P(A,\eta)$ for $\eta \in \sigma(B)$ has an inverse $P^{-1}(A,\eta) \in L(E)$ which is uniformly bounded on $\sigma(B)$ and take $\alpha = \beta$ for simplicity. Let σ_ν be an increasing sequence of bounded Borel sets with $E(\cup\sigma_\nu) = I$ and note that for example

(6.50) $$\lim_{\nu\to\infty} P(A \otimes E(\sigma(B)\cap\sigma_\nu),\ I \otimes BE(\sigma(B)\cap\sigma_\nu))u$$
$$= \lim_{\nu\to\infty} [\sum c_{jk}A^j \otimes (BE(\sigma(B)\cap\sigma_\nu))^k]u$$
$$= \lim_{\nu\to\infty} \int_{\sigma(B)\cap\sigma_\nu} [P(A,\eta) \otimes E(d\eta)]u$$
$$= \sum c_{jk}[A^j \otimes \int_{\sigma(B)} \eta^k E(d\eta)]u = P(A \otimes I,\ I \otimes B)u$$

for $u \in D(A^m) \otimes D(B^n)$. The idea now is to represent $\tilde{P}^{-1}(A \otimes I, I \otimes B) = \tilde{P}^{-1}$ ($= \bar{P}^{-1}$ under circumstances indicated above) in terms of the integral, for $v \in E \otimes G$,

(6.51) $$\tilde{P}^{-1}v = \int_{\sigma(B)} [P^{-1}(A,\eta) \otimes E(d\eta)]v.$$

The details are straightforward using expressions such as (6.50) and one has in particular

Theorem 6.14 Assume the projections $E(d\eta)$ arising from the resolution of the identity for B have the property G as above for $E = F$ and $\alpha = \beta$ (uniform, reasonable, faithful). Let $P(A,\eta)$ satisfy the condition (P) indicated after (6.49) while $\eta^k A^j P^{-1}(A,\eta) \in L(E)$ and is uniformly bounded on $\sigma(B)$ for suitable (j,k). Then $\tilde{P}(A \otimes I, I \otimes B) = (\Sigma c_{jk}(A^j \hat{\otimes}_\alpha B^k))^{\sim}$ and $\tilde{P}$ has a continuous inverse given by (6.51) mapping $E \hat{\otimes}_\alpha G$ into $D(\tilde{P}) \subset \cap D(A^j \hat{\otimes}_\alpha B^k)$ with norm $\Sigma \| (A^j \hat{\otimes}_\alpha B^k)u\|_\alpha$.

For polynomials as in Theorem 6.12, or even larger classes, there are spectral mapping results and closure theorems (cf Ichinose [3]). There are obvious relations here between formulas such as (6.51) and some of the development in Section 1.4 (see e.g. (1.4.63)) and we will expand upon this further in a Hilbert space context following Berezanskij [1]. First let us recall some constructions involving so called negative norm spaces and generalized eigenvectors. Thus let $H_+ \subset H_o$ be two Hilbert spaces with H_+ dense and the injection continuous in the form $\|u\|_o \le \|u\|_+$ (all Hilbert spaces will be assumed separable and of course

$\|u\|_o \le c\|u\|_+$ can be handled by renorming). For $f \in H_o$ we have $\ell_f \in H'_+$ as before with $\ell_f(u) = (f,u)_o$ and $\|\ell_f\| \le \|f\|_o$. The completion of H_o in the norm $\|f\|_- = \|\ell_f\|$ is denoted by H_-. Let $O : H_+ \to H_o$ be the injection and write $(f,Ou)_o = (If,u)_+$ so that $I = O^* : H_o \to H_+$. Then H_- can be viewed as the completion of H_o in the scalar product $(f,q)_- = (If, Ig)_+$ and $\|f\|_- \le \|f\|_o$. A scalar product $(\alpha,u)_o$ is defined for $\alpha \in H_-$ and $u \in H_+$ coinciding with the H_o scalar product for $\alpha \in H_o$. The map I extends to a map $\mathbb{I} : H_- \to H_+$ onto and one has $(f,u)_o = (If,u)_+$ with $(\alpha,g)_- = (\mathbb{I}\alpha,g)_o$. Let $\hat{I} = OI : H_o \to H_o$ and set $K = (\hat{I}^{-1})^{\frac{1}{2}}$; then K is a positive self adjoint operator in H_o with $D(K) = H_+$ and $R(K) = H_o$ while $(u,v)_+ = (Ku,Kv)_o$. Consider K as a map in H_o to H_- and form the closure in the space H_- to get an operator $\mathbb{K}$ with $(f,g)_o = (\mathbb{K}f,\mathbb{K}g)_-$ and $D(\mathbb{K}) = H_o$. One has $\mathbb{I}^{-1} = \mathbb{K}K$ and $(f,Ku)_o = (\mathbb{K}f, u)_o$ (note

$$(f,Ku)_o = (\mathbb{K}f,\mathbb{K}Ku)_- = (\mathbb{K}f,\mathbb{I}^{-1}u)_- = (\mathbb{K}f, u)_o$$

and, for $f \in H_o$, $K^{-1}\mathbb{K}^{-1} f = K^{-1}K^{-1}f = \hat{I}f$ while $\hat{I}f = If$). One sets $J = K^{-1}$ and $\mathbb{J} = \mathbb{K}^{-1}$ so that $\mathbb{I} = J\mathbb{J}$. A further bit of notation involves operators such as K mapping say $H_+ \to H_o$ with $(Ku,f)_o = (u,K^*f)_+$ so that $(f,Ku)_o = (\mathbb{K}f, u)_o$ as above; it follows that, setting I_o = identity, $(\alpha,Ku)_o = (I_o\alpha,Ku)_o = (K^*I_o\alpha,u)_+ = (I^{-1}K^*I_o\alpha,u)_o = (K^+\alpha,u)_o$ (for $u \in H_+$, $\alpha \in H_o$) and K^+ can be regarded as a map $H_o \to H_-$ by extension (replace I by $\mathbb{I}$). Consequently one writes $\mathbb{K} = K^+$ and $\mathbb{J} = J^+$; thus e.g. $\mathbb{K}$ extends $K^* = K$ taken in H_o.

We recall next that an operator $A : H_1 \to H_2$ is Hilbert-Schmidt if for (any) orthonormal basis e_i in H_1 the series $\Sigma\|Ae_i\|^2_{H_2} < \infty$. A nonnegative operator A has a finite trace if $\mathrm{tr}(A) = \Sigma(Ae_i,e_i) < \infty$, which again is independent of the choice of orthonormal basis e_i. Let A be a self adjoint operator in H with $E(\Delta)$ the corresponding resolution of the identity. Let $H_+ \subset H \subset H_-$ be as above with corresponding operators K, $\mathbb{K}$, etc.; one can also look at a chain $\mathcal{D} \subset H_+ \subset H_o \subset H_- \subset \mathcal{D}'$ where $\mathcal{D}$ is a dense nuclear space (for example $\mathcal{D} \subset H^1 \subset L^2 \subset \mathcal{D}'$ with $\mathcal{D}$ the standard Schwartz space). The continuous operator K^{-1} :

$H_o \to H_o$ is Hilbert-Schmidt if we assume the embedding $H_+ \to H_o$ is quasinuclear (by construction $\| u\|_+^2 = \| Ku\|_o^2$ and $H_+ \to H_o$ quasinuclear means $\Sigma\| e_j\|_o^2 < \infty$ for e_j an orthonormal basis in H_+). Consider $\theta(\Delta) = K^{*-1}E(\Delta)K^{-1}$ (* in H_o so $K^* = K$) and $\rho(\Delta) = \mathrm{tr}\theta(\Delta)$ for Borel sets $\Delta \subset (-\infty,\infty)$. There exists an operator function $\psi(\lambda)$ such that $\Theta(\Delta) = \int_\Delta \psi(\lambda)d\rho(\lambda)$ and for $f,g \in H_+ = D(K)$

$$(6.52) \qquad (E(\Delta)f,g) = (\Theta(\Delta)Kf,Kg) = \int_\Delta (\psi(\lambda)Kf,Kg)d\rho(\lambda).$$

Roughly the operator $K^*\psi(\lambda)K = K\,\psi(\lambda)K$ is a projection onto an eigenmanifold corresponding to λ. Thus for any countable set of $u \in D(A) \cap H_+$ with $Au \in H_+$ one has $\psi(\lambda)K(A - \lambda I)u = 0$ a.e. for $d\rho(\lambda)$. Note here that if Δ_ν is a sequence of intervals contracting to λ with u as above then for $f \in H$

$$(6.53) \qquad (\psi(\lambda)K(A - \lambda I)u,\ f) =$$

$$\lim_{\nu\to\infty} \left(\frac{\Theta(\Delta_\nu)}{\rho(\Delta_\nu)} K(A - \lambda I)u,\ f\right) =$$

$$\lim_{\nu\to\infty} \frac{1}{\rho(\Delta_\nu)} (E(\Delta_\nu)(A - \lambda I)u,\ K^{-1}f) =$$

$$\lim_{\nu\to\infty} \frac{1}{\rho(\Delta_\nu)} \int_{\Delta_\nu} (\mu - \lambda)d(E_\mu u,\ K^{-1}f).$$

The measure $\omega(\Delta) = (E(\Delta)u,\ K^{-1}f)$ is absolutely continuous with respect to $\rho(\Delta)$ and a little argument for f in a countable dense set yields the remark above since $\rho(\lambda)$ a.e. the limit in (6.53) is zero.

To extend the treatment here to generalized eigenfunctions consider for $u,v \in H_+$

$$(6.54) \qquad (\psi(\lambda)Ku,\ Kv)_o = (\mathbb{K}\psi(\lambda)Ku,\ v)_o$$

where $P(\lambda) = \mathbb{K}\psi(\lambda)K$ is defined from H_+ to H_- and will be Hilbert Schmidt. Thus for $u,v \in H_+$ (6.52) becomes

$$(6.55) \qquad (E(\Delta)u,\ v) = \int_\Delta (P(\lambda)u,\ v)_o d\rho(\lambda).$$

Suppose $\mathcal{D} \subset H_+$ is dense, $\mathcal{D} \subset D(A^*)$ (A^* in H_o), and $A^* : \mathcal{D} \to H_+$ is continuous. An element $\phi \in H_-$ is a generalized eigenvector of A for λ if for $u \in \mathcal{D}$

$$(6.56) \qquad (\phi, (A^* - \bar{\lambda}I)u)_o = 0.$$

For symmetric A, $A^* \supset A$ and A^* can be replaced by A in (6.56) when $A : \mathcal{D} \to H_+$ is continuous. Now for any continuous $C : \mathcal{D} \to H_+$ there is a continuous $C^+ : H_- \to \mathcal{D}'$ defined by $(C^+\alpha, u)_o = (\alpha, Cu)_o$ as before and one defines $\tilde{A} = A^{*+}$. On $D(A)$ we have $(\tilde{A}f,u)_o = (A^{*+}f,u)_o = (f,A^*u)_o = (Af,u)_o$ for $u \in \mathcal{D} \subset D(A^*)$ so that $A \subset \tilde{A}$. Thus from (6.56) a generalized eigenvector is a solution of $\tilde{A}\phi = \lambda\phi$. Further one can take for $\mathcal{D}$ the space $D(A^*) = \mathcal{D}_+$ with scalar product $((u,v)) = (u,v)_+ + (A^*u, A^*v)_+$ so that $\mathcal{D}_+$ is a separable prehilbert space. In any event every element $\phi \in R(P(\lambda))$ satisfies $\tilde{A}\phi = \lambda\phi$ since (cf. (6.53) and remarks there)

$$(6.57) \qquad (\phi, (A - \lambda I)u)_o = (\mathbb{K}\psi(\lambda)Kv, (A - \lambda I)u)_o$$

$$= (\psi(\lambda)Kv, K(A - \lambda I)u)_o = (Kv, \psi(\lambda)K(A - \lambda I)u)_o = 0$$

($\phi = P(\lambda)v,u$ as in (6.53)). Summarizing some of this we state

<u>Theorem 6.15</u> Let $\mathcal{D} \subset H_+ \subset H_o \subset H_- \subset \mathcal{D}'$ as indicated ($H_+ \to H_o$ quasinuclear) and A a self adjoint operator in H_o with $\mathcal{D} \subset D(A)$ and $A : \mathcal{D} \to H_+$ continuous. There is a nonnegative finite measure $d\rho(\lambda)$ and an almost everywhere defined Hilbert Schmidt valued operator function $P(\lambda) : H_+ \to H_-$ such that (6.55) holds and $R(P(\lambda))$ consists of generalized eigenvectors of A (i.e. $\tilde{A}\phi = \lambda\phi$). One can find a system of such eigenvectors, orthogonal in H_- such that $\Sigma \| \phi_\alpha(\lambda) \|_-^2 = 1$ ($\alpha = 1,\ldots,N_\lambda$) and for $u,v \in H_+$

$$(6.58) \qquad (E(\Delta)u,v)_o = \int_\Delta \sum_1^{N_\lambda} \overline{u_\alpha(\lambda)}\, v_\alpha(\lambda)d\rho(\lambda)$$

where $u_\alpha(\lambda) = (\phi_\alpha(\lambda),u)_o$. Thus $(P(\lambda)u,v)_o = \Sigma\, \overline{u_\alpha(\lambda)}v_\alpha(\lambda)$ or formally one can say

(6.59) $P(\lambda)u = \Sigma\ \overline{(\phi_\alpha(\lambda),u)_o}\phi_\alpha(\lambda)$

(6.60) $u = \int_{-\infty}^{\infty} P(\lambda)ud\rho(\lambda) = \sum \int \overline{(\phi_\alpha(\lambda),u)_o}\phi_\alpha(\lambda)d\rho(\lambda)$

(6.61) $F(A)u = \int_{-\infty}^{\infty} F(\lambda)P(\lambda)ud\rho(\lambda).$

Going now to an operator $C = A \otimes I + I \otimes B$ with A (resp B) acting in a Hilbert space E (resp F) we can produce a resolution of the identity for C in terms of resolutions $E(d\lambda)$ and $F(d\lambda)$ for A and B respectively. Thus assume A and B are self adjoint for simplicity. Then C is symmetric and one has

Theorem 6.16 A resolution of the identity in $E \otimes F$ for $C = A \otimes I + I \otimes B$ is provided by

(6.62) $(G_\lambda(e \otimes f), (g \otimes h)) =$

$$\int_{\infty}^{\infty} (E_{\lambda-\mu}e, g)d_\mu(F_\mu f, h).$$

We note in this respect that as a weak integral (6.62) can be written

(6.63) $G_\lambda = \int_{-\infty}^{\infty} (E_{\lambda-\mu} \otimes I)d_\mu(I \otimes F_\mu)$

and part of the proof of (6.62) is contained in the formula for suitable e, f

(6.64) $(C(e \otimes f), (g \otimes h)) =$

$$\int_{-\infty}^{\infty} \lambda d_\lambda(G_\lambda(e \otimes f), (g \otimes h))$$

which formally follows from

$$(6.65)\qquad \int_{-\infty}^{\infty} \lambda d_\lambda \left[\int_{-\infty}^{\infty} (E_{\lambda-\mu}e,\ g) d_\mu (F_\mu f,\ h) \right] =$$

$$\int_{-\infty}^{\infty} \left[\int_{-\infty}^{\infty} \lambda d_\lambda (E_{\lambda-\mu}e,\ g) \right] d_\mu (F_\mu f,\ h) =$$

$$\int_{-\infty}^{\infty} \left[\int_{-\infty}^{\infty} (\mu + \nu) d_\nu (E_\nu e,\ g) \right] d_\mu (F_\mu f,\ h) =$$

$$\int_{-\infty}^{\infty} \mu \left[\int_{-\infty}^{\infty} d_\nu (E_\nu e,\ g) \right] d_\mu (F_\mu f,\ h) +$$

$$\int_{-\infty}^{\infty} \left[\int_{-\infty}^{\infty} \nu d_\nu (E_\nu e,\ g) \right] d_\mu (F_\mu f,\ h) =$$

$$(e,g)(Bf,\ h) + (Ae,\ g)(f,h) = ((A \otimes I + I \otimes B)(e \otimes f),\ g \otimes h).$$

If $f(\lambda)$ is now a bounded Borel function for example one has

$$(6.66)\qquad f(C) = \int_{-\infty}^{\infty} f(\lambda) dG_\lambda = \int_{-\infty}^{\infty} (f(A + \mu I) \otimes I) d_\mu (I \otimes F_\mu)$$

since $f(A + \mu I) = \int_{-\infty}^{\infty} f(\lambda) dE_{\lambda-\mu}$. Applying this to the resolvant $R_z(A)$, $\mathrm{Im} z \neq 0$, a little argument shows that the closure of C is self adjoint and has the resolution of the identity given by (6.63).

Let now $E_+ \subset E \subset E_-$ and $F_+ \subset F \subset F_-$ be chains as before with quasinuclear embeddings $E_+ \to E$ and $F_+ \to F$. One obtains spectral measures $d\rho(\lambda)$ and $d\sigma(\lambda)$ relative to A and B and the spectral measure $\zeta(\lambda)$ for $\bar{C}$ is given by

$$(6.67)\qquad \zeta(\lambda) = \int_{\infty}^{\infty} \rho(\lambda - \mu) d\sigma(\mu) = (\rho * \mu)(\lambda).$$

Recall here $\rho(\Delta) = \mathrm{tr}(\hat{J}E(\Delta)\hat{J})$ for example where we write $K^{-1} = OJ = \hat{J}$ when considered as an operator in E_o. Then for the chain $E_+ \otimes F_+ \subset E_o \otimes F_o \subset E_- \otimes F_-$ one has maps $J \otimes J'$ etc. and from (6.62)

$$(6.68)\qquad \zeta(\lambda) = \sum \int_{-\infty}^{\infty} (E_{\lambda-\mu}\hat{J}e_i,\ \hat{J}e_i) d_\mu (F_\mu \hat{J}'f_k,\ \hat{J}'f_k)$$

$$= \sum_k \int_{-\infty}^{\infty} \rho(\lambda - \mu) d\mu (F_\mu \hat{J}'f_k,\ \hat{J}'f_k) = \int_{-\infty}^{\infty} \rho(\lambda - \mu) d\sigma(\mu).$$

Similarly there is a generalized projection operator $P^{\otimes}(\lambda)$ for $\bar{C}$ obtained as follows. First by writing out $(G_\lambda(e \otimes f), e' \otimes f')$ from (6.62) and using (6.55) one obtains

$$G_\lambda = \int_{-\infty}^{\infty} \int_{-\infty}^{\lambda-\mu} P(\nu) \otimes P'(\mu) d\rho(\nu) d\sigma(\mu) \tag{6.69}$$

$$= \int_{\nu+\mu<\lambda} P(\nu) \otimes P'(\mu) d\rho(\nu) d\sigma(\mu)$$

where $P(\nu)$ (resp. $P'(\mu)$) is associated with A (resp. B). Hence

$$P^{\otimes}(\lambda) = \frac{d}{d\zeta(\lambda)} \int_{\nu+\mu<\lambda} P(\nu) \otimes P'(\mu) d\rho(\nu) d\sigma(\mu). \tag{6.70}$$

<u>Theorem 6.17</u> For A and B self adjoint the self adjoint $\bar{C}$ has spectral measure given by (6.67) and projection operators given by (6.70). Further

$$f(\bar{C}) = \int_{-\infty}^{\infty} \int_{-\infty}^{\infty} f(\nu + \mu) P(\nu) \otimes P'(\mu) d\rho(\nu) d\sigma(\mu) \tag{6.71}$$

and (6.70) can be expressed in terms of a Borel measure $d\rho_\lambda(\nu,\mu)$ concentrated on the line $\nu + \mu = \lambda$ in the form (cf also (6.85))

$$P^{\otimes}(\lambda) = \int_{\nu+\mu=\lambda} P(\nu) \otimes P'(\mu) d\rho_\lambda(\nu,\mu). \tag{6.72}$$

The material sketched here on tensor products and spectral theory is clearly very relevant to the further study of transmutation operators. It has been organized and included here with that purpose in mind.

In this spirit let us indicate some essentially formal relations which suggest further study. We consider operators P and Q as in Section 2.3 with $P(D_x)\phi(x,y) = Q(D_y)\phi(x,y)$ and $\phi(x,0) = f(x)$ with $\phi(0,y) = (Bf)(y)$. Recall that under suitable hypotheses $\phi(x,y) = \langle\beta(y,\xi, U(x,\xi)\rangle$ where $\beta(y,x) = \langle\Omega(x,\lambda), \theta(y,\lambda)\rangle$ and $U(x,\xi) = T_x^\xi f(x)$ so that (cf. (3.78)) $(f * g)(x) = \langle U(x,\xi), g(\xi)\rangle$ is the generalized convolution associated with P and T. Similarly let $V(y,\eta)$ satisfy $Q(D_y)V = Q(D_\eta)V$ with $V(y,0) = (Bf)(y)$ and suitable other conditions for uniqueness $(V(0,\eta) = (Bf)(\eta))$. Denote by S_η^y the generalized translation operator associated with Q so that $V(y,\eta) = S^y_\eta(Bf)(\eta)$ and set, for $Bf = h$, $h \circledast g(y) = \langle V(y,\eta), g(\eta)\rangle$. Further recall that $P(D_x)h(x,\lambda) = -\lambda^2 h(x,\lambda)$ and $Q(D_y)\theta(y,\lambda) = -\lambda^2\theta(y,\lambda)$. For uniformity of notation therefore let us set $-\lambda^2 = \mu$ and write $H(x,\mu)$ (resp. $\Theta(y,\mu)$) for the solutions of $P(D_x)H = \mu H$ (resp $Q(D_y)\Theta = \mu\Theta$) with the same initial values; thus $H(x,-\lambda^2) = h(x,\lambda)$ and $\Theta(y,-\lambda^2) = \theta(y,\lambda)$. For convenience here we think of P and Q as suitably self adjoint with $\sigma(P) = \sigma(Q)$ so as in Remark 3.18 $T^x = H(x,P)$ and $S^y = \Theta(y,Q)$ and for spectral resolutions dE_λ and dF_λ of P and Q respectively one has

$$(6.73) \qquad T^x = \int H(x,\mu)dE_\mu; \quad S^y = \int \Theta(y,\mu)dF_\mu .$$

Thus $\mathcal{P}f(\lambda) = \langle h(x,\lambda), f(x)\rangle$ corresponds to $\hat{f}(\mu) = \mathcal{H}f(\mu) = \langle H(x,\mu), f(x)\rangle$ (i.e. $\mathcal{H}f(-\lambda^2) = \mathcal{P}f(\lambda))$ and $\mathcal{Q}f(\lambda) = \langle\theta(y,\lambda), f(y)\rangle$ corresponds to $f(\mu) = \mathcal{T}f(\mu) = \langle\Theta(y,\mu), f(y)\rangle$ (i.e. $\mathcal{T}f(-\lambda^2) = \mathcal{Q}f(\lambda)$). Then (cf (3.72) - (3.73)) corresponding to inversions such as $\mathbb{P} = \mathcal{P}^{-1}$ of Theorem 3.5 one writes

$$(6.74) \qquad f(x) = \int \hat{f}(\mu)H(x,\mu)d\sigma(\mu);$$

$$h(y) = \int \tilde{h}(\nu)\Theta(y,\nu)d\rho(\nu).$$

Formally one has $T^x = H(x,P)$ and $S^y = \Theta(y,Q)$ with $\widehat{T^x f} = H(x,\mu)\hat{f}(\mu)$ and $\widetilde{S^y h} = \Theta(y,\nu)\tilde{h}(\nu)$ (thus eg. $H(x,\mu) = \widehat{T^x\delta}$ formally). We recall from Theorem 3.7 that $\phi(x,y)$ can be written as $\phi(x,y) = \langle\theta(y,\lambda), F(\lambda)h(x,\lambda)\rangle$ where $F(\lambda) = \langle\Omega(\xi,\lambda), f(\xi)\rangle = \langle h(\xi,\lambda), f(\xi)\rangle$ since $\Omega = h$ here (up to a constant). This corresponds to a formula

$$(6.75) \qquad \phi(x,y) = \int \Theta(y,\mu)H(x,\mu)\hat{f}(\mu)d\sigma(\mu).$$

Clearly ϕ given by (6.75) satisfies formally $P(D_x)\phi = Q(D_y)\phi$ and $\phi(x,0) = f(x)$. Similarly consider (cf (6.86) - (6.87))

$$(6.76) \qquad \psi(x,y) = \int \Theta(y,\nu)H(x,\nu)\widetilde{Bf}(\nu)d\rho(\nu).$$

Again $P(D_x)\psi = Q(D_y)\psi$ and $\psi(0,y) = (Bf)(y)$. Hence by definitions and assuming uniqueness with $Bf(y)$ prescribed we have

Theorem 6.18(F) Under the hypotheses indicated the unique solution $\phi(x,y)$ of $P(D_x)\phi = Q(D_y)\phi$, $\phi(x,0) = f(x)$, $\phi(0,y) = (Bf)(y)$ is given by either (6.75) or (6.76). In this event the kernels K and $\tilde{K}$ below are inverses with $Bf(y) = \langle K(y,x), f(x)\rangle$ and $f(x) = \langle\tilde{K}(x,y), Bf(y)\rangle$.

$$(6.77)\qquad K(y,x) = \int \Theta(y,\mu)H(x,\mu)d\sigma(\mu);$$

$$\tilde{K}(x,y) = \int \Theta(y,\nu)H(x,\nu)d\rho(\nu).$$

Formally one can think of the formulas (6.77) in the form (cf (6.81))

$$(6.78)\qquad K(y,x) = \Theta(y,P_x);\quad \tilde{K}(x,y) = H(x,Q_y)$$

and one is generally motivated to express B in terms of T and S. It is instructive to recall here that if $P = -D_x^2$ in $L^2(0,\infty)$ with $D(P) = \{u \in L^2;\ u'' \in L^2;\ u'(0) = 0\}$ then $H(x,\mu) = \mathrm{Cos}\sqrt{\mu}x$ is a generalized eigenfunction and $T_x^y f(x) = \frac{1}{2}[\tilde{f}(x+y) + \tilde{f}(x-y)]$ where $\tilde{f}$ is the even extension of f (thus $Bf(y) = f(y)$). Note that $\mathrm{Cos}\sqrt{\mu}y\ \hat{f}(\mu) = \frac{1}{2}\int_0^\infty [\tilde{f}(x+y) + \tilde{f}(x-y)]\mathrm{Cos}\sqrt{\mu}x\,dx$ and recall that $f * g = \int_0^\infty T_x^y f(x)g(y)dy$ (cf (3.75)). All functions are extended to be even here and clearly

$$\delta * g = \langle T_x^y\delta(x), g(y)\rangle = \int_0^\infty T_x^y\delta(x)g(y)dy$$
$$= \frac{1}{2}\int_0^\infty [\delta(x+y) + \delta(x-y)]g(y)dy$$
$$= \frac{1}{2}[g(x) + g(-x)] = g(x).$$

It is interesting to look at ϕ given by (6.75) for $P = Q = -D^2$ in $L^2(0,\infty)$ as indicated since then

$$(6.79)\qquad \phi(x,y) = U(x,y) = \int \mathrm{Cos}\sqrt{\mu}y\ \mathrm{Cos}\sqrt{\mu}x\ \hat{f}(\mu)d\sigma(\mu)$$
$$= \frac{1}{2}\int [\mathrm{Cos}\sqrt{\mu}(x+y) + \mathrm{Cos}\sqrt{\mu}(x-y)]\hat{f}(\mu)d\sigma(\mu)$$
$$= \frac{1}{2}[\tilde{f}(x+y) + \tilde{f}(x-y)] = T_x^y f(x).$$

Note also that $(T^y\delta * f)(x) = T_x^y f(x)$ by a simple calculation and define

$$(6.80)\qquad \langle T_x^y, f(x)\rangle = \langle T_x^y\delta(x), f(x)\rangle = \frac{1}{2}[\tilde{f}(y) + \tilde{f}(-y)]$$

so that (6.78) is to be interpreted as

$$(6.81)\qquad K(y,x) = \Theta(y,P_x)\delta(x);\quad \tilde{K}(x,y) = H(x,Q_y)\delta(y)$$

and then $(Bf)(y) = \langle K(y,x), f(x)\rangle = \langle T_x^y\delta(x), f(x)\rangle$ is correct.

Let us observe next a way of looking at kernels such as K based on Theorem 6.17. Thus think of $P = A$ and $\mathbb{B} = -Q$ and look at ϕ as an element of the null space $N(\bar{C})$ where $C = A \otimes I + I \otimes \mathbb{B}$ acting in $H_o = E_o \otimes F_o$. Let $E_+ \subset E_o \subset E_-$ and $F_+ \subset F_o \subset F_-$ as above and note (cf Maurin [2], Maurin-Maurin [3]) that from (6.72) the generalized eigenspace $H(\lambda)$ for $\bar{C}$ can be written as

$$(6.82)\qquad H(\lambda) = \int_{\nu+\mu=\lambda} E(\nu)\,\hat{\otimes}\,F(\mu)d\rho_\lambda(\nu,\mu)$$

where $E(\nu)$ (resp $F(\mu)$) are the eigenspaces for A (resp $\mathbb{B}$) corresponding to values ν (resp μ). Suppose $\dim E(\nu) = \dim F(\mu) = 1$; then the elements of $H(\lambda)$ have the form

$$(6.83)\qquad h(\lambda) = \int_{\nu+\mu=\lambda} c^h(\mu,\nu)[e(\nu)\otimes f(\mu)]d\rho_\lambda(\nu,\mu).$$

We remark that (6.72) can be written in the form (cf. Berezanskij [1])

$$(6.84)\qquad P^{\otimes}(\lambda) = \int_{-\infty}^{\infty} P(\lambda-\mu)\otimes P'(\mu)d_\mu\rho_\lambda(\lambda-\mu,\mu)$$

where $d_\mu\rho_\lambda(\lambda-\mu,\mu)$ is defined by

$$(6.85)\qquad \rho_\lambda(\lambda-\tau,\tau) - \rho_\lambda(\lambda,0) = \operatorname{sgn}\tau\,\frac{d}{d\zeta(\lambda)}\int_{\substack{\nu+\mu<\lambda\\ \mu\in[0,\tau)}} d\rho(\nu)d\sigma(\mu)$$

(for $\tau < 0$ $[0,\tau)$ means $[\tau,0)$). Further it should be noted that (6.69) - (6.72) are valid with any spectral measures since terms like $P(\nu)d\rho(\nu)$ etc.

will be invariant under changes of measure $d\rho(\nu) = M(\nu)d\hat{\rho}(\nu)$ etc. Thus when the measures are absolutely continuous relative to Lebesgue measure one can simply use $d\nu$ etc. to make some calculations, in particular to compute $d\rho_\lambda(\nu,\mu)$ or equivalently $d_\mu\rho_\lambda(\lambda - \mu,\mu)$. Thus for example if $d\rho(\lambda) = d\sigma(\lambda) = d\zeta(\lambda) = d\lambda$ on $(0,\infty)$ one finds easily from (6.85) that $d_\mu\rho_\lambda(\lambda - \mu,\mu) = d\mu$ for $\mu \in (0,\lambda)$ and is zero for $\mu \notin (0,\lambda)$. If $d\rho(\lambda) = d\lambda$ on $(0,\infty)$ with $d\sigma(\lambda) = d\lambda$ on $(-\infty,0)$ then $d\zeta(\lambda) = d\lambda$ on $(-\infty,\infty)$ and from (6.85) again $d_\mu\rho_\lambda(\lambda - \mu,\mu) = d\mu$ for $\mu \in (-\infty, \min(\lambda,0))$ and is zero for $\mu \in [\min(\lambda,0), \infty)$.

Now for $\nu + \mu = 0$ we consider $Ae(\nu) = \nu e(\nu)$ with $A = P$ so $e(\nu) = H(x,\nu)$ while $\mathbb{B}f(\mu) = -Qf(\mu) = \mu f(\mu)$ gives $f(\mu) = \Theta(y,-\mu) = \Theta(y,\nu)$. The null space $N(\bar{C})$ will thus be associated with generalized eigenvectors $H(x,\nu) \otimes \Theta(y,\nu) = H(x,\nu)\Theta(y,\nu)$. Take $\sigma(A) = (0,\infty)$ and $\sigma(\mathbb{B}) = (-\infty,0)$ (e.g. $A = P = -D_x^2$ and $\mathbb{B} = -Q = D_y^2$) so $\rho \sim A$ and $\sigma \sim \mathbb{B}$ above. For $\lambda = 0$ $d_\mu\rho_\lambda(\lambda - \mu,\mu) = d\mu$ on $(-\infty,0)$ and (6.83) becomes, for $d\sigma(\mu) = M(\mu)d\mu$

$$(6.86) \qquad h = \int_{-\infty}^{0} c_h(\mu)H(x,\mu)\Theta(y,\mu)d\mu$$

$$= \int_{\sigma(\mathbb{B})} (c_h(\mu)/M(\mu))H(x,\mu)\Theta(y,\mu)d\sigma(\mu).$$

Note here that there is a shift in notation $\rho \leftrightarrow \sigma$ from (6.75) = (6.76)! Now for $c_h(\mu) = M(\mu)(\tilde{B}f)(\mu)$ (B is the transmutation operator) the h of (6.86) corresponds to the ϕ of (6.76). On the other hand setting $\nu = -\mu$ in (6.86) we have, for $d\rho(\nu) = N(\nu)d\nu$,

$$(6.87) \qquad h = \int_{0}^{\infty} c_h(-\nu)H(x,-\nu)\Theta(y,-\nu)d\nu$$

$$= \int_{\sigma(A)} (c_h(-\nu)/N(\nu))H(x,-\nu)\Theta(y,-\nu)d\rho(\nu)$$

which corresponds to (6.75) with $c_h(-\nu) = N(\nu)\hat{f}(\nu)$ since μ could be replaced by $-\mu$ in the $H\Theta$ of (6.75). This suggests the following assertion which we state as a formal theorem without attempting here a complete proof.

Theorem 6.19(F) In the self adjoint case with $\sigma(P) = \sigma(Q)$ unique solutions ϕ

of $P\phi = Q\phi$, $\phi(x,0) = f(x)$, $\phi(0,y) = (Bf)(y)$, can be obtained from (6.8.3) and written in the form (6.86) - (6.87).

Remark 6.20 Let us note that (6.78) and (6.81) are in keeping with the requirement in Section 2.3 that $\langle\Omega(x,\lambda), 1\rangle = \delta(x)$ since $\Omega = h$ here and in writing for example

(6.88) $$\delta(x) = \int H(x,\mu)d\sigma(\mu)$$

one obtains $\Theta(y,P)\delta = \int\Theta(y,P)H(x,\mu)d\sigma(\mu) = \int\Theta(y,\mu)H(x,\mu)d\sigma(\mu) = K(y,x)$. In this spirit one can then write

(6.89) $$(Bf)(y) = \langle\Theta(y,P_x)\delta(x), f(x)\rangle.$$

Another way of writing (6.78) which displays S and T is

(6.90) $$K(y,x) = \int \widetilde{S^y\delta}\ \widehat{T^x\delta}\ d\sigma;$$

$$\tilde{K}(x,y) = \int \widetilde{S^y\delta}\ \widehat{T^x\delta}\ d\rho.$$

Remark 6.21 The formulations of section 2.3 are further combined with the preceeding spectral considerations in Carroll [39] and a transform theory is developed. Thus given $PH = \mu H$, $Q\Theta = \mu\Theta$, $H(0,\mu) = \Theta(0,\mu) = 1$, and $H'(0,\mu) = \Theta'(0,\mu) = 0$ as above with $P^*\Omega = \mu\Omega$ and $Q^*\hat{\Omega} = \mu\hat{\Omega}$ satisfying $\langle\Omega(x,\mu), 1\rangle_\sigma = \delta(x)$ and $\langle\hat{\Omega}(y,\mu), 1\rangle_\rho = \delta(y)$ we write

(6.91) $$\mathcal{P}f(\mu) = \langle\Omega(x,\mu), f(x)\rangle$$

(6.92) $$Pf(\mu) = \langle H(x,\mu), f(x)\rangle$$

(6.93) $$\mathcal{B}F(x) = \langle F(\mu), H(x,\mu)\rangle_\sigma$$

(6.94) $$\mathbb{P}F(x) = \langle F(\mu), H(x,\mu)\rangle_\rho$$

(6.95) $$\mathbb{P}F(x) = \langle F(\mu), \Omega(x,\mu)\rangle_\sigma$$

(6.96) $$\mathcal{A}g(\nu) = \langle\hat{\Omega}(y,\nu), g(y)\rangle$$

(6.97) $$\mathcal{Q}g(\nu) = \langle\Theta(y,\nu), g(y)\rangle$$

(6.98) $\mathcal{A}G(y) = \langle G(\nu), \Theta(y,\nu)\rangle_\rho$

(6.99) $\mathcal{Q}G(y) = \langle G(\nu), \Theta(y,\nu)\rangle_\sigma$

(6.100) $\mathbb{Q}G(y) = \langle G(\nu), \hat{\Omega}(y,\nu)\rangle_\rho$

where the subscripts σ or ρ indicates different spectral pairings in conformity with our notation above. Then if everything fits together one has

Theorem 6.22 $\mathcal{B} = \mathcal{B}^{-1}$, $\mathcal{A} = \mathcal{A}^{-1}$, $\mathbb{P} = P^{-1}$, $\mathbb{Q} = Q^{-1}$, $B = \mathcal{Q}\mathcal{B}$, and $B = B^{-1} = \mathbb{P}\mathcal{A}$. This implies also $\mathcal{Q}^{-1} = \mathcal{B}\mathbb{P}\mathcal{A}$ and $\mathbb{P}^{-1} = \mathcal{A}\mathcal{Q}\mathcal{B}$.

Observe that in Section 2.3 (equation (3.31)) we used the symbol $\mathbb{P}$ for $\mathcal{B}$ and $\mathcal{Q}$ for $\mathcal{A}$; the particular nature of the kernels there made this permissable. For further results on the transform theory see Appendix 2. For nonselfadjoint operators see also Marčenko [4] and Gasymov [6].

APPENDIX 1

We will outline here someof the basic ideas from functional analysis which are used in the text. The proofs are all standard and can be found for example in Bourbaki [1; 2], Carroll [4], Dunford-Schwartz [1], Garsoux [1], Horváth [1], Köthe [1], Marinescu [1], Nachbin[1], Reed-Simon [1], Schwartz [1], Treves [1], and Yosida [1] (cf. also Bjorck [1], Ehrenpreis [2], Rosinger [1]). The material can be thought of as a kind of minimal working background. First assuming known the concept of a vector space we define a seminorm in a vector space F over $\mathbb{C}$ to be a function $p : F \to \mathbb{R}_+$ satisfying

$$(1.1) \qquad p(x + y) \leq p(x) + p(y); \quad p(\alpha x) = |\alpha| p(x).$$

If in addition $p(x) = 0$ implies $x = 0$ then p is called a norm. Now a topology on F consists of the prescription of a family of open sets $\mathbb{O}$ satisfying

$$\cup O_\alpha \in \mathbb{O} \text{ (any index set)},$$

$$\cap O_\alpha \in \mathbb{O} \text{ (finite intersection)}, \quad F \in \mathbb{O}, \quad \text{and} \quad \Phi \in \mathbb{O}$$

(Φ being the empty set); F is a topological vector space (TVS) if the vector space operations $(\alpha,x) \to \alpha x : \mathbb{C} \times F \to F$, $(x,y) \to x + y : F \times F \to F$, and $x \to -x : F \to F$ are continuous with this topology (see below). A neighborhood (nbh) of a point $x \in F$ is any set containing an open set containing x. Thus a set is open if and only if it is a nbh of each of its points and we denote by $N(x)$ the nbhs of x. In a linear vector space F one has $N(x) = x + N(0)$ so the topology is completely determined by $N(0)$. Here, given a family $N(x)$ satisfying (a) every set in F containing a set in $N(x)$ itself belongs to $N(x)$ (b) $N_i \in N(x)$ implies $\cap N_i \in N(x)$ (finite intersection) (c) $x \in N$ if $N \in N(x)$, and (d) if $V \in N(x)$ there exists $W \in N(x)$ such that $V \in N(y)$ for all $y \in W$, it follows that there is a unique topology on F such that $N(x)$ is the family of nbhs of

x. Indeed one simply picks as open sets those 0 such that $0 \in N(x)$ for each $x \in 0$. A family $B(x) \subset N(x)$ is called a base of nbhs or a fundamental system of nbhs at x if every $N \in N(x)$ contains a set $B \in B(x)$. $N(x)$ is then recovered as the family of sets containing a set in $B(x)$. A family $B(x)$, prescribed for each $x \in F$, qualifies as a fsn, if for example (a) each $B \in B(x)$ contains x (b) $B_1, B_2 \in B(x)$ implies $B_1 \cap B_2 \supset B_3 \in B(x)$ and (c) axiom (d) above holds for $B(x)$.

Definition 1.1 A locally convex TVS (abbreviated LCS) is a topological vector space F whose topology is defined by a family of seminorms. Thus if $V_\alpha(\varepsilon) = \{x \in F;\ p_\alpha(x) \leq \varepsilon\}$ then a fsn $B(0)$ is taken to be the family of finite intersections of the $V_\alpha(\varepsilon)$ with $B(x) = x + B(0)$.

We recall also that closed sets are the complements of open sets and the closure $\bar{B}$ of a set B, defined as the smallest closed set containing B, can be characterized as the set of x such that every nbh of x intersects B. Further, directing the nbhs U of x by inclusion, if $x \in \bar{B}$ then picking $x_U \in B \cap U$ one can produce a net $x_U \to x$. Recall here that a net in F is a pair $(S_\alpha, \geq)$ where $S_\alpha \in F$ and $\geq$ directs the set of α (i.e. $\alpha \geq \beta$, $\beta \geq \gamma \Rightarrow \alpha \geq \gamma$; $\alpha \geq \alpha$; and given $\alpha, \beta\ \exists\ \gamma$ with $\gamma \geq \alpha$ and $\gamma \geq \beta$). A net S_α is eventually in a set B if $\exists\ \beta$ such that $\alpha \geq \beta \Rightarrow S_\alpha \in B$; S_α converges to $x \Leftrightarrow S_\alpha$ is eventually in every nbh of x. Thus in general a set $B \subset F$ is closed if and only if no net in B converges to a point in $F - B$. Hence closed sets, and consequently open sets, are completely characterized by convergence properties. Therefore the topology is completely described in terms of convergence. We remark also that a topological space F is separated or Hausdorff if for any $x, y \in F\ \exists$ open sets $U \ni x$ and $V \ni y$ such that $U \cap V = \Phi$; all TVS will be assumed separated unless otherwise stated.

Now to construct the Schwartz distributions in $\mathbb{R}^n$ let $K_m \subset K_{m+1}$ be an increasing sequence of compact sets with $\cup K_m = \mathbb{R}^n$. Let $\mathcal{D}_m$ be the vector space of C^∞ complex valued functions with support in K_m. The support of ϕ, written supp ϕ, is the smallest closed set outside of which $\phi \equiv 0$; equivalently it is the closure of the set of x where $f(x) \not\equiv 0$. Define seminorms in $\mathcal{D}_m$ by

$$(1.2) \qquad N_p(\phi) = \sup_{x \in K_m} |D^p \phi|$$

where $D^p = D_1^{p_1} \ldots D_n^{p_n}$ with $D_k = \partial/\partial x_k$. Define now, with $|p| = \Sigma p_i$,

$$(1.3) \qquad V_p(K_m, \varepsilon) = \{\phi \in \mathcal{D}_m;\ N_p(\phi) \leq \varepsilon\}$$

$$(1.4) \qquad V(K_m, \varepsilon, s) = \{\phi \in \mathcal{D}_m;\ N_p(\phi) \leq \varepsilon \text{ for } |p| \leq s\}.$$

Evidently $V(K_m, \frac{1}{n}, s) = \cap V_p(K_m, \frac{1}{n})$ for $|p| \leq s$, and this particular set of finite intersections for $n = 1,2,\ldots$ and $s = 0,1,\ldots$ suffices to define $B(0)$ in $\mathcal{D}_m$ as indicated in Definition 1.1. To verify that the vector space operations $(\alpha,\phi) \to \alpha\phi$, $(\phi,\psi) \to \phi + \psi$, and $\phi \to -\phi$ are continuous we recall

Definition 1.2 A map $f : F \to G$ between two topological spaces is continuous at x if for any $W \in N(f(x))$ $\exists$ $V \in N(x)$ such that $f(V) \subset W$. Equivalently $f^{-1}(U)$ should be open for any open U.

For linear maps between TVS one need only check continuity at 0 (since $N(x) = x + N(0)$) and this is routine now for $\mathcal{D}_m$. We note also that $N(0)$ in $\mathcal{D}_m$ has a countable fsn $B(0)$ as defined and thus closure and convergence can be discussed in terms of sequences (nets are not necessary). Such a LCS will be metrizable which means there can be defined on $F = \mathcal{D}_m$ a metric function $d : F \times F \to \mathbb{R}_+$ satisfying

$$(1.5) \qquad d(x,y) \geq 0; \quad d(x,y) = d(y,x);$$

$$d(x,z) \leq d(x,y) + d(y,z)$$

and $d(x,y) = 0 \iff x = y$ such that the sets $V_n(x) = \{y \in F;\ d(x,y) \leq \frac{1}{n}\}$ form a fsn at x. Such a metric can easily be constructed but we omit details (see Carroll [4], Köthe [1]). We recall next that a net S_α in a TVS is Cauchy if given any $V \in N(0)$ $\exists$ β such that $S_\alpha - S_\gamma \in V$ for $\alpha,\gamma \geq \beta$. Such a space F

is complete if every Cauchy net converges. When F is metrizable it is complete if every Cauchy sequence converges. (Unless otherwise stated all spaces will be complete.) A complete metrizable LCS is called a Frechet space and a complete normed space is a Banach space. The construction of $\mathcal{D}_m$ now yields, with a routine argument for completeness, the result that $\mathcal{D}_m$ is a Frechet space.

Let now $\mathcal{D}$ be the vector space of all C^∞ functions with compact support in $\mathbb{R}^n$. Clearly $\mathcal{D}_m \subset \mathcal{D}$ and we denote by i_m the injection $i_m : \mathcal{D}_m \to \mathcal{D}$. We shall put on $\mathcal{D}$ the finest locally convex topology such that all i_m are continuous. Here finest means strongest in the sense of having the largest collection of open sets or nbhs. We shall construct the nbh system explicitly and recall first that a set B is disced if $x \in B$ and $|\alpha| \leq 1$ implies $\alpha x \in B$. A set B in a TVS over $\mathbb{R}$ is convex if $x,y \in B$ and $0 < \lambda < 1$ implies $\lambda x + (1 - \lambda)y \in B$. If F is a TVS over $\mathbb{C}$ let F_o be the same space over $\mathbb{R}$ with multiplication by i considered as an automorphism of F_o. Then $B \subset F$ is convex if it is convex in F_o. A set B is called absorbing if for $x \in F$ $x \in \lambda B$ when $|\lambda| \geq \lambda_o$. Given a convex disced absorbing set $B \subset F$ one defines the gauge of B or the Minkowsk functional as $p_B(x) = \inf\{\lambda;\ x \in \lambda B\}$ for $\lambda > 0$. Then p_B is a seminorm and $\bar{B} = \{x \in F;\ p_B(x) \leq 1\}$. Thus if one has a TVS with a fsn consisting of convex disced absorbing sets then the topology can be defined by a family of seminorms as in Definition 1.1 so the space is locally convex. Now for $\mathcal{D}$ let $V_m \in N(0)$ in $\mathcal{D}_m$ and write $\Gamma(\cup V_m)$ for the convex disced envelope in $\mathcal{D}$ of $\cup V_m$; this is the smallest convex disced set in $\mathcal{D}$ containing $B = \cup V_m$. Let Λ be the collection of such $\Gamma(\cup V_m)$ and take it as a fsn for a topology T on $\mathcal{D}$. Since $V_m \subset \Gamma(\cup V_m)$ with $i_m(V_m) = V_m$ we see that i_m is continuous. On the other hand if i_m is continuous for a locally convex topology T' on $\mathcal{D}$ and V is a convex disced nbh of 0 for T' then $V \supset \Gamma(\cup(V \cap \mathcal{D}_m))$ with $V \cap \mathcal{D}_m \in N(0)$ in $\mathcal{D}_m$ since $i_m^{-1}(V) = V \cap \mathcal{D}_m$. Hence $V \in N(0)$ for T and T is stronger than T'. The sets $B = \Gamma(\cup V_m)$ are absorbing since they are taken as nbhs of 0 and $(\gamma,x) \to \gamma x$ is continuous; hence the topology T is locally convex and indeed the finest locally convex topology on $\mathcal{D}$ such that the i_m are all continuous.

The space $\mathcal{D}$ with this topology is called the strict inductive limit of the $\mathcal{D}_m$ and one easily checks that it is independent of the defining sequence $K_m \subset K_{m+1}$ with $\cup K_m = \mathbb{R}^n$. We remark also that if $\Omega \subset \mathbb{R}^n$ is any open set and $K_m \subset K_{m+1}$ is a sequence of compact sets in Ω with $\cup K_m = \Omega$ then one defines $\mathcal{D}(\Omega)$ in the same manner.

There are other equivalent ways of defining the topology of $\mathcal{D}$ (cf. Schwartz [1], Hörmander [1]) but the above method is perhpas the most revealing. A crucial property of this topology, which allows one usually to replace net arguments by sequence arguments, is determined by

Theorem 1.3 A linear map $f : \mathcal{D} \to F$, F a LCS, is continuous if and only if each restriction $f_m = f|_{\mathcal{D}_m} : \mathcal{D}_m \to F$ is continuous.

To prove this simply note that f is continuous $\iff f^{-1}(V) \in N(0)$ in $\mathcal{D}$ for $V \in B(0)$ in F. Take V convex and disced so that $f^{-1}(V)$ is also and then $f^{-1}(V) \in N(0)$ in $\mathcal{D} \iff f^{-1}(V) \cap \mathcal{D}_m \in N(0)$ in $\mathcal{D}_m$ which is equivalent to saying f_m is continuous. We now define the Schwartz distributions in $\mathbb{R}^n$ as the space $\mathcal{D}'$ of continuous linear maps $\mathcal{D} \to \mathbb{C}$; this is called the (topological) dual of $\mathcal{D}$ (similarly one obtaind $\mathcal{D}'(\Omega)$ for $\Omega \subset \mathbb{R}^n$ open). That it is not a trivial space follows from the Hahn-Banach theorem below but one can also produce elements of $\mathcal{D}'_n$ directly. For example the δ "function" is defined by $(\phi \in \mathcal{D})$

$$(1.6) \qquad \delta(\phi) = \langle \delta, \phi \rangle = \phi(0).$$

If $f \in L^1_{loc}$ then f determines an element $f \in \mathcal{D}'$ by the rule

$$(1.7) \qquad \langle f, \phi \rangle = \int_{-\infty}^{\infty} f(x)\phi(x)dx.$$

The idea in constructing distributions is partly to have a space where one can differentiate the objects at will so that a theory of differential operators can be developed. That this is possible and natural follows now from the fact that if $f \in C^1(\mathbb{R})$ for example then

$$<f, \phi'> = \int f\phi' dx = -\int f'\phi dx = -<f', \phi>.$$

Then if $T \in \mathcal{D}'(\mathbb{R}^n)$ we define

(1.8) $$<D_k T, \phi> = -<T, D_k\phi>$$

where $D_k = \partial/\partial x_k$. This makes sense since the map $\phi \to D_k\phi : \mathcal{D} \to \mathcal{D}$ is obviously continuous so that $\phi \to <T, D_k\phi> : \mathcal{D} \to \mathbb{C}$ is continuous and defines therefore a distribution.

We will define some further spaces of distributions now and then give some additional general information about functional analysis which should be of use in the text. First for $T \in \mathcal{D}'$ one says $T = 0$ in an open set $\Omega \subset \mathbb{R}^n$ if $<T, \phi> = 0$ for all $\phi \in \mathcal{D}(\Omega)$ where $\mathcal{D}(\Omega)$ is the space of C^∞ functions with compact support in Ω. $\cup\Omega$ where $T = 0$ is open and its complement is defined to be the support of T (written supp T). Thus $x \in \text{supp } T$ if $T \neq 0$ in any open nbh of x. Let now $\mathcal{E} = C^\infty(\mathbb{R}^n)$ (no restriction on supports) with topology defined as follows. Let $K_m \subset K_{m+1}$ be a sequence of compact sets as before with $\cup K_m = \mathbb{R}^n$ and define $N_p(\phi)$, $V_p(K_m,\varepsilon)$, and $V(K_m,\varepsilon,s)$ as before but referred to functions $\phi \in \mathcal{E}$. Then take the sets $V(K_m, \frac{1}{n}, s)$, $m = 1,2,\ldots$, $n = 1,2,\ldots$, $s = 0,.,\ldots$, as a fsn at 0 for $\mathcal{E}$. This is a countable family and $\mathcal{E}$ will be metrizable; since it is also easily seen to be complete $\mathcal{E}$ is a Frechet space. Next let $\mathcal{S}$ be the space of C^∞ functions on $\mathbb{R}^n$ decreasing as $|x| \to \infty$, along with all derivatives, faster than any power of $1/|x|$. Thus if $P(x)$ is any polynomial and $Q(D)$ any polynomial in the D_k then $P(x)Q(D)\phi$ is a continuous bounded function on $\mathbb{R}^n$ when $\phi \in \mathcal{S}$. As a fsn of 0 in $\mathcal{S}$ we take the sets $(k = 0,1,\ldots;\ q = 1,2,\ldots;\ m = 0,1,\ldots)$

(1.9) $$W(m,k,q) = \{\phi \in \mathcal{S};\ (1 + |x|^2)^k |D^p\phi| \leq \frac{1}{q}\}$$

for $|p| \leq m$ and $x \in \mathbb{R}^n$. Thus one is dealing with seminorms $N_{m,k} = \sup n_{p,k}$ for $|p| \leq m$ where $n_{p,k}(\phi) = \sup\ (1 + |x|^2)^k |D^p\phi|$ for $x \in \mathbb{R}^n$. Again we have a countable fsn and $\mathcal{S}$ is easily seen to be complete; hence $\mathcal{S}$ is also a

Frechet space.

The two spaces $\mathcal{E}'$ and $\mathcal{S}'$ paly an important role in the theory of distributions (recall F' is the space of continuous linear maps $F \to \mathbb{C}$). One must show of course that $\mathcal{E}' \subset \mathcal{D}'$ and $\mathcal{S}' \subset \mathcal{D}'$ (it does require proof) and then one can prove that $\mathcal{E}'$ is the space of distributions with compact support. The space $\mathcal{S}'$ on the other hand provides a natural setting for the Fourier transform defined as follows. For $\phi \in \mathcal{S}$ set

$$(1.10) \qquad \mathcal{F}(\phi) = \hat{\phi}(s) = \int_{-\infty}^{\infty} \phi(x) \exp i\langle s,x\rangle dx$$

where $\langle s,x\rangle = \Sigma x_i s_i$. The inversion formula is (cf Titchmarsh [1] for proof)

$$(1.11) \qquad \bar{\mathcal{F}}(\hat{\phi}) = \phi(x) = \frac{1}{(2\pi)^n} \int_{-\infty}^{\infty} \hat{\phi}(s) \exp(-i\langle s,x\rangle) dx.$$

It is easily seen that $\mathcal{F} : \mathcal{S} \to \mathcal{S}$ is an isomorphism of LCS. Next we recall the Parseval formula ($\langle\ ,\ \rangle$ is simply integration)

$$(1.12) \qquad \langle \mathcal{F}\phi, \psi\rangle = \langle \phi, \mathcal{F}\psi\rangle$$

valid for functions in $\mathcal{S}$. Let $T \in \mathcal{S}'$ and consider the linear map $\psi \to \langle T, \mathcal{F}\psi\rangle$ $: \mathcal{S} \to \mathbb{C}$ where $\langle\ ,\ \rangle$ denotes $\mathcal{S} - \mathcal{S}'$ duality. Since $\mathcal{F} : \mathcal{S} \to \mathcal{S}$ is continuous this determines an element of $\mathcal{S}'$ which we denote by $\mathcal{F}T$ so that $\langle \mathcal{F}T, \psi\rangle = \langle T, \mathcal{F}\psi\rangle$. This extends $\mathcal{F}$ to a map $\mathcal{S}' \to \mathcal{S}'$ which will be an algebraic isomorphism.

In order to define convolution, and for other applications in Chapter 1, we need the idea of tensor product. Thus let F and G be LCS and define $\mathcal{B}(F,G; \mathbb{C})$ to be the space of bilinear forms $F \times G \to \mathbb{C}$. (The constructions are algebraic until otherwise indicated.) Let $\Lambda = \Lambda(F, G)$ be the set of all <u>formal</u> finite linear combinations $\Sigma\alpha_j(f_j,g_j)$ with $\alpha_j \in \mathbb{C}$. Then let us define a vector space structure on Λ by $\alpha\Sigma\alpha_j(f_j, g_j) = \Sigma\alpha\alpha_j(f_j, g_j)$ and $\Sigma\alpha_j(f_j, g_j) + \Sigma\beta_k(f_k, g_k) = \Sigma(\alpha_\ell + \beta_\ell)(f_\ell, g_\ell)$ where the ℓ indices include the j and k indices. Let Λ_o be the subvector space of finite linear combinations with coefficients in $\mathbb{C}$ of elements of the form $(f_1 + f_2, g) - (f_1, g) - (f_2, g)$, $(f, g_1 + g_2) - (f, g_1) - (f, g_2)$, $(\alpha f, g) - \alpha(f, g)$, and $(f, \alpha g) - \alpha(f, g)$. Then the tensor product $F \otimes G$ is defined to be the vector space Λ/Λ_o. If $b \in \mathcal{B}(F, G; \mathbb{C})$ we can extend it to be a linear map $b : \Lambda \to \mathbb{C}$ by the rule $b(\Sigma\alpha_j(f_j, g_j)) = \Sigma\alpha_j b(f_j, g_j)$ and since b vanishes on Λ_o we can factor it through the quotient Λ/Λ_o to define a linear map $\dot{b} : F \otimes G \to \mathbb{C}$. Similarly if $\dot{b}$ is a linear map $F \otimes G \to \mathbb{C}$ it defines a bilinear map $b(f, g) = \dot{b}(f \otimes g)$ from $F \times G$ to $\mathbb{C}$. Thus one has a 1-1 correspondence (algebraic) between $\mathcal{B}(F, G; \mathbb{C})$ and $\mathcal{L}(F \otimes G; \mathbb{C})$ where $\mathcal{L}(H; K)$ denotes linear maps $H \to K$. The tensor product can also be characterized in terms of a universal factorization property (see e.g. Carroll [4]). Next let us recall that if E is a TVS there exists a unique (up to isomorphism) complete TVS denoted by $\hat{E}$ such that E is isomorphic as a TVS to a dense subspace of $\hat{E}$. There are various natural topologies on $F \otimes G$ for LCS F and G due to Grothendieck [1]. The π topology is characterized by the property that $L(F \hat{\otimes}_\pi G; \mathbb{C}) = B(F, G; \mathbb{C})$ where L and B refer to continuous maps; we will not have occasion to use the π topology. The ε topology on $F \otimes G$ is the topology of uniform convergence on products of (convex, disced) equicontinuous sets in $F' \times G'$. Here a set $H \subset L(E; \mathbb{C}) = E'$ is equicontinuous if given ε the set $\cap h'^{-1}\{|z| < \varepsilon\}$ for $h' \in H$ is a nbh of 0 in E; equivalently $\exists$ $V \in N(0)$ in E such that $h'(V) \subset \{|z| < \varepsilon\}$ for $h' \in H$. Thus a net $e_\alpha \to 0$ in $F \otimes_\varepsilon G$ if $\langle e_\alpha, (f', g')\rangle \to 0$ uniformly for $f' \in A$ and $g' \in B$ when $A \subset F'$ and $B \subset G'$ are convex, disced, equicontinuous sets; the completion in the ε topology is denoted by $F \hat{\otimes}_\varepsilon G$. In general the π topology is stronger than the ε topology and when E is a so called nuclear space the canonical injection $E \hat{\otimes}_\pi F \to E \hat{\otimes}_\varepsilon F$ is an isomorphism into, for every LCS F. Most of the distribution spaces are nuclear but we will not explicitly use this property. Now for $S \in \mathcal{D}'_x$ and $T \in \mathcal{D}'_y$ one can identify $S_x \otimes T_y$ with the distribution $W \in \mathcal{D}'_{x,y}$ defined by $\langle W, \phi(x,y)\rangle = \langle S_x, \langle T_y, \phi(x,y)\rangle\rangle = \langle T_y, \langle S_x, \phi(x,y)\rangle\rangle$ for $\phi \in \mathcal{D}_{x,y}$. Here one notes that there is a natural map $\mathcal{D}'_x \otimes \mathcal{D}'_y$ into $(\mathcal{D}_x \otimes \mathcal{D}_y)^* = \mathcal{L}(\mathcal{D}_x \otimes \mathcal{D}_y; \mathbb{C})$ defined by extending the map $\langle S \otimes T, \phi \otimes \psi\rangle = \langle S, \phi\rangle\langle T, \psi\rangle$ and thus W acts like $S_x \otimes T_y$ in $(\mathcal{D}_x \otimes \mathcal{D}_y)^*$. But W can easily be shown to be continuous $\mathcal{D}_{x,y} \to \mathbb{C}$ and since $\mathcal{D}_x \otimes \mathcal{D}_y$ is clearly dense in $\mathcal{D}_{x,y}$ we have $W = S_x \otimes T_y$ in $\mathcal{D}'_{x,y}$ which gives an algebraic map $\mathcal{D}'_x \otimes \mathcal{D}'_y$ into $\mathcal{D}'_{x,y}$. The Schwartz kernel theorem, which we do not need, states in fact that $\mathcal{D}'_x \hat{\otimes}_\varepsilon \mathcal{D}'_y = \mathcal{D}'_{x,y} = \mathcal{D}'_x(\mathcal{D}'_y)$ algebraically and topologically (for suitably defined topologies). Using the tensor product one can now define the convolution $S * T$, when it makes sense, by

$$(1.13) \qquad \langle S * T,\phi\rangle = \langle S \otimes T,\phi(\xi + \eta)\rangle$$

for $\phi \in \mathcal{D}_x$, $S \in \mathcal{D}'_x$, and $T \in \mathcal{D}'_x$. One must be careful with supports here since $(\xi,\eta) \to (\xi + \eta)$ does not have compact support in general. It can be easily demonstrated that if $S \in \mathcal{E}'_x$ and $T \in \mathcal{D}'_x$ however then $S * T \in \mathcal{D}'_x$ is well defined. For suitable functions S and T of course

$$(S * T)(x) = \int S(x-y)T(y)dy.$$

Further it is easily verified that $D_k\delta * T = D_kT$ and also $D_k(S * T) = D_k\delta * S * T = D_kS * T = S * D_kT$ assuming $S * T$ makes sense. We mention two more spaces which play a role in convolution. Let $\mathcal{O}_M$ be the space of C^∞ functions on $\mathbb{R}^n$ such that $D^p\phi$ is bounded by a polynomial (depending on p). Thus $fD^p\phi$ is bounded for $f \in \mathcal{S}$ and $\phi \in \mathcal{O}_M$ and we say $\phi_\alpha \to 0$ in $\mathcal{O}_M$ if for any p and any $f \in \mathcal{S}$ one has $fD^p\phi_\alpha \to 0$ uniformly in $\mathbb{R}^n$. Define further $\mathcal{O}'_C = F\mathcal{O}_M$ ($\mathcal{O}_M \subset \mathcal{S}'$ evidently) and stipulate that $S_\alpha \in \mathcal{O}'_C$ converges $\Longleftrightarrow$ FS_α converges in $\mathcal{O}_M$. Then $\mathcal{O}_M$ is the space of multipliers $\mathcal{S}' \to \mathcal{S}'$ and $\mathcal{O}'_C$ maps $\mathcal{S}' \to \mathcal{S}'$ under convolution.

It will be useful to define also the idea of the order of a distribution. Let $\mathcal{D}^m$ be the space of C^m functions with compact support (i.e. $D^p\phi \in C^0$ for $|p| \leq m$). Clearly $\mathcal{D}^m = \cup\, \mathcal{D}^m(K_n)$ where $K_n \subset K_{n+1}$ are compact with $\cup\, K_n = \mathbb{R}^n$. Topologize $\mathcal{D}^m(K_n)$ in the obvious manner and put on $\mathcal{D}^m$ the finest locally convex topology such that all $i_n : \mathcal{D}^m(K_n) \to \mathcal{D}^m$ are continuous. The same construction also applies for $\mathcal{D}^m(\Omega)$ when $\Omega \subset \mathbb{R}^n$ is open. One has a sequence of continuous maps $\mathcal{D}^{m+1} \to \mathcal{D}^m$ and injections $\mathcal{D}'^m \to \mathcal{D}'^{m+1}$ where $\mathcal{D}'^m = L(\mathcal{D}^m;\mathbb{C})$. A distribution is said to be of order m if it belongs to $\mathcal{D}'^m$ but not to $\mathcal{D}'^{m-1}$. It can be shown that every $T \in \mathcal{E}'$ is of finite order for example. If one writes $\mathcal{K}$ for the strict inductive limit of $C_0^0(K_n) = \{\phi \in C^0;\ \text{supp}\,\phi \subset K_n\}$ then $\mathcal{K}'$ is defined to be the spaceof Radon measures on $\mathbb{R}^n (\mathcal{K}' = \mathcal{D}'^0)$. If $T \in \mathcal{D}'^m$ then $\exists$ a finite family μ_p of measures such that $T = \Sigma\, D^p \mu_p$ $(|p| \leq m)$ where the derivatives are in $\mathcal{D}'$. If $T \in \mathcal{E}'^m = \mathcal{D}'^m \cap \mathcal{E}'$ and $\phi \in \mathcal{E}^m$ is such that $D^p\phi = 0$ for $x \in \text{supp}\, T$ and $|p| \leq m$ then $\langle T,\phi\rangle = 0$.

We mention also an elementary but useful fact about convergence of vector valued functions. If $t \in \mathbb{R}$ and $f(t) \in F$ where F is e.g. a TVS then $f(\cdot)$ has a limit ℓ as $t \to t_0$ if for any sequence $t_n \to t_0$ $f(t_n) \to \ell$. Another standard fact is that if E and F are LCS with F complete and f is a continuous linear map from a dense subspace $A \subset E$ to F then f can be extended by continuity to a unique continuous linear map $\tilde{f} : E \to F$. Now let us turn to some observations about duality theory.

Theorem 1.4. (Hahn-Banach) Let p be a continuous seminorm in a complex LCS E and suppose given a linear form ℓ defined on a closed subspace $F \subset E$ such that $|\ell(x)| \leq p(x)$ for $x \in F$. Then ℓ may be extended to a continuous linear form $\tilde{\ell}$ on E with $|\tilde{\ell}(x)| \leq p(x)$ for $x \in E$.

In particular if $x_0 \in E$ define ℓ on $\{x_0\} = \{\alpha x_0\ ;\ \alpha \in \mathbb{C}\}$ by $\ell(\alpha x_0) = \alpha p(x_0)$ and extend ℓ to E. This gives the existence of nontrivial elements of E' so one has nontrivial duals E' for E a LCS since continuous seminorms exist. Now let us define a set B in a TVS to be bounded if it is absorbed by any nbh of 0 or equivalently by any $V \in B(0)$; this means $B \subset \lambda V$ for $|\lambda| \geq \lambda_0$. One notes that Cauchy sequences are bounded. Two TVS F and G are said to be in duality if there is a bilinear form $\langle\cdot,\cdot\rangle$ on $F \times G$ with the properties (1) for any $x \neq 0$ in F there is a $y \in G$ such that $\langle x,y\rangle \neq 0$ (2) for any $y \neq 0$ in G there is an $x \in F$ such that $\langle x,y\rangle \neq 0$. Clearly if F is an LCS and F' its dual then the natural pairing $\langle\ ,\ \rangle$ establishes a duality between F and

F' Let F' be dual to F with F an LCS. Then the strong topology $\beta(F',F)$ on F' is the topology of uniform convergence on bounded sets of F; seminorms of the form $p_B(f') = \sup |<f',f>|$ for $f \in B$ bounded can be used. One can use here as a fsn the polars of closed bounded disced sets $B \subset F$ where the polar B^0 of a disced B is

$$B^0 = \{f' \in F';\ |<f',f>| \leq 1 \text{ for } f \in B\}.$$

If F is a normed space then the strong topology in F' has the norm determined by $||f'|| = \sup |<f',f>|$ for $||f|| \leq 1$. The weak topology $\sigma(F',F)$ on F' is the coarsest topology on F' making all the linear maps $f' \to <f',f> : F' \to \mathbb{C}$ continuous. Thus $F' \subset \mathbb{C}^F = \Pi_F \mathbb{C}$ and $\sigma(F',F)$ is the induced product topology with a fsn of 0 consisting of finite intersections of sets $\{f' \in F';$ $|<f',f>| \leq \varepsilon\}$. A barrel in a TVS is a closed, convex, disced, absorbing set and a space is called barreled if every barrel is a nbh of 0. A Baire-space is a topological space such that every countable union of closed sets without interior points itself has no interior point.

<u>Theorem 1.5</u>. A complete metrizable space is Baire. Every Baire LCS is barreled.

As a corollary we see that every Frechet or Banach space is barreled and a further elementary argument shows that $\mathcal{D}$ is also barreled. Now let E and F be LCS and consider the vector space $L(E;F)$ of continuous linear maps $E \to F$. Let V run over a fsn of 0 in F composed of convex closed disced sets and B run over all bounded convex disced sets in E. The strong topology in $L(E;F)$ has a fsn of 0 composed of sets $N(B,V) = \{u \in L(E;F);\ u(B) \subset V\}$. The weak topology or topology of simple convergence has a fsn of 0 composed of finite intersections of sets $N(x,V) = \{u \in L(E;F);\ u(x) \subset V;\ x \in E\}$. A set $H \subset L(E;F)$ is then strongly bounded if $H \subset \lambda N(B,V)$ for $|\lambda| \geq \lambda_0$ or equivalently $\cup\, y'(B) \subset \lambda V$ for $y' \in H$ or equivalently $B \subset \lambda \cap y'^{-1}(V)$ for $y' \in H$. Thus it is required that $\cap\, y'^{-1}(V)$ for $y' \in H$ absorb every bounded set. Similarly $H \subset L(E;F)$ is weakly bounded $\Longleftrightarrow \cap\, y'^{-1}(V)$ for $y' \in H$ absorbs every point. A set $H \subset L(E;F)$

is equicontinuous if $\cap y'^{-1}(V)$ for $y' \in H$ is a nbh of 0 in E. Obviously if $H \subset L(E;F)$ is equicontinuous then it is both weakly and strongly bounded. On the other hand it is easy to see that if E is barreled then every weakly bounded $H \subset L(E;F)$ is both strongly bounded and equicontinuous. Now for V a convex disced nbh of 0 in a LCS F one says a set $A \subset F$ is small of order V if for any $x, y \in A$ it follows that $x - y \in V$. A set $B \subset F$ is said to be precompact if for any V as indicated there is a finite covering of B by sets $x_i + V$ which are small of order V. One knows that B is compact $\Longleftrightarrow$ B is precompact and complete. It is easy to show now that if $H \subset L(E;F)$ is equicontinuous for LCS E and F then the weak topology on H is equivalent to the topology of uniform convergence on precompact subsets of E and this leads to

<u>Theorem 1.6.</u> (Banach-Steinhaus) Let E be barreled and $U_n \in L(E;F)$ a sequence converging simply to a map $U_0 : E \to F$. Then $U_0 \in L(E;F)$ and $U_n \to U_0$ uniformly on precompact subsets of E. If $U_\alpha \in L(E;F)$ is a simply counded net converging simply on a dense subset $\tilde{E} \subset E$ to a map $U_0 : E \to F$ then $U_0 \in L(E;F)$ and $U_\alpha \to U_0$ uniformly on precompact subsets of E.

Recall next that a set A in a TVS E is said to be relatively compact if $\bar{A}$ is compact and compact means of course that any covering by open sets has a finite subcovering. A topological space is locally compact if every point has a compact nbh. $C(E;F)$ means continuous maps $E \to F$, not necessarily linear; we cite then

<u>Theorem 1.7.</u> (Ascoli) Let E be locally compact and F a TVS (both Hausdorff spaces). Then $H \subset C(E;F)$ is relatively compact for the topology of uniform convergence on compact subsets of E $\Longleftrightarrow$ H is equicontinuous and for all $x \in E$ the set $H(x) = \{h(x); h \in H\}$ is relatively compact in F.

One can show that $B \subset \mathcal{D}$ is bounded $\Longleftrightarrow B \subset \mathcal{D}_m$ for some m and is bounded there. Using the above theorems it then follows that bounded sets in $\mathcal{D}$ or $\mathcal{E}$ are relatively compact. A LCS which is barreled and in which every bounded set is relatively compact is called a Montel space so that $\mathcal{D}$ and $\mathcal{E}$ are Montel spaces (as also is $\mathcal{S}$). These facts allow one to assert that if a sequence $T_j \in \mathcal{D}'$ is weakly convergent then it is strongly convergent; also $D_k : \mathcal{D}' \to \mathcal{D}'$ is continuous in the strong topology. We mention also in passing that $\mathcal{D}$ and $\mathcal{D}'$ are complete and $\mathcal{D}'$, $\mathcal{E}'$, and $\mathcal{S}'$ are Montel since the strong dual of a Montel space is Montel; $\mathcal{E}'$ and $\mathcal{S}'$ are also complete. If E is a TVS with strong dual E' then $e \in E$ determines a continuous linear map $e' \to \langle e,e' \rangle : E' \to \mathbb{C}$ embedding E into E''. If this map is onto E is called semireflexive and if in addition the strong topology of E'' coincides with the topology of E then E is called reflexive. The spaces $\mathcal{D}$, $\mathcal{S}$, and $\mathcal{E}$ are all reflexive.

Next let $u : E \to F$ be a continuous linear map where E and F are LCS. Then u is continuous from $\sigma(E,E')$ to $\sigma(F,F')$ and its transpose ${}^tu : F' \to E'$ is continuous from $F' \to E'$ and from $\sigma(F',F)$ to $\sigma(E',E)$. Here one writes $\langle u(e),f' \rangle = \langle e, {}^tu(f') \rangle$ and $\sigma(E,E')$ for example is the induced topology from $\mathbb{C}^{E'} = \Pi_{E'}\mathbb{C}$. In fact any linear u continuous for $\sigma(E,E')$ and $\sigma(F,F')$ can be represented as $\langle u(e),f' \rangle = \langle e, {}^tu(f') \rangle$ where ${}^tu$ is continuous for $\sigma(F',F)$ and $\sigma(E',E)$. The Mackey topology $\tau(E,E')$ on E is the topology of uniform convergence on closed disced $\sigma(E',E)$ compact sets in E'.

Theorem 1.8. (Mackey-Arens) A topology T on E is compatible with the E - E' duality (i.e. yields E' as dual) if and only if T is a topology of uniform convergence on sets covering E' which are convex disced and $\sigma(E',E)$ compact.

Thus a topology T on E yields E' as dual $\Longleftrightarrow \sigma(E,E') \subset T \subset \tau(E,E')$. Further if a linear $u : E \to F$ is continuous for $\sigma(E,E')$ and $\sigma(F,F')$ then it is continuous for $\tau(E,E')$ and $\tau(F,F')$. If E is barreled its initial topology must be $\tau(E,E')$ and in particular, using the Tichonov theorem, the strongly closed unit ball in the dual E' of a Banach space E is weakly compact. A geometric version of the Hahn-Banach theorem asserts that if A is a closed convex set in a TVS E and $x \notin A$ then they are separated by a closed hyperplane. As a consequence every closed convex $A \subset E$ is the intersection of the closed halfspaces containing it. Since closed hyperplanes are determined by E' the closed convex

convex sets in E are the same for all topologies compatible with $E - E'$ duality. In fact the bounded sets in E are the same for all such topologies. Next we mention

Theorem 1.9. (Theorem of bipolars) Let $A \subset E$ with E a LCS. Then A^{oo} is the closed convex envelope for $\sigma(E,E')$ of A and 0.

A few facts and definitions about vector valued functions and distributions will also be useful. The space K was defined earlier in discussing the order of a distribution and its dual K' is the space of Radon measures. Denote $\mathbb{R}^n$ or $\mathbb{R}^1$ or any open $\Omega \subset \mathbb{R}^n$ by E and define $K(E)$ in the same way; if one has a compact E use $C^o(E)$ instead with uniform convergence and measures are elements of the dual. Let $\mu \in K'$ and $z' \in F'$ (F is a LCS over $\mathbb{R}$) and let $f : E \to F$ be continuous with compact support (we speak then of $f \in K(E,F)$). Then $x \to \langle f(x), z' \rangle : E \to \mathbb{R}$ is continuous with compact support and one sets $\phi(z') = \int \langle f(x), z' \rangle \, d\mu$. This defines a linear form on F' and hence an element of F'^* ($*$ denotes algebraic dual) which we call $\mu(f)$ or $\int f d\mu$ so that $\langle \int f d\mu, z' \rangle = \langle \int f(x), z' \rangle \, d\mu$. We note that $\sigma(F'^*, F')$ induces on F the topology $\sigma(F,F')$. If $h : E \to [0,1]$ is continuous with compact support, $h = 1$ on $K = \text{supp } f$, and C is the weakly closed convex envelope of $f(E)$ in F'^* then $\int f d\mu \in \mu(h) C \subset F'^*$ for any positive μ. If the closed convex envelope in F of a compact set is compact then $\mu(f) \in F$ and this holds whenever F is complete for example or in the weak topology of the dual of a Banach space.

Now for $\phi \in K(E)$ and $1 \leq p < \infty$ write $N_p(\phi) = (\int_E |\phi(x)|^p d\mu)^{1/p}$. For $h \geq 0$ lower semicontinuous define $\int^* h d\mu = \sup \int \phi d\mu$ for $\phi \leq h$. For any $g \geq 0$ $\int^* g d\mu = \inf \int^* h d\mu$ for $h \geq g$ lower semicontinuous. Let F be a Banach space now and $F^p(E,F)$ be the space of all functions $E \to F$, defined everywhere on E, such that $N_p(f) = (\int^* ||f(x)||^p d\mu)^{1/p} < \infty$. With N_p as seminorm $F^p(E,F)$ is complete but it is not Hausdorff. Let $N^p(E,F)$ be the adherence of 0 in $F^p(E,F)$ and $L^p(E,F)$ the closure of $K(E,F)$ in $F^p(E,F)$; then $\mathrm{L}^p(E,F) = L^p(E,F)/N^p(E,F)$. We know if $f \in K(E,F)$ and F is Banach then $\int f d\mu \in F$;

further $||\int f d\mu|| \leq \int ||f(x)|| d\mu = N_1(f)$. Hence $f \to \int f d\mu : K(E,F) \to F$ is continuous when $K(E,F)$ has the topology of $L^1(E,F)$. Since $K(E,F)$ is dense we can extend this map to $L^1(E,F)$ to define $\int f d\mu \in F$. Since any real measure can be written $\mu = \mu^+ - \mu^-$ with $\mu^+ \geq 0$, $\mu^- \geq 0$ one can speak of integrating with respect to any measure; complex measures can also be handled routinely. A function $f : E \to F$ is said to be scalarly integrable if for any $z' \in F'$ $x \to \langle f(x),z'\rangle$ is integrable. If for example F is reflexive Banach and $f(K)$ is bounded for K compact then a scalarly integrable f satisfies $\int f d\mu \in F$. For F Banach $f : E \to F$ is μ-measurable if for any compact $K \subset E$ and $\varepsilon > 0$ there is a compact $K' \subset K$ with $\mu(K \cap K') \leq \varepsilon$ and f continuous on K'. Equivalently f is μ-measurable $\Leftrightarrow$ a) for every $z' \in F'$ $x \to \langle f(x),z'\rangle$ is measurable and b) for any compact $K \subset E$ there is a denumerable $H \subset F$ such that $f(x) \in \overline{H}$ for μ almost all $x \in K$.

Define now $\mathcal{D}'(F) = L(\mathcal{D};F)$, F a LCS; no topology will be needed here but it is instructive to think of $\mathcal{D}'(F) = \mathcal{D}' \otimes_\varepsilon F$. Let $D_k = \partial/\partial x_k$ and define D_k in $\mathcal{D}'(F)$ as the extension by continuity of $D_k \otimes I$ in $\mathcal{D}' \otimes_\varepsilon F$. This is continuous in the ε topology since if $U \subset \mathcal{D}'' = \mathcal{D}$ and $V \subset F'$ are equicontinuous, $\phi \in U$, $f' \in V$ then

$$\langle \Sigma\, D_k T_\alpha \otimes f_\alpha, \phi \otimes f'\rangle = - \langle \Sigma\, T_\alpha \otimes f_\alpha, D_k \phi \otimes f'\rangle$$

and $D_k U$ is equicontinuous with U. Clearly one has

$$\langle \Sigma\, D_k T_\alpha \otimes f_\alpha, \phi \otimes f'\rangle = \langle \Sigma \langle D_k T_\alpha, \phi\rangle f_\alpha, f'\rangle = \langle \Sigma \langle f_\alpha, f'\rangle D_k T_\alpha, \phi\rangle$$

and this leads to the determination of $D_k T$, $T \in \mathcal{D}'(F)$, by either of the rules $\langle D_k T, \phi\rangle = - \langle T, D_k \phi\rangle$ in F or $\langle D_k T, f'\rangle = D_k \langle T, f'\rangle$ in $\mathcal{D}'$.

A few more general facts should be mentioned. Let $L : E \to F$ be a linear operator defined on a dense linear set $D(L) \subset E$ (E and F can be general LCS). The graph $G(L)$ of L is the set of pairs (e,Le) $E \times F$ for $e \in D(L)$. L is closed if $G(L)$ is closed and L is closable if $\overline{G(L)}$ is a graph (i.e. $\overline{G(L)}$ contains no elements (o,f)); $\overline{L}$ is the operator with graph $\overline{G(L)}$. $D(L^*)$ is the set of $f' \in F'$ such that $e \to \langle Le,f'\rangle : D(L) \to \mathbb{C}$ is continuous when $D(L)$ has the topology of E. Extending by continuity we get an $e' = L^*f' \in E'$ with $\langle L^*f',e\rangle = \langle Le,f'\rangle$. L is closable if and only if L^* is densely defined and then $\overline{L} = L^{**}$. A family $T(t) \in L(F)$, F Banach, is called a strongly contin- semigroup if $T(t+\tau) = T(t)T(\tau)$, $T(0) = I$, and $t \to T(t) \in C^0(L_s(F))$, $0 \leq t < \infty$; other kinds of groups and semigroups are defined in Chapter 1 and we need not deal with them here. The infinitesimal generator A of $T(t)$ is the linear operator defined for $x \in D(A)$ by $Ax = \lim (T(t)x - x)/t$ as $t \to 0$. $D(A)$ is dense, A is closed and for $t > 0$, $D_t T(t)x = AT(t)x = T(t)Ax$ for $x \in D(A)$. The resolvant $R_\lambda(A) = (\lambda I - A)^{-1}$ is defined for $\operatorname{Re} \lambda > \omega_0$ where $\omega_0 = \lim \log ||T(t)||/t$ as $t \to \infty$ and

$$(1.14) \qquad R_\lambda(A)x = \int_0^\infty \exp(-\lambda t)T(t)x\,dt$$

<u>Theorem 1.10.</u> (Hille-Yosida-Phillips-Miyadera) A necessary and sufficient condition for a closed linear operator A with dense domain to generate a strongly continuous semigroup $T(t)$ is that there exist $m > 0$ and ω_0 such that $||(\lambda I - A)^{-n}|| \leq m/(\lambda - \omega_0)$ for real $\lambda > \omega_0$ and all integers n. In this event $||T(t)|| \leq m \exp \omega_0 t$.

A corollary is the classical Hille-Yosida theorem which states that a necessary and sufficient condition for a closed densely defined linear A to generate a strongly continuous semigroup of contractions is that $||(\lambda I - A)^{-1}|| \leq 1/\lambda$ for $\lambda > 0$. In a Hilbert space A generates a strongly continuous semigroup of contractions if and only if it is a closed maximal dissipative operator with dense domain. A dissipative means $\operatorname{Re}(Ax,x) \leq 0$ for $x \in D(A)$ and maximal dissipative means A is not the proper restriction of another linear dissipative operator. A map $A : E \to F$, E and F Banach, is called open if it maps open sets to open sets or equivalently if $||e|| \leq c||Ae||$ for $e \in D(A)$. If A is a closed densely defined linear map $E \to F$ then the following are equivalent: (a) $R(A)$ is closed (b) $R(A^*)$ is closed (c) A is open (d) A^* is open (e) $R(A) = \{f; \langle f,f'\rangle = 0 \text{ for } f' \in N(A^*)\}$ (f) $R(A^*) = \{e'; \langle e',e\rangle = 0 \text{ for } e \in N(A)\}$

(here N and R denote nullspace and range respectively).

Theorem 1.11. (Banach) Let E and F be complete metrizable TVS. Then every continuous linear map $u : E \to F$, onto F, is open.

A corollary is the closed graph theorem stating that a linear map $u : E \to F$ (into) is continuous if and only if $G(u) = \text{graph } u$ is closed (E and F complete metrizable). We recall next that a Banach algebra B is a Banach space which is an algebra over $\mathbb{C}$ with $||xy|| \leq ||x||\,||y||$. We assume B is commutative and has an identity e with $||e|| = 1$. An element x is regular if x^{-1} exists and otherwise singular. The group G of regular elements in B is open with $x \to x^{-1}$ continuous. An ideal $I \subset B$ is maximal if $I \subset B$ properly and is not contained in any other ideal. B/I is a field if and only if I is maximal and in fact $B/I = \mathbb{C}$; for $x \in B$ one writes $\dot{x}$ for its image in B/I and one can write $\dot{x} = \lambda\dot{e}$ with $||\dot{x}|| = |\lambda|$. This identifies $\dot{x}$ with λ. The spectrum $\sigma(x)$ of x is the set of λ such that $x = \lambda e$ is singular and the resolvant set $\rho(x)$ is the set of λ where $x - \lambda e$ is regular. The resolvant is again $R_\lambda(x) = (\lambda e - x)^{-1}$ defined in $\rho(x)$. $\sigma(x)$ is a nonvoid closed bounded set and $R_\lambda(x)$ is analytic in $\rho(x)$, vanishing of ∞, and satisfying

$$R_\lambda(x) - R_\mu(x) = (\mu - \lambda)R_\lambda(x)R_\mu(x).$$

If $\phi : B \to \mathbb{C}$ is a continuous homomorphism then $\ker \phi = I$ is a maximal ideal. The converse also holds and the set Φ_B of maximal ideals is called the carrier space. Φ_B is a weakly closed subset of the unit ball in B' and one writes $\phi(x) = \hat{x}(\phi)$. Φ_B is thus (weakly) compact and the map $x \to \hat{x} : B \to C(\Phi_B)$ from B into the space of continuous functions on a compact Hausdorff space is called the Gelfand map. One sees that x^{-1} exists if and only if $\hat{x}$ does not vanish on Φ_B. Further $\sigma(x)$ is the set of values taken on by $\hat{x}(\phi)$ for $\phi \in \Phi_B$.

Let now $f(z)$ be an analytic function in a nbh of $\sigma(x)$ and Γ be a smooth closed curve enclosing $\sigma(x)$. Then consider

(1.15) $$F(x) = \frac{1}{2\pi i}\int_\Gamma (\lambda e - x)^{-1} f(\lambda)d\lambda$$

Clearly $F(x) \in B$ and $\widehat{F(x)}(\phi) = f(\hat{x}(\phi))$. One has

Theorem 1.12. Let Ξ be a closed bounded set in $\mathbb{C}$ and $H(\Xi)$ the algebra of functions f analytic in some open ubh Ω_f of Ξ. Let $D \subset B$ be the set of elements $x \in B$ with $\sigma(x) \subset \Xi$ and for $x \in D$ define $J_x : H(\Xi) \to B$ by $J_x(f) = F(x)$ where F is defined by (1.15) with suitable Γ_f. Then (a) J_x is an algebra homomorphism (b) $f(\lambda) = 1$ implies $F(x) = e$ (c) $f(\lambda) = \lambda$ implies $F(x) = x$ (d) If $f_n(\lambda)$ converges uniformly on a compact nbh K of Ξ, f_n analytic in an open nbh of K, then $F_n(x)$ converges in norm.

In general the spectrum $\sigma(L)$ of a linear operator $L \in L(F)$, F Banach, is divided into three parts: (1) the set of $\lambda \in \sigma(L)$ such that $\lambda I - L$ is not $1 - 1$ is called the point spectrum $\sigma_p(L)$ (2) the set of λ for which $\lambda I - L$ is $1 - 1$ with $(\lambda I - L)F$ dense in F but not equal to F is called the continuous spectrum $\sigma_c(L)$ (3) the set of λ such that $(\lambda I - L)$ is $1 - 1$ but $(\lambda I = L)F$ is not dense in F is the residual spectrum $\sigma_r(L)$. Similar considerations apply to closed densely defined linear unbounded L. For such L $\sigma(L)$ may be bounded, unbounded, void, or all of $\mathbb{C}$. We assume $\rho(L)$ is not void however in general. Then if $H(L)$ is the set of functions analytic in some nbh of $\sigma(L)$ and at ∞ and Γ is a suitable smooth curve enclosing $\sigma(L)$ one can write

(1.16) $$f(L) = f(\infty)I + \frac{1}{2\pi i}\int f(\lambda)R(\lambda,L)d\lambda$$

where $R(\lambda,L) = R_\lambda(L) = (\lambda I - L)^{-1}$. Similar conclusions to those of Theorem 1.12 hold for such L, namely $(fg)(L) = f(L)g(L)$, $\sigma(f(L)) = f(\sigma(L) \cup \{\infty\})$, etc. A subset $\Delta \subset \sigma(L)$ which is both open and closed in $\sigma(L)$ is called a spectral set. For such Δ there is a function $f \in H(L)$ equal to 1 on Δ and 0 elsewhere on $\sigma(L)$; one writes $E(\Delta) = E(\Delta,L) = f(L)$. The map $\Delta \to E(\Delta)$ is an isomorphism from the Boolean algebra of spectral sets to the algebra of projections $E(\Delta)$ and

(1.17) $$E(\Delta) = \frac{1}{2\pi i}\int_\Gamma (\lambda I - L)^{-1} d\lambda$$

where Γ surrounds Δ but no other point of $\sigma(L)$. Let us now go to a Hilbert space H and let B be a commutative * algebra in $L(H)$ (i.e. B is a commutative Banach algebra of bounded operators with involution * which is simply $A \to A^*$ here). Adjoin a unit e, the identity operator, if necessary. Then B is isometrically * isomorphic to $C(\Phi_B)$ $(A^* \to \bar{\hat{a}})$ and we write $\Gamma : \Lambda = C(\Phi_B) \to B$ (this result is due to Gelfand-Naimark). By a theorem of Riesz there is a measure μ on Λ such that

$$(\Gamma(f)x,y) = \int f(\lambda)d\mu(\lambda,x,y).$$

Further $d\mu(\lambda,x,y) = \overline{d\mu(\lambda,y,x)}$ and for each Borel set $\sigma \in \mathcal{B}$ there is a selfadjoint spectral measure $E(\sigma)$ with $\mu(\delta,x,y) = (E(\delta)x,y)$. One has $E(\sigma)A = AE(\sigma)$ for $A \in B$ and $\Gamma(f) = \int f(\lambda)dE(\lambda)$ for $f \in C(\Lambda)$ (we write $dE(\lambda) = E(d\lambda)$ occasionally). Applied to a bounded normal operator L $(LL^* = L^*L)$ these results yield a spectral family of projections $E(\sigma)$ such that E vanishes on $\rho(L)$ while $f(L) = \int f(\lambda)dE(\lambda)$ for $f \in C(\sigma(L))$. This is called the spectral resolution of L. In fact for each complex bounded Borel function f on $\sigma(L)$ set $f(L) = \int f(\lambda)E(d\lambda)$ and $f \to f(L)$ is then a continuous * homomorphism of the algebra of bounded Borel functions (sup norm) or $\sigma(L)$ into $L(H)$ with $1 \to I$ and $\lambda \to L$. For L selfadjoint one has

$$E(a,b) = \lim_{\substack{\varepsilon\to 0\\ \delta\to 0}} \frac{1}{2\pi i}\int_{a+\delta}^{b-\delta} [R(\mu-i\varepsilon,L) - R(\mu+i\varepsilon,L)]d\mu \tag{1.18}$$

Let us spell out a few more details for the selfadjoint case where $\sigma(L)$ is a subset of $(-\infty,\infty)$ and L may not be bounded. Let $E(d\lambda)$ be the resolution of the identity or spectral resolution for L. Then for f a Borel function on $\sigma(L)$,

$$D(f(L)) = \{x \in H; \int |f(\lambda)|^2 (E(d\lambda)x,x) < \infty\}$$

and $(f(L)x,y) = \int f(\lambda)(E(d\lambda)x,y)$. Further

$$R(\lambda,L) = \int_{-\infty}^{\infty} \frac{E(d\mu)}{\lambda-\mu}\ ; \quad \lambda \in \rho(L) \tag{1.19}$$

one can also proceed somewhat differently and, taking H to be separable, it will follow that there is a measure space (M,μ), μ finite, a unitary operator $U : H \to L^2(M,d\mu)$, and a real function f on M finite a.e., such that $x \in D(L) \Longleftrightarrow f(\cdot)(Ux)(\cdot) \in L^2(M,d\mu)$ and if $\phi \in U(D(L)$ then $ULU^{-1}\phi(\lambda) = f(\lambda)\phi(\lambda)$. If h is a bounded Borel function on $\mathbb{R}$ define $h(L) = U^{-1}T_{h(f)}U$ where $T_{h(f)}$ is the operator on $L^2(M,d\mu)$ acting as multiplication by $h(f(\lambda))$. Then the map $\hat{\phi}(h \to h(L))$ from bounded Borel functions into $L(H)$ is a (unique) algebraic * homomorphism, norm continuous, and satisfies (a) if $Lx = \zeta x$ then $\phi(h)x = h(\zeta)x$ (b) $h \geq 0 \Rightarrow \hat{\phi}(h) \geq 0$ (c) if $h_n(\lambda) \to h(\lambda)$ pointwise with $||h_n||_\infty \leq M$ then $\hat{\phi}(h_n) \to \hat{\phi}(h)$ strongly. Clearly $h(\lambda) = \lambda$ corresponds to $\hat{\phi}(h) = L$ but λ has to be obtained here as a limit of suitable bounded functions $h_n(\lambda)$. A vector $x \in H$ is cyclic for L if $\{h(L)x;\ h \in C_\infty(\mathbb{R}) = C^\infty$ functions vanishing at $\infty\}$ is dense in H and if x is cyclic $H = L^2(\mathbb{R},d\mu_x)$ where $d\mu_x$ is determined by $\int h(\lambda)d\mu_x(\lambda) = (x,h(L)x)$ and L becomes multiplication by λ. In general H decomposes into a direct sum of cyclic subspaces so that M above is a union of copies of $\mathbb{R}$ and the function $f(\lambda)$ above arises in selecting a finite μ and inserting Radon-Nikodym derivatives. The spectral projections arise in another common notation as $P_\Omega = \chi_\Omega(L)$ where χ_Ω is the characteristic function of Ω. For $x \in H$ $(x,P_\Omega x)$ is a Borel measure on $\mathbb{R}$ denoted by $d(x,P_\lambda x)$ and a bounded Borel function h determines $h(L)$ by $(x,h(L)x) = \int h(\lambda)d(x,P_\lambda x)$; this satisfies the properties (a) - (c) etc. above and by uniqueness of $\hat{\phi}$ must coincide with $h(L)$ defined there as $U^{-1}T_{h(f)}U$. If h is an unbounded Borel function define then

$$D_h = D(h(L)) = \{x \in H;\ \int |h(\lambda)|^2\ d(x,P_\lambda x) < \infty\}$$

and D_h is dense in H. Symbolically $h(L) = \int h(\lambda)dP_\lambda$ and clearly $dP_\lambda = dE(\lambda)$. Formula (1.18) holds also for unbounded L.

Still another point of view involves representing H as a direct integral of Hilbert spaces $\int^\oplus H(\lambda)d\nu(\lambda)$ which diagonalizes L (cf Carroll [4], Dixmier [1], Maurin [1]). This is similar to writing H as $L^2(M,d\mu)$ above where the

Radon-Nikodym derivatives represented by $f(\lambda)$ are incorporated into $d\nu$. Generally let Z be a locally compact Hausdorff space and ν a positive measure on Z. A family or field of Hilbert spaces on Z is a map $\zeta \to H(\zeta)$, $H(\zeta)$ Hilbert, and one looks at $H = \Pi H(\zeta)$ with elements $\zeta \to x(\zeta) \in H(\zeta)$. The field $H(\zeta)$ is measurable if there is a vector subspace $E \subset H$ such that (a) for $x \in E$ $\zeta \to ||x(\zeta)||_\zeta$ is measurable ($|| \; ||_\zeta$ denotes the norm in $H(\zeta)$) (b) if $y \in H$ is such that $\zeta \to (x(\zeta),y(\zeta))_\zeta$ is measurable for all $x \in E$ then $y \in E$ (c) there is a (fundamental) sequence $x_k \in E$ such that for each $\zeta \in Z$ the closed subspace of $H(\zeta)$ generated by the $x_k(\zeta)$ equals $H(\zeta)$. In particular $H(\zeta)$ is separable. The elements of E are called measurable vectors and we let K be the set of such vectors with $||x||^2 = \int ||x(\zeta)||_\zeta^2 d\nu < \infty$. This gives a prehilbert structure on K with $x = 0$ meaning $x(\zeta) = 0$ a.e. The associated Hilbert space will be denoted by $\mathcal{H} = \int^{\oplus} H(\zeta)d\nu$. A family $A(\zeta)$ of linear operators in $H(\zeta)$ is called measurable if $\zeta \to A(\zeta)x(\zeta)$ is measurable for $x \in \mathcal{H}$. An operator A in $\mathcal{H}$ is diagonalizable if it is represented by functions $A(\zeta)$ in $H(\zeta)$. If $\mathcal{A}$ is a (suitable) commutative $*$ algebra of operators in a separable Hilbert space H with spectrum $\Lambda \sim \Phi_{\mathcal{A}}$ then there is a measurable field of Hilbert spaces $H(\zeta)$ on Λ relative to a (basic) measure ν such that $H = \mathcal{H} = \int^{\oplus} H(\zeta)d\nu$ and the Gelfand map becomes an isomorphism $L^\infty(\Lambda)$ onto the algebra of diagonalizable operators. Applied to a selfadjoint operator A in H, possibly unbounded, one obtains a direct integral over a possibly infinite interval such that A corresonds to multiplication by ζ.

It is appropriate here to indicate also the construction of generalized eigenfunctions in the context of rigged Hilbert spaces (cf. Berezanskij [1], Gelfand-Šilov [3], Šmulyan [1]). There are various ways to approach this and we simply choose one. Thus let H be a Hilbert space and $\Phi \subset H$ a dense nuclear space with continuous injection (everything separable) and think of $\mathcal{D} \subset L^2$ as a model situation. We will not dwell here on nuclearity; a main fact needed is that in Φ bounded sets are relatively compact. For $h \in H$ the map $\phi \to (h,\phi) : \Phi \to \mathbb{C}$ is continuous and determines an element $f_h \in \Phi'$ written $\langle f_h,\phi\rangle$ so that we have

$\Phi \subset H \subset \Phi'$; Φ' will be the antidual or conjugate linear dual. If dE_λ is a spectral family of projections in H for example and $\sigma(\lambda) = (E_\lambda e, e)$ for $e \in H$ fixed one can show the existence of $\chi_\lambda \in \Phi'$ such that $\langle \chi_\lambda, \phi \rangle = d(E_\lambda e, \phi)/d\sigma$ in the sense of a Radon-Nikodym derivative. Indeed $e_\lambda = E_\lambda e$ considered as a functional on Φ is of strongly bounded variation since Φ is nuclear and one writes $\chi_\lambda = dE_\lambda e/d\sigma$. Now let $H(e)$ be the subspace of H generated by the $e_\lambda = E_\lambda e$. The χ_λ associated to e are orthonormal and complete in the sense that for $\phi \in H(e) \cap \Phi$

$$\phi = \int_{-\infty}^{\infty} \overline{\langle \chi_\lambda, \phi \rangle} \chi_\lambda \, d\sigma(\lambda) \tag{1.20}$$

$$||\phi||^2 = \int_{-\infty}^{\infty} |\langle \chi_\lambda, \phi \rangle|^2 d\sigma(\lambda) \tag{1.21}$$

and if $\langle \chi_\lambda \ \phi \rangle = 0$ for all λ then $\phi = 0$. We note here that $\langle \chi_\lambda, \phi \rangle = \overline{\langle \phi, \chi_\lambda \rangle}$. Further

$$E(\Delta)e = \int_\Delta \chi_\lambda d\phi(\lambda). \tag{1.22}$$

The space H can then be decomposed as an orthogonal direct sum of spaces $H(e_\alpha)$ and one has associated elements $\chi_\lambda^\alpha \in \Phi'$. Then for any $\phi \in \Phi$ there results

$$\phi = \sum_\alpha \int_{-\infty}^{\infty} \langle \phi, \chi_\lambda^\alpha \rangle \chi_\lambda^\alpha d\sigma_\alpha(\lambda). \tag{1.23}$$

Let now A be a continuous linear operator $A : \Phi \to \Phi$ with $(A\phi, \psi) = (\phi, A\psi)$ for $\phi, \psi \in \Phi$. Then $A^* : \Phi' \to \Phi'$ is continuous and extends A on Φ. If A admits a selfadjoint extension to H then it has a complete system of eigenelements χ_λ in Φ'. Indeed one starts from the associated resolution of the identity $E_\lambda = E(\Delta_{-\infty}^\lambda)$ and constructs χ_λ as above. Then

$$\begin{aligned} \langle A\chi_\lambda, \phi \rangle &= \langle \chi_\lambda, A\phi \rangle = \lim \left(\frac{E(\Delta)e}{\sigma(\Delta)}, A\phi\right) \\ &= \lim \frac{1}{\sigma(\Delta)} \left(E(\Delta)e, \int_{\infty}^{\infty} \lambda dE_\lambda \phi\right) \\ &= \lim \frac{1}{\sigma(\Delta)} \left(e, \int_{\infty}^{\infty} \lambda dE(\Delta)E_\lambda \phi\right) = \lim \frac{1}{\sigma(\Delta)} \left(e, \int_\Delta \lambda dE_\lambda \phi\right) \\ &= \lim \frac{1}{\sigma(\Delta)} \left(\int_\Delta \lambda dE_\lambda e, \phi\right) = \lim \left(\int_\Delta \frac{\lambda dE_\lambda e}{\sigma(\Delta)}, \phi\right) = \langle \lambda \chi_\lambda, \phi \rangle \end{aligned} \tag{1.24}$$

so that $A\chi_\lambda = \lambda\chi_\lambda$. For example if one takes for A the Sturm-Liouville operator $Ay = -y'' + qy$ for $q \in C^\infty$ real with say $y'(0) = \theta y(0)$, θ real, then the χ_λ satisfy $-\chi''_\lambda + q\chi_\lambda = \lambda\chi_\lambda$. If e_α is a system of generators for $H = L^2(0,\infty)$ relative to E_λ set $\sigma_\alpha(\lambda) = (E_\lambda e_\alpha, e_\alpha)$ and $\chi_\alpha(\lambda) = b_\alpha(\lambda)y(x,\lambda)$ where $y(x,\lambda)$ is the classical solution of $Ay = \lambda y$ (take here $\Phi = \mathcal{D}$). Set $d\sigma(\lambda) = \Sigma\, b_\alpha^2(\lambda)d\sigma_\alpha(\lambda)$ and $F(\lambda) = \int_{-\infty}^{\infty} \overline{y(\xi,\lambda)}\phi(\xi)d\xi$. Then

$$\phi(x) = \int_{-\infty}^{\infty} y(x,\lambda)F(\lambda)d\sigma(\lambda) \tag{1.25}$$

and this can be extended to an isomorphism between $L^2(0,\infty)$ and L^2_σ since $||\phi||^2 = \int_0^\infty |\phi(x)|^2 dx = \int_0^\infty |F(\lambda)|^2 d\sigma(\lambda)$.

APPENDIX 2

Notation of (2.6.91) - (2.6.100) with a few minor changes in expanding the framework somewhat to fit the model situation of $P(D) = L_m(D)$ and $Q(D) = D^2$ (cf. Section 2.3). Thus let in (2.6.91) and (2.6.96) (cf. Carroll [39]; Remark 2.6.21)

(2.1) $\mathfrak{P} : E \to \hat{E}^{\sigma}; \quad \mathfrak{Q} : F \to \tilde{F}^{\rho}$

with inverses $\mathfrak{P}$ and $\mathfrak{Q}$ given by (2.6.93) and (2.6.98). Let in (2.6.92) and (2.6.97)

(2.2) $P : \mathbb{E} \to \hat{\mathbb{E}}^{\sigma'}; \quad Q : \mathbb{F} \to \tilde{\mathbb{F}}^{\rho'}$

with inverses $\mathbb{P}$ and $\mathbb{Q}$ given by (2.6.95) and (2.6.100) (with notation changed to σ' and ρ'). The "twisting" transforms $\mathcal{P}$ and $\mathcal{Q}$ given by (2.6.94) and (2.6.99) will in general be unbounded operators (cf. below)

(2.3) $\mathcal{P} : \tilde{F}^{\rho} \to E; \quad \mathcal{Q} : \hat{E}^{\sigma} \to F.$

Let us use the model indicated above to put together a realistic picture. Thus

(2.4) $H(x,\mu) = \hat{R}^m(x,\lambda) = 2^m\Gamma(m+1)(\lambda x)^{-m}J_m(\lambda x)$

(2.5) $\Omega(x,\mu) = 2^{-2m}\Gamma(m+1)^{-2}(\lambda x)^{2m+1}\hat{R}^m(x,\lambda)$

(2.6) $\Theta(y,\mu) = \text{Cos}\ \lambda y; \quad \hat{\Omega}(y,\mu) = \frac{2}{\pi}\ \text{Cos}\ \lambda y$

for $\mu = -\lambda^2$. Then

(2.7) $\mathfrak{P} f(\mu) = \dfrac{\lambda^{m+\frac{1}{2}}}{2^m\Gamma(m+1)}\ \mathbb{H}_m[x^{m+\frac{1}{2}}f(x)] = \hat{f}(\mu)$

(2.8) $\mathfrak{P} F(x) = \dfrac{2^m\Gamma(m+1)}{x^{m+\frac{1}{2}}}\ \mathbb{H}_m[\lambda^{-m-\frac{1}{2}}F(\lambda)]$

and similarly for P and $\mathbb{P}$, $x^{-m-\frac{1}{2}}f(x) \leftrightarrow \lambda^{m+\frac{1}{2}}\hat{\hat{f}}(\mu)$ under Hankel transformation. Now for suitable m, $\mathbb{H}_m$ will be an isometric isomorphism $L^2(0,\infty) \to L^2(0,\infty)$ and we choose $E = \{f;\ x^{m+\frac{1}{2}}f(x) \in L^2\}$ with $\hat{E}^{\sigma} = \{\hat{f};\ \lambda^{-m-\frac{1}{2}}\hat{f}(\mu) \in L^2_{\lambda}\} = L^2(-\infty,0;\ d\sigma)$ where $d\sigma = (-\mu)^{-m-\frac{1}{2}}d\mu/2\sqrt{-\mu}$. Take now $E' = \{f;\ x^{-m-\frac{1}{2}}f(x) \in L^2\} = \mathbb{E}$ (thus we do not identify E and E') and take $\tilde{\mathbb{E}}^{\sigma'} = (\hat{E}^{\sigma})' = \{\tilde{f};\ \lambda^{m+\frac{1}{2}}\tilde{f}(\mu) \in L^2_{\lambda}\} = L^2(-\infty,0;\ d\sigma')$ where $d\sigma' = (-\mu)^{m+\frac{1}{2}}d\mu/2\sqrt{-\mu}$. Note that the duality $\hat{E}^{\sigma} - (\hat{E}^{\sigma})'$ for example is expressed in terms of the "basic" measure $d\lambda = d\mu/2\sqrt{-\mu}$ by the rule $\langle\hat{f},\tilde{g}\rangle = \int \hat{f}(\mu)\tilde{g}(\mu)d\lambda$. For Q we take $F = L^2(0,\infty) = \mathbb{F}$ and $\tilde{F} = \tilde{\mathbb{F}} = L^2(0,\infty;\ d\lambda)$, in identifying these spaces with their duals. Then the following diagram displays the spaces and maps in a clear manner.

(2.9)

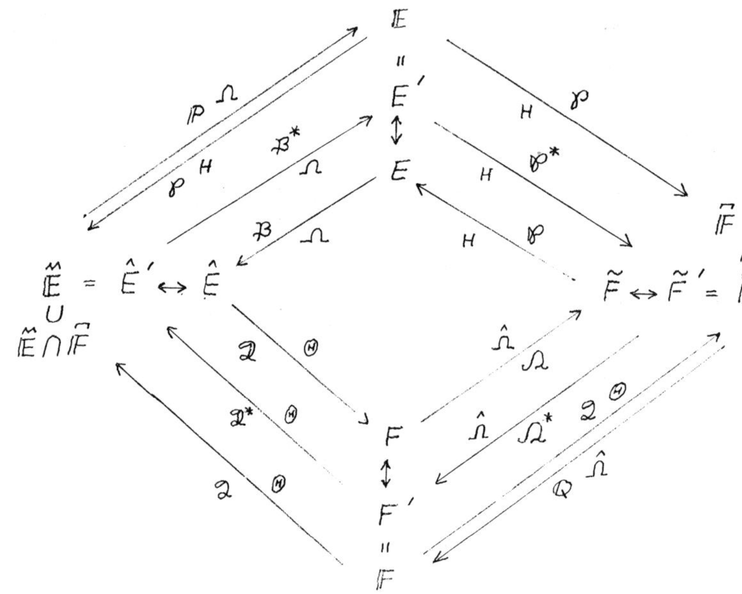

<u>Theorem 2.1</u>. The diagram (2.9) is a typical model of a transform theory based on transmutation and in addition to the formulas of Theorem 2.6.22 displays the identifications $\mathfrak{P}^* = \mathbb{P}$, $\Omega^* = \mathbb{Q}$, $\mathbb{P}^* = P$, and $\mathfrak{Q}^* = Q$ which yield $B^* = (\mathfrak{Q}\mathfrak{P})^* = \mathbb{P}Q$ and $\mathcal{B}^* = (\mathbb{P}\Omega)^* = \mathbb{Q}P$.

A few further remarks are appropriate here and we refer to Carroll [39] for an extensive treatment. First from $Bf(y) = \langle\beta(y,x),f(x)\rangle$ and β given by one of its determinations (2.3.45) we see that f should have n derivatives when $-\frac{1}{2} < m < n - \frac{1}{2}$ so B will not be defined on all of E. Another way to see this and to specify $D(\mathfrak{Q})$ is to write

$$(2.10) \qquad \mathcal{QB}f(y) = \int_0^\infty \lambda^{m+\frac{1}{2}} F(\lambda) \operatorname{Cos} \lambda y d\lambda$$

where $\hat{f}(\mu) = \lambda^{m+\frac{1}{2}} F(\lambda)$ from (2.7) with $F \in L^2_\lambda$. In order for $\mathcal{QB}f \in L^2 = F$ we want also $\hat{f}(\mu) \in L^2_\lambda$ so $D(\mathcal{Q}) = \hat{E}^\sigma \cap \tilde{F}$. Similarly for $\mathcal{B} = \mathcal{P}\Omega$ we have $G = \Omega g \in \tilde{F} = L^2_\lambda$ and $\mathcal{P}G$ is given by the formula (2.8); thus for $\mathcal{P}G$ to lie in E we want $\lambda^{-m-\frac{1}{2}} G(\lambda) \in L^2_\lambda$ or $G \in \hat{E}^\sigma$. Hence $D(\mathcal{P}) = \tilde{F} \cap \hat{E}^\sigma$. We note also that writing $\mathcal{B}g(x) = \langle \gamma(x,y), g(y) \rangle$ with $\gamma(x,y) = \langle H(x,\mu), \hat{\Omega}(y,\mu) \rangle$ one obtains

$$(2.11) \qquad \gamma(x,y) = R_+^m(x,y) = \frac{\Gamma(m+1)x^{-2m}}{\Gamma(\frac{1}{2})\Gamma(m+\frac{1}{2})} (x^2-y^2)_+^{m-\frac{1}{2}}$$

(cf. (2.3.26)). A further study of (more general) formulas such as (2.6.90) is also conducted in Carroll [39].

REFERENCES

Abdukadyrov, E.
1. The Green's function of a Sturm-Liouville equation with operator coefficients, Dokl. Akad. Nauk SSSR, 195 (1970), 519-522.

Abdulkerimov, L.
1. On the solvability of the Cauchy problem in Banach space..., Izv. Akad. Nauk Azerb. SSR, 1 (1966), 48-55.

Agranovič, Z. and Marčenko, V.
1. The inverse problem of scattering theory, transl. Gordon-Breach, N. Y., 1963.

Akhiezer, N. and Glazman, I.
1. The theory of linear operators in Hilbert space, Moscow, 1950.

Allakhverdiev, D. and Gasanov, E.
1. A theorem on the completeness of a system of characteristic and adjoined elements of rational operator pencils in a Banach space, Funkts. Anal. Prilož., 8 (1974), 76-78.

Audroščuk, A.
1. On the transmutation operator for a differential equation of second order with operator coefficients, Ukrain. Mat. Žur., 23 (1971), 52-55.

Atkinson, F.
1. Multiparameter spectral theory, Bull. Amer. Math. Soc., 74 (1968), 1-27.

2. Multiparameter eigenvalue problems, Vol. 1, Academic Press, N. Y., 1972.

Aubin, J.
1. Approximation of elliptic boundary value problems, Wiley-Interscience, N. Y., 1972.

2. Abstract boundary value operators and their adjoints, Rend. Sem. Mat. Univ. Padova, 43 (1970), 1-33.

Babolola, V.
1. Semigroups of operators on locally convex spaces, Trans. Amer. Math. Soc., 199 (1974), 163-179.

Baiocchi, C.
1. Regolarità e unicità della soluzione di una equazione diferenziale astratta, Rend. Sem. Mat. Univ. Padova, 35 (1965), 380-417.

2. Sul problema misto per l'equazione parabolica del tipo del calore, Rend. Sem. Mat. Univ. Padova, 36 (1966), 80-121.

3. Soluzioni ordinarie e generalizzate del problema di Cauchy per equazioni differenziali astratte lineari del secondo ordine in spazi di Hilbert, Richerche Mat., 16 (1967), 27-95.

Bairamogly, M. and Khalilova, R.
1. On a theorem of Phragmen-Lindeloff type for operator differential equations, Izv. Akad. Nauk Azerb. SSR, 2 (1973), 83-86.

Bakalrishnan, A.
1. An operational calculus for infinitesimal generators of semigroups, Trans. Amer. Math. Soc., 91 (1959), 330-353.

2. Fractional powers of closed operators and the semigroups generated by them, Pacific Jour. Math., 10 (1960), 419-437.

Barantzev, R.
1. Transvukovoi gazodinamike, Izd. Leningrad. Univ., 1965.

Barbu, V.
1. Semigrupuri de contractii neliniare in spatii Banach, Bucharest, 1974.

Bardos, C. and Brézis H.
1. Sur une classe de problèmes d' évolution nonlineáries, Jour. Diff. Eqs., 6 (1969), 345-394.

Bardos, C. and Cooper J.
2. A nonlinear wave equation in a time dependent domain, Jour. Math. Anal. Appl., 42 (1973), 29-60.

Beals, R.
1. On the abstract Cauchy problem, Jour. Ful. Anal, 10 (1972), 281-299.

2. Semigroups and abstract Gevrey spaces, Jour. Ful. Anal. 10 (1972), 300-308.

Berezanskij, Yu.
1. The expansion of selfadjoint operators in terms of eigenfunctions, Izd. Nauk Dumka, Kiev, 1965.

Berezin, I.
1. On Cauchy's problem for linear equations of the second order with initial conditions on a parabolic line, Mat. Sbornik, 24 (1949), 301-320.

Berg, L.
1. Introduction to the operational calculus, North Holland, Amsterdam, 1967.

Berger, M.
1. Nonlinearity and functional analysis, Academic Press, N. Y., 1977.

Bernardi, M.
1. Sulla regolarità delle soluzioni di equazioni differenziali lineari astratte del primo ordine in domini variabili, Boll. U. M. I., 10 (1974), 182-201.

Bernardi, M. and Brezzi, F.
2. Alcune osservazioni sugli operatori di proiezione in spazi di Hilbert, Boll. U. M. I., 9 (1974), 495-512.

Bers, L.
1. Mathematical aspects of subsonic and transonic gas dynamics, Wiley, N. Y., 1958.

Bitsadze, A.
1. Equations of mixed type, Izd. Akad. Nauk SSSR, Moscow, 1959.

Björck, G.
1. Linear partial differential operators and generalized distributions, Arkiv Mat., 6 (1966), 351-407.

Bobisud, L. and Hersh, R.

1. Perturbation and approximation theory for higher order abstract Cauchy problems, Rocky Mount. Jour. Math., 2 (1972), 57-73.

Bochner, S.

1. Sturm Lionville and heat equations whose eigenfunctions are ultraspherical polynomials or associated Bessel Functions, Proc. Conf. Diff. Eqs., Univ. Maryland, 1955, pp. 23-48.

2. Vorlesungen über Fouriersche Intergrale, Chelsea, N. Y., 1948.

Bourbaki, N.

1. Espaces vectoriels topologiques, Chaps. 1-2, Hermann, Paris, 1953.

2. Espaces vectoriels topologiques, Chaps. 3-5, Hermann, Paris, 1955.

3. Intégration, Chaps. 1-4, Hermann, Paris, 1952.

4. Intégration des measures, Hermann, Paris, 1956.

5. Intégration vectorielle, Hermann, Paris, 1959.

Božinov, N.

1. Operational calculus for the general linear partial differential operation of the first order, Computes Rendus Acad. Bulgare Sci., 29 (1976), 1261-1264.

Božinov, N. and Dimovski, I.

2. Operational calculus for general linear partial differential operators of second order, Comptes Rendus Acad. Bulgare Sci, 28 (1975), 727-730.

Braaksma, B.

1. A singluar Cauchy problem and generalized translations, Int. Conf. Diff. Eqs., Academic Press, 1975, pp. 40-52.

Braaksma, B. and de Snoo, H.

2. Generalized translation operators associated with a singluar differential operator, Proc. Conf. Theory Ord. and Part. Diff. Eqs., Springer Lecture Notes.

Braaksma, B. and Schnitman, A.

3. Some classes of Watson transforms and related integral equations for generalized functions, SIAM Jour. Math. Anal., 6 (1976), 771-798.

Bragg, L.

1. Linear evolution equations that involve products of commutative operators, SIAM Jour. Math. Anal., 5 (1974), 327-335.

2. Related nonhomogeneous partial differential equations, Applicable Anal., 4 (1974), 161-189.

3. Hypergeometric operator series and related partial differential equations, Trans. Amer. Math. Soc., 143 (1969), 319-336.

4. Singular nonhomogeneous abstract Cauchy and Dirichlet problems related by a generalized Stieltjes transform, Indiana Univ. Math. Jour., 24 (1974), 183-195.

5. The Riemann-Liouville integral and parameter shifting in a class of linear abstract problems, SIAM Jour. Math. Anal., 7 (1976), 1-12.

Bragg, L. and Dettman, J.

6. An operator calculus for related partial differential equations, Jour. Math. Anal. Appl., 22 (1968), 261-271.

7. Related partial differential equations and their applications, SIAM Jour. Appl. Math., 16 (1968), 459-467.

8. A class of related Dirichlet and initial value problems, Proc. Amer. Math. Soc., 21 (1969), 50-56.

Brézis, H.

1. Opérateurs maximaux monotones ef semigroupes de contraction dans les espaces de Hilbert, North-Holland, Amsterdam, 1973.

2. On some degenerate nonlinear parabolic problems, Proc. Symp. Pure Math., Amer. Math. Soc., 18, Pt. 1, 1970, pp. 28-38.

3. Equations et inéquations nonlinéaires dans les espaces vectoriels en dualité, Ann. Inst. Fourier, 18 (1968), 115-175.

Browder, F.

1. Nonlinear operators and nonlinear equations of evolution in Banach spaces, Proc. Symp. Pure Math., 28, Pt. 2, Amer. Math. Soc., 1976.

2. Functional analysis and partial differential equations, Math. Annalen, 138 (1959), 55-79.

3. Nonlinear initial value problems, Annals Math., 82 (1965), 51-87.

4. Nonlinear equations of evolution, Annals Math., 80 (1964), 485-523.

5. Nonlocal elliptic boundary value problems, Amer. Jour. Math., 86 (1964), 735-750.

6. Analyticity and partial differential equations, Amer. Jour. Math., 84 (1962), 666-710.

7. Families of linear operators depending upon a parameter, Amer. Jour. Math., 87 (1965), 752-758.

8. Nonlinear maximal monotone operators in Banach spaces, Math. Annalen, 175 (1968), 89-113.

Browne, P.

1. Multiparameter spectral theory, Indiana Univ. Math. Jour., 24 (1974), 249-257.

Bruk, V.

1. Some questions of the spectral theory of second order differential equations with variable unbounded operator coefficients, Mat. Zametki, 16 (1974), 813-822.

2. Generalized resolvants of a second order differential operator with unbounded operator coefficients, Fnl. Anal., Ulyanovsk. Univ., 2 (1973), pp. 3-11.

3. Generalized resolvants and spectral functions of odd order differential operators in a space of vector valued functions, Fnl. Anal., Ulyanovsk. Univ., 3 (1974), pp. 44-50.

Butzer, P. and Trebels, W.

1. Hilbert transformation, gebrochene Integration und Differentiation, Westdeutscher Verlag, Köln, 1968.

Calderón, A.

1. Integrales singulares y sus aplicaciones a ecuaciones diferenciales hiperbolicas, Sem., Buenos Aires, 1960.

Calkin, J.

1. Abstract symmetric boundary conditions, Trans. Amer. Math. Soc., 45 (1939), 369-442.

Carroll, R.

1. L'équation d'Euler-Poisson-Darboux et les distributions sousharmoniques, C. R. Acad. Sci. Paris, 246 (1958), 2560-2562.

2. On some generalized Cauchy problems and the convexity of their solutions, AFOSR-TN-59-649, Univ. Maryland, 1959.

3. Some singular Cauchy problems, Ann. Mat. Pura Appl., 56 (1961), 1-31.

4. Abstract methods in partial differential equations, Harper-Row, N. Y., 1969.

5. On some hyperbolic equations with operator coefficients, Proc. Japan Acad., 49 (1973), 233-238.

6. On a class of canonical singular Cauchy problems, Proc. Colloq. Anal., Univ. Fed Rio de Janeiro, 1972; Anal. Fonct. Appl., Act. Sci. Ind., Hermann, Paris, 1975, pp. 71-90.

7. Some remarks on kernels, recovery formulas, and evolution equations, Rocky Mount. Jour. Math., 5 (1975), 61-74.

Carroll, R. and Showalter, R.

8. Singular and degenerate Cauchy problems, Academic Press, N. Y., 1976.

Carroll, R. and Silver, H.

9. Suites canomiques de problèmes de Cauchy singuliers, C. R. Acad. Sci. Paris, 273 (1971), 979-981.

10. Growth properties of solutions of certain canonical hyperbolic equations with subharmonic initial data, Proc. Symp. Pure Math., Amer. Math. Soc., 23 (1973), pp. 97-104.

11. Canonical sequences of singular Cauchy problems, Applicable Anal., 3 (1973), 247-266.

Carroll, R.

12. A uniqueness theorem for EPD type equations in general spaces, Applicable Anal., 7 (1977), 15-18.

13. Some remarks on degenerate Cauchy problems in general spaces, Osaka Math. Jour., 14 (1977), 609-618.

14. On some general abstract degenerate Cauchy problems, Anais Acad. Brasil. Ciencias, 49(1977), 209-211.

15. Algunos resultados sobre ecuaciones diferenciales abstractas relacionadas, Lecture CIMAS 1975, unpublished.

16. Uniqueness theorems for systems of abstract differential equations in general spaces, Applicable Anal., 7 (1978), 147-151.

17. Systems of abstract differential equations in general spaces, Ricerche de Mat., to appear.

18. On the singluar Cauchy problem, J. Math. Mech., 12 (1963), 69-102.

19. On some singular quasilinear Cauchy problems, Math. Zeit., 81 (1963), 135-154.

20. On the spectral determination of the Green's operator, J. Math. Mech., 15 (1966), 1003-1018.

21. Some differential problems related to spectral theory in several operators, Rend. Accad. Lincei, 39 (1965), 170-174.

22. Quelques problèmes differentiels abstraits, Eqs. aux Derivées Part., Univ. Montréal, 1966, pp. 9-46.

Carroll, R. and Neuwirth, J.

23. Some uniqueness theorems for differential equations with operator coefficients, Trans. Amer. Math. Soc., 110 (1964), 459-472.

Carroll, R. and Cooper, J.

24. Remarks on some variable domain problems in abstract evolution equations, Math. Annalen, 188 (1970), 143-164.

Carroll, R. and State, E.

25. Existence theorems for some weak abstract variable domain hyperbolic problems, Canad. Jour. Math., 23 (1971), 611-626.

Carroll, R.

26. Some degenerate Cauchy problems with operator coefficients, Pacific Jour. Math., 13 (1963), 471-485.

27. Problems in linked operators, I and II, Math. Annalen, 151 (1963), 272-282 and 160 (1965), 233-256.

28. On the structure of the Green's operator, Proc. Amer. Math. Soc., 15 (1964), 225-230.

29. On the degenerate Cauchy problem, Canad. Jour. Math., 17 (1965), 245-256.

30. On the structure of some abstract boundary value problems, I and II, Ann. Math. Pura Appl., 72 (1966), 305-318 and 81 (1969), 93-110.

31. Some remarks on the propagator equation, Jour. London Math. Soc., 42 (1967), 737-743.

Carroll, R. and Mazumdar T.

32. Solutions of some possibly noncoercive evolution problems with regular data, Applicable Anal., 1 (1972), 381-395.

Carroll, R.
33. Singular Cauchy problems in symmetric spaces, Jour. Math. Anal. Appl., 56 (1976), 41-54.

34. The group theoretic nature of certain recursion relations for singular Cauchy problems, Jour. Math. Anal. Appl., 63 (1978), 156-167.

35. Transmutation and separation of variables, Applicable Anal., to appear.

36. On the propagator equation, Illinois Jour. Math., 11 (1967), 506-527.

37. On the nature of weak solutions and some abstract Cauchy problems, Bull. Amer. Math. Soc., 72 (1966), 1068-1072.

38. Some variable domain problems in abstract evolution equations, Proc. Conf. Eqs. Evolution and Nonlinear Semigroups, Univ. Kentucky, 1969, 11-24.

39. Transmutation and generalized translation, to appear.

Chazarain, J.
1. Problèmes de Cauchy abstracts et applications à quelques problèmes mixtes, Jour. Fnl. Anal., 7 (1971), 386-446.

Chebli, H.
1. Opérateurs de translation généralisées et semigroupes de convolution, Théorie de potential et analyse harmonique, Springer Lect. Notes 404, 1974, pp. 35-59.

Churchill, R.
1. Modern operational mathematics in engineering, McGraw-Hill, 1944.

Coddington, E. and Levinson, N.
1. Theory of ordinary differential equations, McGraw-Hill, N. Y., 1955.

Colojoară, I. and Foiaş, C.
1. Theory of generalized spectral operators, Gordon-Breach, N. Y., 1968.

Cooper, J.
1. Evolution equations in Banach space with variable domain, Jour. Math. Anal. Appl., 36 (1971), 151-171.

2. Two point problems for abstract evolution equations, Jour. Diff. Eqs., 9 (1971), 453-495.

3. Local decay of solutions of the wave equation in the exterior of a moving body, Jour. Math. Anal. Appl., 49 (1975), 130-153.

4. Scattering of plane waves by a moving obstacle, Arch. Rat. Mech. Anal., to appear.

5. An approximation for one dimensional scattering by a periodically moving boundary, to appear.

6. The inverse scattering problem for moving obstacles, to appear.

Cooper, J. and Medeiros, L.
7. A nonlinear wave equation in a domain with moving boundary, Ann. Sc. Norm. Sup. Pisa, 26 (1972), 829-838.

Cooper, J. and Strass, W.

8. Energy boundedness and decay of waves reflecting off a moving obstacle, Indiana Univ. Jour. Math., 25 (1976), 671-690.

9. Representations of the scattering operator for moving obstacles, to appear.

Copson, E. and Erdelyi, A.

1. On a partial differential equation with two singular lines, Arch. Rat. Mech. Anal., 2 (1958), 76-86.

Cordes, H.

1. On maximal first order partial differential operators, Amer. Jour. Math., 82 (1960), 63-91.

2. Uber die Spekfralzerleguug von hypermaximalen Operatoren die durch Separation der Variablen Zerfallen, I, Math. Annalen, 128 (1954), 257-289; II, Math Annalen, (1955), 373-411.

Cotlar, M.

1. Condiciones de continuidad de operadores potenciales y de Hilbert, Univ. Nac. Buenos Aires, 1959.

Courant, R. and Hilbert, D.

1. Methods of mathematical physics, Vols 1-2, Interscience, N. Y., 1953-1962.

Crandall, M.

1. Semigroups of nonlinear transformations in Banach spaces, Conf. Nonlin. Ful. Anal., Academic Press, N. Y., 1971, pp. 157-179.

Crandall, M. and Pazy, A.

2. Nonlinear evolution equations in Banach spaces, Israel Jour. Math., 11 (1972), 57-94.

Crandall, M. and Liggett, T.

3. Generation of semigroups of nonlinear transformations in general Banach spaces, Amer. Jour. Math., 93 (1971), 157-179.

Crandall, M.

4. Boundary value problems for symmetric positive differential operators of odd order, Jour. Math Mech., 18 (1968), 155-172.

Daleckij, Yu. and Krein, M.

1. Stability of solutions of differential equations in Banach space, Izd. Nauka, Moscow, 1970.

Da Prato, G.

1. Somma di generatori infinitesimali di semigruppi di contrazioni e equazioni di evoluzione in spazi di Banach, Ann. Mat. Pura Appl., 78 (1968), 131-158.

2. Somme d'applications non-linéaires dans les cônes et équations d'évolution dans des espaces d'opérateurs, Jour. Math. Pure Appl., 49 (1970), 289-348.

Da Prato, G. and Grisvard, P.

3. Sommes d'opérateurs linéaires et équations differéntielles opérationnelles, Jour. Math. Pures Appl., 54 (1975), 305-387.

Da Prato, G.

4. Equations d'évolution dans des algèbres d'opérateurs et application à des équations quasi-linéaires, Jour. Math Pure Appl., 48 (1969), 59-107.

Da Prato, G. and Giusti, E.
5. Une caratterizzazione dei generatori di funzioni coseno astratte, Boll. U. M. I, 22 (1967), 357-362.

Delsarte, J.
1. Sur certains transformations fonctionnelles relatives aux équations linéairesaux derivées partielles du second ordre, C. R. Acad. Sci. Paris, 206 (1938), 1780-1782.

Delsarte, J. and Lions, J.
2. Transmutations d'opérateurs differentiels dans le domaine complexe, Comm. Math. Helv., 32 (1957), 113-128.

3. Moyennes généralisées, Comm. Math. Helv., 33 (1959), 59-69.

Dembart, B.
1. On the theory of semigroups of operators on locally convex spaces, to appear.

Dettman, J.
1. Initial-boundary value problems related through the Stieltjes transform, Jour. Math. Anal. Appl., 15 (1969), 341-349.

2. Related semigroups and the abstract Hilbert transform, in Function theoretic methods in differential equation, Pittman Press, London, 1976, pp. 94-108.

3. The wave, Laplace, and heat equations and related transforms, Glasgow Math. Jour., 11 (1970), 117-125.

4. Related singular problems and the generalized Hilbert transform, Proc. Royal Soc. Edinburgh, 79 (1977), 173-182.

Dezin, A.
1. Existence and uniqueness theorems..., Usp. Mat. Nauk, 14 (1959), 21-73.

Dijksma, A. and de Snoo, H.
1. Distributional Watson transforms, SIAM Jour. Math. Anal., 5 (1974), 888-892.

Dikopolov, G.
1. On boundary values for differential equations with constant coefficients in a half space, Mat. Sbornik, 59 (1962), 215-228.

Dikopolov, G. and Šilov, G.
2. On correct boundary values for partial differential equations in a half space, Izv. Akad Nauk SSSR, 24 (1960), 369-380.

3. On correct boundary values in a half space for partial differential equations with a right hand side, Sibirsk. Mat. Žur., 1 (1960), 45-61.

Ditkin, V. and Prudnikov, A.
1. Integral transforms and operational calculus, Moscow, 1961.

Dixmier, J.
1. Les algèbres d'opérateurs dans l'espace Hilbertien, Gauthier-Villars, Paris, 1957.

Djedour, M.
1. Existence and approximation of weak solutions of abstract differential equations, Ann. Sc. Norm. Sup. Pisa, 26 (1972), 463-478.

Doetsch, G.
1. Theorie und Andwendung der Laplace-Transformation, Dover, N. Y., 1943.

Donaldson, J.
1. A singluar abstract Cauchy problem, Proc. Nat'l. Acad. Sci., 66 (1970), 269-274.

2. New integral representations for solutions of Cauchy's problem for abstract parabolic equations, Proc. Nat'l. Acad. Sci., 68 (1971), 2025-2027.

3. An operational calculus for a class of abstract operator equations, Jour. Math. Anal. Appl., 37 (1972), 167-184.

4. The Cauchy problem for a first order system of abstract operator equations, Bull. Amer. Math. Soc., 81 (1975), 576-578.

5. The abstract Cauchy problem and Banach space valued distributions, Howard Univ. Math. Rept., 13, 1975.

Donaldson, J. and Hersh, R.
6. A perturbation series for Cauchy's problem for higher order abstract parabolic equations, Proc. Nat'l. Acad. Sci., 67 (1970), 41-44.

Donaldson, J.
7. The abstract Cauchy problem, Jour. Diff. Eqs., 25 (1977), 400-409.

8. A uniqueness class for two improperly posed problems arising in mathematical physics, Improperly posed boundary value problems, Pitman Press, London, 1975.

Donaldson, T.
1. A laplace transform calculus for partial differential operators, Mem. Amer. Math. Soc., 143, 1974.

Dorroh, J.
1. A simplified proof of a theorem of Kato on nonlinear evolution equations, Jour. Math. Soc. Japan, 27 (1975), 474-478.

Dube, L. and Pandey, J.
1. An inversion of the Hankel potential transform of generalized functions, Tohoku Math. Jour., 29 (1977), 189-201.

2. On the Hankel transform of distributions, Tohoku Math. Jour., 27 (1975), 337-354.

Dubinskij, Yu.
1. On an abstract theorem and its application to boundary value problems for nonclassical equations, Mat. Sbornik, 79 (1969), 91-117.

2. On some differential operational equations of arbitrary order, Mat. Sbornik, 90 (1973), 1-22.

Du Chateau, P.
1. The Cauchy-Goursat problem, Mem. Amer. Math. Soc., 118, 1972.

Dunford, N. and Schwarts, J.
1. Linear operators, Vols. 1-3, Wiley-Interscience, N. Y., 1958, 1963, 1971.

Durovin, B., Matveev, V., and Novikov, S.
1. Nonlinear equations of Kortewe -de Vries type, finite zoned linear operators, and algebraic varieties, Uspekhi Mat. Nauk, 31 (1976), 55-136.

Duvaut, G. and Lions, J.
1. Les inéquations en mecanique et en physique, Dunod, Paris, 1972.

Ehrenpreis, L.
1. Fourier analysis in several complex variables, Interscience-Wiley, N. Y., 1970.

2. Analytic functions and the Fourier transform of distributions, I, Annals Math., 63 (1956), 129-159; II, Trans. Amer. Math. Soc., 89 (1958), 450-483.

3. Theory of distributions for locally compact spaces, Mem. Amer. Math. Soc., 21, 1956.

Ekeland, I. and Teman, R.
1. Analyse convexe et problèmes variationnels, Dunod, Paris, 1974.

Emami-Rad, H. A.
1. Semi-groupe distribution engendré par - A^{α}, Thèse, Dijon, 1973.

Erdelyi, A., et. al.
1. Bateman manuscript project, Tables of integral transforms, Vols. 1,2, McGraw-Hill, N. Y., 1954.

2. Operational calculus and generalized functions, Holt, Rinehart, Winston, N. Y., 1962.

Eringen, A.
1. The finite Sturm-Lionville transform, Quart. Jour. Math., Oxford, 5 (1954), 120-129.

2. A transform technique for boundary value problems in fourth order partial differential equations, Quart. Jour. Math., Oxford, 6 (1955), 241-249.

Faddeev, L.
1. Expansion in eigenfunctions of the Laplace operator on the fundamental domain of a discrete group on the Lobačevskij plane, Trudy Mosk. Mat. Obšč., 17 (1967), 323-350.

Fattorini, H.
1. Ordinary differential equations in linear topological spaces, I and II, Jour. Diff. Eqs., 5 (1968), 72-105 and 6 (1969), 50-70.

2. On a class of differential equations for vector valued distributions, Pacific Jour. Math., 32 (1970), 79-104.

3. The underdetermined Cauchy problem in Banach spaces, Math. Annalen, 200 (1973), 103-112.

Fattorini, H. and Radnitz, A.
4. The Cauchy problem with incomplete initial data in Banach spaces, Mich. Math. Jour., 18 (1971), 291-320.

Feller, W.
1. The parabolic differential equations and the associated semigroups of transformations, Annals Math., 55 (1952), 468-519.

Foiaş, C., Gussi, G. and Poenaru, V.
1. L'étude de l'équation u' = A(t)u pour certaines classes d'opérateurs non-bornés dans l'espace de Hilbert, Trans. Amer. Math. Soc., 86 (1957), 335-347.

Foiaş, C. and Sz.Nagy, B.
2. Analyse harmonique des opérateurs de l'espace de Hilbert, Masson, Paris, 1967.

Friedman, A.
1. Generalized functions and partial differential equations, Prentice-Hall, Englewood Cliff, N. J., 1963.

2. Partial differential equations of parabolic type, Prentice-Hall, 1964.

3. Partial differential equations, Holt-Rinehart-Winston, N. Y., 1969.

Friedman, A. and Schuss, Z.
4. Degenerate evolution equations in Hilbert space, Trans. Amer. Math. Soc., 161 (1971), 401-427.

Friedman, B.
1. An abstract formulation of the methods of separation of variables, Proc. Conf. Diff. Eqs., Univ. Maryland, 1956, pp. 209-226.

2. Principles and techniques of applied mathematics, Wiley, N. Y., 1956.

Friedrichs, K.
1. Symmetric positive linear differential equations, Comm. Pure Appl. Math., 11 (1958), 333-418.

Galpern, S.
1. The Cauchy problem for Sobolev equations, Sibirsk. Mat. Žur., 4 (1963), 758-774.

Garding, L.
1. Application of the theory of direct integrals of Hilbert spaces to some integral and differential operators, Lectures, Univ. Maryland, 1954.

Garnir, H.
1. Les problèmes aux limites de la physique mathematique, Birkhauser, Basel, 1958.

Garsoux, J.
1. Espaces vectoriels topologiques et distributions, Dunod, Paris, 1963.

Gasymov, M.
1. On the theory of polynomial operator pencils, Dokl. Akad. Nauk SSSR, 199 (1974), 747-750.

2. On the multiply complete part of eigenvectors and associated vectors of polynomial operator pencils, Izv. Akad. Nauk Armyan. SSR, 6 (1971), 131-147.

3. On the theory of evolution equations of regular type, Dokl. Akad Nauk SSSR, 200 (1971), 13-16.

Gasymov, M. and Maksudov, F.
4. The principal part of the resolvant of nonselfadjoint operators in a neighborhood of spectral singularities, Funkts. Anal. Prilož., 6 (1972), 16-24.

Gasymov, M. and Magerramov, A.
5. On the existence of transformation operators for higher order differential equations polynomially dependent on a parameter, Dokl. Akad. Nauk SSSR, 235 (1977), 259-262.

Gasymov, M.
6. Expansion in eigenfunctions of a nonselfadjoint differential operator of second order with a singularity at zero, Trudy Let. Sk. Spekt. Teor. Oper. ..., Izd. Elm, Baku, 1975, pp. 20-45.

Gearhart, L.
1. Weighted translation semigroups and a fine dependent Cauchy problem, to appear.

Gelfand, I. and Šilov, G.
1. Generalized functions, Vol. 1, Generalized functions and operations on them, Moscow, 1958.

2. Generalized functions, Vol. 2, Spaces of basic and generalized functions, Moscow, 1958.

3. Generalized functions, Vol. 3, Some questions in the theory of differential equations, Moscow, 1958.

Gelfand, I. and Levitan, B.
4. On the determination of a differential equation from its spectral function, Izv. Akad. Nauk SSSR, 15 (1951), 309-360.

Gil de Lamadrid, J.
1. Uniform crossnorms and tensor products of Banach algebras, Duke Math. Jour., 32 (1965), 359-369.

Gohberg, I. and Krein M.
1. Introduction to the theory of linear nonselfadjoint operators in Hilbert space, Moscow, 1965.

2. The theory of Volterra operators in Hilbert space and applications, Izd. Nauka, Moscow, 1967.

Goldberg, S.
1. Unbounded linear operators, McGraw-Hill, New York, 1966.

Goldstein, J.
1. Abstract evolution equations, Trans. Amer. Math. Soc., 141 (1969), 159-185.

2. Time dependent hyperbolic equations, Jour. Fnl. Anal., 4 (1969), 31-49.

3. Semigroups and second order differential equations, Jour. Fnl. Anal., 4 (1969), 50-70.

4. Semigroups of operators and abstract Cauchy problems, Tulane Univ. Press, 1970.

5. On the absence of necessary conditions for linear evolution operators, Proc. Amer. Math. Soc., 64 (1977), 77-80.

6. On a connection between first and second order differential equations in Banach spaces, J. Math. Anal. Appl., 30 (1970), 246-251.

Goldstein, J. and Sandefur, J.
7. Asymptotic equipartition of energy for differential equations in Hilbert space, Trans. Amer. Math. Soc., 219 (1976), 397-406.

8. Abstract equipartition of energy theorems, to appear.

Gorbačuk, M.
1. Selfadjoint boundary problems for a second order differential equation with unbounded operator coefficient, Funkts. Anal. Prilož., 5 (1971), 10-21.

Gorbačuk, V. and Gorbačuk, M.
2. Certain questions of the spectral theory of a second order linear differential equation with unbounded operator coefficients, Ukrain. Mat. Žur., 23 (1971), 3-14.

Gorbačuk, M. and Kočubei, A.
3. Selfadjoint boundary value problems for certain classes of differential operator equations of high order, Dokl. Acad. Nauk SSSR, 201 (1971), 1029-1032.

Gorbačuk, M. and Vainerman, L.
4. The selfadjointness of semibounded abstract differential operators, Ukrain, Mat. Žur., 22 (1970), 806-808.

Gorbačuk, V. and Gorbačuk, M.
5. On operator functions of distribution for Sturm-Liouville equations with operator coefficients, Trudy Let. Sk. Spekt. Teor. Oper. ..., Izd. Elm, Baku, 1975, pp. 90-129.

Grabovskaya, R. and Krein, S.
1. Second order differential equations with operators generating a Lie algebra representation, Math. Nachrichten, 75 (1976), 9-29.

Grisvard, P.
1. Equations différentielles abstraites, Ann. Sci. Ecole Norm. Sup., 2 (1969), 311-395.

2. Equations opérationnelles abstraites et problèmes aux limites, Ann. Sc. Norm. Sup. Pisa, 21 (1967), 308-347.

Groetsch, C.
1. Generalized inverses of linear operators, Dekker, N. Y., 1977.

Grothendieck, A.
1. Produits tensoriels topologiques et espaces nucléaires, Mem. Amer. Math. Soc., 16 (1955).

Hackman, M.
1. The abstract time dependent Cauchy problem, Trans. Amer. Math. Soc., 133 (1968), 1-50.

Helgason, S.
1. Fundamental solutions of invariant differential operators on symmetric spaces, Am. Jour. Math., 86 (1964), 565-601.

2. Differential geometry and symmetric spaces, Academic Press, N. Y., 1962.

Hersh, R.
1. Explicit solution of a class of higher order abstract Cauchy problems, Jour. Diff. Eqs., 8 (1970), 570-579.

2. Direct solution of general one dimensional linear parabolic equation via an abstract Plarcherel formula, Proc. Nat'l. Acad. Sci., 63 (1969), 648-654.

3. The method of transmutations, Part. Diff. Eqs. and Related Topics, Springer Lect. Notes 446, N. Y., 1975, pp. 264-282.

4. Mixed problems in several variables, Jour. Math. Mech., 12 (1963), 317-334.

5. Boundary conditions for equations of evolution, Arch. Rat. Mech. Anal., 16 (1964), 243-264.

6. On the general theory of mixed problems, Battelle Rencontres 1968, Springer, Berlin, 1970, pp. 85-95.

Hersh, R. and Steinberg, S.
7. Abstract hyperbolic equations with non commuting coefficients, Tech. Rep. 356, Univ. New Mexico, 1978.

Heywood, P.
1. On a modification of the Hilbert transform, Jour. London Math. Soc., 42 (1967), 641-645.

Hille, E. and Phillips, R.
1. Functional analysis and semigroups, Amer. Math. Soc. Colloq. pub., Vol. 31, 1957.

Hille, E.
2. The abstract Cauchy problem, Jour. Anal. Math., 3 (1953-54), 81-196.

3. Lectures on ordinary differentions, Addison-Wesley, Reading, Mass., 1969.

Hirshman, I. and Widder, D.
1. The convolution transform, Princeton Univ. Press, 1955.

Hörmander, L.
1. Linear partial differential operators, Springer, Berlin, 1963.

Horváth, J.
1. Topological vector spaces and distribution, Addison-Wesley, Reading, Mass., 1966.

2. Composition of hypersingular integral operators, Applicable Anal., 7 (1978), 171-190.

Hutson, V. and Pym, J.
1. General translations associated with an operator, Math Annalen, 187 (1970), 241-258.

2. Generalized translations associated with a differential operator, Proc. London Math. Soc., 24 (1972), 548-576.

Ichinose, T.
1. On the spectra of tensor products of linear operators in Banach spaces, Jour. Reine Angew. Math., 244 (1970), 119-153.

2. Operators on tensor products of Banach spaces, Trans. Amer. Math. Soc., 170 (1972), 197-219.

3. Tensor products of linear operators and the method of separation of variables, Hokkaido Math Jour., 4 (1975), 161-189.

4. Operational calculus for tensor products of linear operators in Banach spaces, Hokkaido Math Jour., 4 (1975), 306-334.

Ince, E.
1. Ordinary differential equations, Dover, N.Y., 1956.

Inoue, A.
1. Sur $\Box u + u^3 = f$ dans un domaine noncylindrique, Jour. Math Anal. Appl., 46 (1973), 777-819.

Jury, E.
1. Theory and application of the Z transform, Wiley, N. Y., 1964.

Källström, A. and Sleeman, B.
1. Multiparameter spectral theory, Ark. Mat., 15 (1977), 93-99.

Kato, T. and Tanabe, H.
1. On the abstract evolution equation, Osaka Math. Jour., 14 (1962), 107-133.

Kato, T.
2. Integration of the equations of evolution in a Banach space, Jour. Math. Soc. Japan, 5 (1953), 208-234.

3. Nonlinear evolution equations in Banach space, Proc. Symp. Appl. Math., Amer. Math. Soc. 17 (1965), 50-67.

4. Perturbation theory for linear operators, Sec. Ed., Springer, N. Y., 1976.

5. Nonlinear semigroups and evolution equations, Jour. Math. Soc. Japan, 19 (1967), 508-520.

6. Linear evolution equations of hyperbolic type, Jour. Fac. Sci. Univ. Tokyo, 17 (1970), 241-258.

7. Linear evolution equations of hyperbolic type, II, Jour. Math. Soc. Japan, 25 (1973), 648-666.

8. On linear differential equations in Banach spaces, Comm. Pure Appl. Math., 9 (1956), 479-486.

Kharitonenko, P. and Yurčuk, N.
1. The Cauchy problem for the abstract Love equation, Diff. Urav., 8 (1972), 1256-1266.

Kisynski, J.
1. Sur les opérateurs de Green des problèmes de Cauchy abstraits, Studia Math., 23 (1964), 285-328.

Kleiman, E.
1. On the Green's function for the Sturm-Liouville equation with a normal operator coefficient, Vest. Mosk. Univ. Mat., 29 (1974), 47-53.

Kober, H.
1. A modification of Hilbert transforms, the Weyl integral, and functional equations, Jour. London Math. Soc., 42 (1967), 42-50.

Komatsu, H.
1. Semigroups of operators in locally convex spaces, Jour. Math. Soc. Japan, 16 (1964), 230-262.

2. Editor, Hyperfunctions and pseudo differential equations, Springer, Berlin, 1973.

Komura, T.
1. Semigroups of operators in locally convex spaces, Jour. Fnl. Anal., 2 (1968), 258-296.

Komura, Y.
1. Nonlinear semigroups in Hilbert space, Jour. Math. Soc. Japan, 19 (1967), 493-507.

Kothe, G.
1. Topologische lineare Raume, Springer, Berlin, 1960.

Krabbe, G.
1. Operational calculus, Springer, N. Y., 1970.

Krasnov, M.
1. Mixed boundary value problems for degenerate linear hyperbolic differential equations of the second order, Mat. Sbornik, 91 (1959), 29-84.

Krein, S.
1. Linear differential equations in Banach spaces, Izd. Nauka, Moscow, 1967.

Krein, S and Šikhvatov, A.
2. Linear differential equations on a Lie group, Funk Anal., 4 (1970), 52-61.

Lacomblez, C.
1. Une équation d'évolution du second ordre en t à coefficients dégénérés ou singuliers, Pub. Math. Bordeaux, Fasc. 4 (1974), 33-64.

Ladas, G. and Lakshmikantham, V.
1. Differential equations in abstract spaces, Academic Press, N. Y., 1972.

Ladyženskaya, O. and Uraltzeva, N.
1. Linear and quasilinear elliptic equations, Moscow, 1964.

Ladyženskaya, O., Solonnikov, V., and Uraltzeva, N.
2. Linear and quasilinear parabolic equations of parabolic type, Moscow, 1967.

Ladyženskaya, O.
3. On the solution of non stationary operational equations, Mat. Sbornik, 39 (1956), 491-524.

4. On nonstationary operational equations..., Mat. Sbornik 45 (1958), 123-158.

Lagnese, J.
1. Perturbations in variational inequalities, Jour. Math. Anal. Appl., to appear.

2. Perturbations in a class of nonlinear abstract equations, SIAM Jour. Math. Anal., 6 (1975), 616-627.

3. The final value problem for Sobolev equations, Proc. Amer. Math. Soc., to appear.

4. Exponential stability of solutions of differential equations of Sobolev type, SIAM Jour. Math. Anal., 3 (1972), 625-636.

5. General boundary value problems for differential equations of Sobolev type, SIAM Jour. Math. Anal., 3 (1972), 105-119.

Lakshmikantham, V. and Leela, S.
1. Differential and integral inequalities, Vols. I and II, Academic Press, N. Y., 1969.

Lang, S.
1. $SL_2(\mathbb{R})$, Addison-Wesley, Reading, Mass., 1975.

Laptev, G.
1. Operational calculus of linear unbounded operators and semigroups, Funkts. Anal. Prilož., 4 (1970), 31-40.

Lax, P. and Phillips, R.
1. Scattering theory, Academic Press, N. Y., 1967.

2. Scattering thoery for automorphic functions, Princeton Univ. Press, 1976.

Leblanc, N.
1. Classification des algèbres de Banach associées aux opérateurs différentiels de Sturm-Liouville, Jour. Ful. Ánal., 2 (1968), 52-72.

Lee, W.
1. On Schwartz's Hankel transformation of certain spaces of distributions, SIAM Jour. Math. Anal., 6 (1975), 427-432.

Leray, J. and Lions, J.
1. Quelques résultats de Višik sur les problèmes elliptiques nonlinéaires par les méthodes de Minty-Browder, Bull. Soc. Math. France, 93, (1965), 97-107.

Levin, B.
1. Transforms of Fourier and Laplace type with the aid of solutions of differential equations of second order, Dokl. Akad. Nauk SSSR, 106 (1956), 187-190.

Levine, H.
1. Some uniqueness and growth theorems in the Cauchy problem for $Pu'' + Mu' + Nu = 0$ in Hilbert space, Math. Zeit., 126 (1972), 345-360.

2. Uniqueness and growth of weak solutions to certain linear differential equations in Hilbert space, Jour. Diff. Eqs., 17 (1975), 73-81.

3. Logarithmic convexity and the Cauchy problem for some abstract second order differential inequalities, Jour. Diff. Eqs., 8 (1970), 34-55.

4. Some nonexistence and instability theorems for solutions to formally parabolic equations of the form $Pu_t = -Au + F(u)$, Arch. Rat'l. Mech. Anal., 51 (1973), 371-386.

Levitan, B.
1. Generalized translation operators and some of their applications, Gos. Izd. Fiz - Mat. Lit., Moscow, 1962.

Levitan, B. and Sangsyan, I.

2. Introduction to spectral theory: Self adjoint ordinary differential operators, Moscow, 1970.

Levitan, B.

3. The application of generalized displacement operators to linear differential equations of the second order, Uspekhi Mat. Nauk, USSR, 4 (1949), 3-112.

4. Investigation of the Green's function for a Sturm-Liouville equation with operator coefficients, Mat. Sbornik, 76 (1968), 239-270.

5. The theory of generalized translation operators, Izd. Nauka, Moscow, 1973.

6. Some questions of spectral analysis for Sturm-Liouville equations with operator coefficients, Trudy Let. Sk. Spekt. Teor. Oper. ..., Izd. Elm. Baku, 1975, pp. 161-169.

Lighthill, J.

1. Introduction to Fourier analysis and generalized functions, Cambridge Univ. Press, 1958.

Lions, J.

1. Sur les semigroupes distributions, Portug. Math., 19 (1960), 141-164.

2. Opérateurs de Delsarte et problèmes mixtes, Bull. Soc. Math. France, 84 (1956), 9-95.

3. Opérateurs de transmutation singuliers et équations d'Euler-Poisson-Darboux généralisées, Rend. Sem. Mat. Fis. Milano, 28 (1959), 3-16.

4. Equations différentielles-opérationnelles, Springer, Berlin, 1961.

5. Quelques méthodes de résolution des problèmes aux limites nonlinéaires, Dunod, Paris, 1969.

6. Contrôle optimal de systemes gourvernés par des équations aux derivées partielles, Dunod-Gauthier-Villars, Paris, 1968.

7. Perturbations singulières dans les problèmes aux limites et en contrôle optimal, Springer, Berlin, 1973.

Lions, J., and Magenes, E.

8. Problèmes aux limites nonhomogenes, Vols. 1-3, Dunod, Paris, 1968-1970.

Lions, J.

9. Une rémarque sur les applications du théorème de Hille-Yosida, Jour. Math. Soc. Japan, 9 (1957), 62-70.

10. Rémarque sur les équations différentielles-opérationnelles, Osaka Jour. Math., 15 (1963), 131-142.

11. Une rémarque sur les problèmes d'évolution nonlinéaires dans des domaines non cylindriques, Rev. Roumaine Math. Pure Appl., 9 (1964), 11-18.

12. Sur les problèmes mixtes pour certains systemes paraboliques dans les ouverts non cylindriques, Annales Inst. Fourier, 7 (1957), 143-182.

Löfström, J. and Peetre, J.

1. Approximation theorems connected with generalized translations, Math. Annalen, 18 (1969), 255-268.

Löftström, J.
2. On certain interpolation spaces related to generalized semigroups, Math. Scand., 16 (1965), 41-54.

Lowndes, J.
1. Note on the generalized Mehler transform, Proc. Camb. Phil. Soc., 60 (1964), 57-59.

Lyubič, Yu.
1. The classical and local Laplace transformation in an abstract Cauchy problem, Uspekhi Mat. Nauk, 21 (1966), 3-51.

Magnus, W., Oberhettinger, F., and Soni, R.
1. Formulas and theorems for the special functions of mathematical physics, Springer, Berlin, 1966.

Marčenko, V.
1. Some questions in the theory of one dimensional linear differential operators of the second order, I and II, Trudy Mosk. Mat. Obsc., 1 (1952), 327-420, and 2 (1953), 3-82.

2. The spectral theory of Sturm-Liouville operators, Izd. Nauk. Dumka, Kiev, 1972.

3. Sturm-Liouville operators and their applications, Izd. Nauk. Dumka, Kiev, 1977.

4. Expansion in eigenfunctions of a nonselfadjoint singular differential operator of second order, Mat. Sbornik, 52 (1966), 739-788.

Marinescu, G.
1. Espaces vectoriels pseudotopoloquiques et théorie des distributions, Deut. Ver. Wiss., Berlin, 1963.

Markus, A. and Mereutsa, I.
1. On the connection between the spectral properties of a polynomial bundle and of the roots of the corresponding operator equation, Funkts. Anal. Priloz., 7 (1973), 77-78.

Martin, R.
1. Nonlinear operators and differential equations in Banach spaces, Wiley-Interscience, 1976.

Maslov, V. and Manin, Yu.
1. Algebraic aspects of nonlinear differential equations and Equations of selfcongruent fields, Sovrem, Prob. Mat., Itogi Nauka Tech., Vol. 2, Ed. R. Gamkrelidze, Moscow, 1978.

Maslov, V.
2. Operator methods, Izd. Nauka, Moscow, 1973.

Massera, J. and Schaeffer, J.
1. Linear differential equations and function spaces, Academic Press, N. Y., 1966.

Maurin, K.
1. Methods of Hilbert spaces, Warsaw, 1967.

2. General eigenfunction expansions and unitary representations of topological groups, Warsaw, 1968.

Maurin, K. and Maurin, L.
3. Spektraltheorie separierbarer Operatoren, Studia Math., 23 (1963), 1-29.

Mazumder, T.
1. Generalized projection theorem with application to linear noncoercive equations and some nonlinear situations, Jour. Math. Anal. Appl., 43 (1973), 72-100.

2. Generalized projection theorem and weak noncoercive evolution problems in Hilbert space, Jour. Math. Anal. Appl., 46 (1974), 143-168.

3. Regularity of solutions of linear coercive evolution equations with variable domain, I, Jour. Math. Anal. Appl., 52 (1975), 615-647.

4. Existence of solutions for linear systems and hyperbolic problems with variable domains, Applicable Anal., to appear.

Mazya, V. and Plamenevskij, B.
1. The asympotics of the solutions of differential equations with operator coefficients, Dokl. Akad. Nauk SSSR, 196 (1971), 512-515.

Mercer, A.
1. On integral transform pairs arising from second order differential equations, Proc. Edinburgh Math. Soc., 13 (1962), 63-68.

2. On integral transform pairs arising from differential equations of any even order, Quart. Jour. Math., Oxford, 14 (1963), 9-15.

3. A theory of integral transforms, Quart. Jour. Math., Oxford, 15 (1964), 149-154.

Mikhailov, V.
1. The first boundary value problem for a class of hypoelliptic equations, Mat. Sbornik, 63 (1964), 238-264.

Mikusinski, J.
1. Rachunek operatorow, Warsaw, 1953.

Millionšc̆ikov, V.
1. On the theory of differential equations in locally convex spaces, Mat. Sbornik, 57 (1961), 385-406.

Milman, D.
1. The formulation and methods of solving a general boundary value problem..., Dokl. Akad. Nauk SSSR, 161 (1965), 1276-1281.

Miyadera, I.
1. Semigroups of operators in Frechet space and applications to partial differential equations, Tohoku Math. Jour., 11 (1959), 162-183.

Mizohata, S.
1. Le problème de Cauchy pour les équations paraboliques, Jour. Math. Soc. Japan, 8 (1956), 269-299.

Mohammed, S. E. A.
1. Separation of variables (an abstract approach), Proc. Roy. Soc. Edinburgh, 72 (1974), 57-78.

Nachbin, L.
1. Lectures on the theory of distributions, Textos do Matematica, 15, Inst. Fis. Mat., Univ. Recife, 1964.

Naimark, M.
1. Linear differential operators, Moscow, 1954.

2. Normed rings, Moscow, 1956.

Nasim, C.
1. An inversion formula for a class of integral transforms, Jour. Math. Anal. Appl., 52 (1975), 525-537.

Oberhettinger, F. and Higgins, T.
1. Tables of Lebedev, Mehler, and generalized Mehler transforms, Math. Note 246, Math. Res. Lab., Boeing Sci. Res. Labs., 1961.

Okikiolu, G.
1. Aspects of the theory of bounded integral operators in L^p spaces, Academic Press, N. Y., 1971.

2. A generalization of the Hilbert transform, Jour. London Math. Soc., 40 (1965), 27-30.

Orton, M.
1. Hilbert transforms, Plemelj relations, and Fourier transforms of distributions, SIAM Jour. Math. Anal., 4 (1973), 656-670.

2. Hilbert boundary value problems-a distributional approach, Proc. Roy. Soc. Edinburgh, 76 (1977), 193-208.

Oucii, S.
1. Semigroups of operators in locally convex spaces, Jour. Math. Soc. Japan, 25 (1973), 265-276.

Ovsyannikov, L.
1. Singular operators in Banach spaces, Dokl. Akad. Nauk SSSR, 163 (1965), 819-822.

Palamodov, V.
1. Linear differential operators with constant coefficients, Izd. Nauka, Moscow, 1967.

2. On correct boundary values for partial differential equations in a half space, Izv. Akad. Nauk SSSR, 24 (1960), 381-386.

Pascali, D.
1. Operatori neliniari, Ed. Acad. Rep. Soc. Romania, Bucharest, 1974.

Paškovskij, V.
1. Related operators and boundary value problems for hyperbolic and degenerate hyperbolic partial differential equations, Diff. Urav., 11 (1975), 127-136.

2. Related operators and boundary value problems for equations of elliptic type, Diff. Urav., 12 (1976), 118-128.

Pavlov, A.
1. On general boundary values for partial differential equations in a half space, Mat. Sbornik, 103 (1977), 367-391.

Phillips, R.
1. Dissipative operators and hyperbolic systems of partial differential equations, Trans. Amer. Math. Soc., 90 (1959), 193-254.

2. Dissipative hyperbolic systems, Trans. Amer. Math. Soc., 86 (1957), 109-173.

vander Pol, B. and Bremmer, H.
1. Operational calculus based on the two sided Laplace transform, Cambridge Univ. Press, 1964.

Polyanskij, E. and Fedoryuk, M.
1. On the relation of boundary value problems for elliptic equations in a semicylinder to mixed problems for the heat and Schrodinger equations, Dokl. Akad. Nauk SSSR, 200 (1971), 560-563.

Poulsen, E.
1. Evolutions-gleichungen in Banach-Räumen, Math. Zeit., 90 (1965), 286-309.

Povzner, A.
1. Differential equations of Sturm-Liouville type on a half axis, Mat. Sbornik, 23 (1948), 3-52.

Protter, M.
1. The Cauchy problem for a hyperbolic second order equation with data on the parabolic line, Canad. Jour. Math., 6 (1954), 542-553.

Rao, V. Gopala
1. Sobdev-Galpern equations of order $n + 2$ in $\mathbb{R}^m \times \mathbb{R}$, $m \geq 2$, Trans. Amer. Math. Soc., 210 (1975), 267-278.

Raskin, V. and Sobolevskij, P.
1. The Cauchy problem for second order differential operators in Banach spaces, Sibirsk. Mat. Žur., 8 (1967), 70-90.

Raskin, V.
2. On the solvability of differential operational equations of arbitrary order, Diff. Urav., 12 (1976), 362-364.

Reed, M. and Simon, B.
1. Methods of modern mathematical physics, Vols. 1 and 2, Academic Press, N. Y., 1975.

2. Methods of modern mathematical physics, Vol. 4, Academic Press, N. Y., 1978.

3. Tensor products of closed operators on Banach spaces, Jour. Ful. Anal., 13 (1973), 107-124.

Roach, G. and Sleeman, B.
1. Coupled operator systems and multiparameter spectral theory, Proc. Roy. Soc. Edinburgh, 80 (1978), 23-34.

Rofe-Beketov, F.
1. The expansion in eigenfunctions of an infinite system of differential equations in the nonselfadjoint and selfadjoint cases, Mat. Sbornik, 51 (1960).

Roetman, E.
1. Some observations about an odd order parabolic equation, Jour. Diff. Eqs., 9 (1971), 335-345.

Rosenbloom, P.
1. Linear equations of parabolic type with constant coefficients, Annals Math. Studies, 33, Princeton, 1954, pp. 191-200.

Rosinger, E.
1. Distribution solutions of nonlinear partial differential equations and stability properties, Tech, Rep. 127, Technion, Haifa, 1978.

Rubel, L.
1. An operational calculus in miniature, Applicable Anal., 6 (1977), 299-304.

Sandefur, J.
1. Higher order abstract Cauchy problems, Jour. Math. Anal. Appl., to appear.

Sansone, G.
1. Orthogonal functions, Wiley-Interscience, N. Y., 1959.

Sarason, L.
1. On boundary value problems for hyperbolic equations, Comm. Pure Appl. Math., 15 (1962), 373-395.

2. On hyperbolic mixed problems, Arch. Rat. Mech. Anal., 18 (1965), 310-334.

Schaeffer, H.
1. Topological vector spaces, MacMillan, N. Y., 1966.

Schechter, M.
1. Spectra of partial differential operators, North Holland, Amsterdam, 1971.

2. On the spectra of operators on tensor products, Jour. Ful. Anal., 4 (1969), 95-99.

Schuss, Z.
1. Backward and degenerate parabolic equations, to appear.

Schwartz, A.
1. An inversion theorem for Hankel transforms, Proc. Amer. Math. Soc., 22 (1969), 713-717.

Schwartz, L.
1. Théorie des distributions,Edition "Papillon", Hermann, Paris, 1966.

2. Théorie des distributions à valeurs vectorielles, Ann. Inst. Fourier, 7 and 8, 1957-58, 1-141 and 1-209.

3. Espaces de fonctions différentiables à valeurs vectorielles, Jour. Anal. Math., 4 (1954-55), 88-148.

4. Lectures on mixed problems in partial differential equations and representations of semigroups, Tata Institute, Bombay, 1957.

5. Les équations d'évolution liées au produit de composition, Ann. Inst. Fourier, 2 (1950), 19-49.

Sebastiao Silva, J.
1. Sur le calcul symbolique des opérateurs différentiels à coefficients variables, Rend. Accad. Naz. Lincei, 27 (1959), 42-47 and 118-122.

2. Le calcul opérationnel pour des opérateurs à spectre non borné, Mem. Accad. Naz. Lincei, 6 (1959), 3-13.

Segal, I.
1. Nonlinear semigroups, Annals Math., 78 (1963), 339-364.

Shapira, P.
1. Théorie des hyperfonctions, Springer, Berlin, 1970.

Shatten, R.
1. A theory of cross spaces, Ann. Math. Studies 26, Princeton Univ. Press, 1950.

Showalter, R.
1. Hilbert space methods for partial differential equations, Pittman Press, London, 1977.
2. Nonlinear degenerate evolution equations and partial differential equations of mixed type, SIAM Jour. Math. Anal., 6 (1975), 25-42.
3. Degenerate evolution equations, Indiana Univ. Math. Jour., 23 (1974), 655-677.
4. Degenerate parabolic initial value problems, Jour. Diff. Eqs., to appear.
5. Initial and final value problems for degenerate parabolic evolution systems, to appear.
6. The Sobolev equation, I and II, Applicable Anal., 5 (1975), 15-22 and 81-99.
7. A nonlinear parabolic-Sobolev equation, Jour. Math. Anal. Appl., 50 (1975), 183-190.
8. Weak solutions of nonlinear evolution equations of Sobolev-Galpern type, Jour. Diff. Eqs., 11 (1972), 252-265.
9. Partial differential equations of Sobolev-Galpern type, Pacific Jour. Math., 31 (1969), 787-793.
10. Existence and representation theorems for a semilinear Sobolev equation in Banach space, SIAM Jour. Math. Anal., 3 (1972), 527-543.

Šilov, G.
1. On boundary values in a quadrant for partial differential equations with constant coefficients, Sibirsk. Mat. Žur., 2 (1961), 144-160.

Simon, J.
1. On the integrability of representations of finite dimensional real Lie algebras, Comm. Math. Physics, 28 (1972), 39-46.

Smirnov, M.
1. Degenerate elliptic and hyperbolic problems, Izd. Nauka, Glav. Red. Fiz-Mat. Lit., Moscow, 1966.
2. Equations of mixed type, Izd. Nauka, Glav. Red. Fiz-Mat. Lit ., Moscow, 1970.

Šmulyan, Yu.
1. Extended resolvants and extended spectral functions of a Hermitean operator, Mat. Sbornik, 84 (1971), 440-455.

Sneddon, I.
1. The use of integral transforms, McGraw-Hill, N. Y., 1972.

Sobolev, S.
1. Functional analysis in mathematical physics, Amer. Math. Soc. Translation, 1963.

2. Some new problems in mathematical physics, Izx. Akad. Nauk SSSR, 18 (1954), 3-50.

Sobolevskij, P.
1. On equations of parabolic type in a Banach space, Trudy Mosk. Mat. Obsc., 10 (1961), 29]-350.

2. On a type of differential equation of second order in a Banach space, Uc. Zap. Azerb. Inst., 3 (1962), 87-106.

3. On differential equations of second order in a Banach space, Dokl. Akad. Nauk SSSR, 146 (1962), 774-777.

Sololev, A.
1. The inverse problem of scattering theory for equations of second order with operator coefficients, Trudy Let. Sk. Spekt. Teor. Oper. ..., Izd. Elm, Baku, 1975, pp. 72-77.

Stein, E.
1. Singular integrals and differentiability properties of functions, Princeton Univ. Press, 1970.

Stein, E. and Weiss, G.
2. Introduction to Fourier analysis on Euclidean spaces, Princeton Univ. Press, 1971.

Steinberg, S. and Treves, F.
1. Pseudo Fokker-Planck equations and hyperdifferential operators, Jour. Diff. Eqs., 8 (1970), 333-366.

Stone, M.
1. Linear transformations in Hilbert space, Amer. Math. Soc. Colloq. Pub., Vol. 15, 1932.

Strauss, W.
1. The energy method in nonlinear partial differential equations, Notas de Mat., 47, Rio de Janeiro, 1969.

2. The initial value problem for certain nonlinear evolution equations, Amer. Jour. Math., 89 (1967), 249-259.

3. Existence of the scattering operator for moving obstacles, Jour. Fnl. Anal., to appear.

Struble, R.
1. An algebraic view of distributions and operators, Studia Math., 37 (1971), 103-109.

2. Operator homomorphisms, Math. Zeit., 130 (1973), 275-285.

3. Linear transformations in the operational calculus, SIAM Jour. Math. Anal., 8 (1977), 258-270.

Tanabe, H.
1. On the equations of evolution in a Banach space, Osaka Math. Jour., 11 (1959), 121-145 and 12 (1960), 145-166 and 363-376.

Thyssen, M.
1. Opérateurs de Delsarte particuliers, Bull. Soc. Roy. Sci. Liege, 26 (1957), 87-96.

2. Sur certains opérateurs de transmutation particuliers, Mem. Soc. Roy. Sci. Liege, 6 (1961), 7-32.

Ting, T.
1. Parabolic and pseudoparabolic partial differential equations, Jour. Math. Soc. Japan, 21 (1969), 440-453.

Titchmarsh, E.
1. Introduction to the theory of Fourier integrals, Oxford Univ. Press, London, 1937.

2. Eigenfunction expansions associated with second order differential equations, Oxford, 1962.

Torelli, G.
1. Un complemento ad un teorema di J. L. Lions sulle equazione diferenziale astratte del secondo ordine, Rend. Sem. Mat. Padova, 34 (1964), 224-241.

Treves, F.
1. Topological vector spaces, distributions, and kernals, Academic Press, N. Y., 1967.

2. Linear partial differential equations with constant coefficeints, Gordon-Breach, N. Y., 1966.

3. Basic linear partial differential equations, Academic Press, N. Y., 1975.

4. Locally convex spaces and linear partial differential equations, Springer, New York, 1967.

5. Ovcyannikov theorem and hyperdifferential operators, Notas de Mat., 46, Inst. Mat. Pura Appl., Rio de Janeiro, 1968.

6. Problèmes de Cauchy et problèmes mixtes en théorie des distributions, Jour. Anal. Math. (1959), 105-187.

7. On the theory of linear partial differential operators with analytic coefficients, Trans. Amer. Math. Soc., 137 (1969), 1-20.

8. Rélations de domination entre opérateurs differéntiels, Acta Math., 101 (1959), 1-139.

9. Problèmes de Cauchy et problèmes mixtes en théorie des distributions, Jour. Anal. Math., 7 (1959), 105-187.

Trione, S.
1. Sopra la trasformata di Hankel distribuzionale, Rend. Lincei, 57 (1974), 316-320.

Ungar, A.
1. A differential transform, SIAM Jour. Math. Anal., 5 (1974), 920-926.

2. An operator related to the inverse Laplace transform, SIAM Jour. Math. Anal., 5 (1974), 367-375.

3. The use of a new operational calculus method, to appear.

Varadarajan, V.
1. Lie groups, Lie algebras, and their representations, Prentice Hall, 1974.

Virozub, A. and Matsaev, V.
1. The spectral properties of a certain class of selfadjoint operator functions, Funkts. Anal. Prilož., 8 (1974), 1-10.

Višik, M.
1. The Cauchy problem for equations with operator coefficients..., Mat. Sbornik, 39 (1956), 51-148.

2. On general boundary value problems for elliptic differential equations, Trudy Mosk. Mat. Obšč., 1 (1952), 187-246.

Višik, M. and Ladyženskaya, O.
3. Boundary value problems for partial differential equations and certain classes of operator equations, Usp. Mat. Nauk, 11 (1956), 41-97.

Voropaeva, G. and Maslov, V.
1. M. V. Keldyš's multiple completeness and the uniqueness of the solution of the corresponding Cauchy problem, Funkts. Anal. Priloz., 4 (1970), 10-17.

Waelbroeck, L.
1. Les semigroupes differéntiables, Deux. Colloq. Anal. Fonct., CBRM (1964), pp. 97-103.

Walker, W.
1. A nonsymmetric singular Cauchy problem, Applicable Anal., 7 (1978), 121-131.

Walton, J.
1. A distributional approach to dual integral equations of Titchmarsh type, SIAM Jour. Math. Anal., 6 (1975), 628-643.

Wang, C.
1. On the degenerate Cauchy problem for linear hyperbolic equations of the second order, Thesis, Rutgers Univ., 1964.

2. A uniqueness theorem on the degenerate Cauchy problem, Canad. Math. Bull., 18 (1975), 417-421.

Weinstein, A.
1. On Tricomi's equation and generalized axially symmetric potential theory, Acad. Roy. Belg., 37 (1951), 348-358.

2. The singular solutions and the Cauchy problem for generalized Tricomi equations, Comm. Pure Appl. Math., 7 (1954), 105-116.

3. Generalized axially symmetric potential theory, Bull. Amer. Math. Soc., 59 (1953), 20-38.

Westphal, U.
1. Ein kalkül für gebrochene Potenzen infinitesmaler Erzeuger von Halbgruppen und Gruppen von Operatoren, Compositio Math., 22 (1970), 67-136.

Widder, D.
1. The Laplace transform, Princeton Univ. Press, Princeton, N. J., 1946.

Wyler, O.
1. Green's operators, Ann. Mat. Pura. Appl., 66 (1964), 251-263.

Yakubov, S.
1. On the Cauchy problem for differential equations of second order in a Banach space, Dokl. Akad. Nauk SSSR, 168 (1966), 759-762.

2. On the solvability of the Cauchy problem for evolution equations, Dokl. Akad. Nauk SSSR, 156 (1964), 1041-1044.

3. The solvability of the Cauchy problem for differential equations in a Banach space, Funks. Anal., Trudy Inst. Mat., Azerb. SSR, 1967, pp. 187-206.

4. The Cauchy problem for differential equations of the second order of parabolic type in a Banach space, Izv. Akad. Nauk Azerb. SSR, 4 (1966), 3-8.

5. Differential equations of higher order with variable unbounded operators in a Banach space, Izv. Akad. Nauk Azerb. SSR, 1 (1966), 20-27.

6. Investigation of the Cauchy problem for evolution equations of hyperbolic type, Dokl. Akad. Nauk Azerb. SSR, 20 (1964), 3-6.

7. The solvability of the Cauchy problem for abstract quasilinear hyperbolic equations of the second order and applications, Trudy Mosk. Mat. Obsc., 23 (1970), 37-60.

Yakubov, S. and Balaev, M.
8. The correct solvability of differential operator equations on the whole line, Dokl. Akad. Nauk SSSR, 229 (1976), 562-565.

Yakubov, S.
9. Solution of mixed problems for partial differential equations by operator methods, Dokl. Akad. Nauk SSSR, 196 (1971), 545-548.

Yosida, K.
1. Functional analysis, Springer, Berlin, 1965.

2. An operator theoretical integration of the wave equation, Jour. Math. Soc. Japan, 8 (1956), 79-92.

Zaidman, S.
1. Equations différentielles abstraites, Sem. Mat. Sup. Univ. Montreal, 1966.

Zarnitskaya, N., Selezneva, F. and Eidelman, S.
1. The mixed problem for Petrovskij correct systems of equations with constant coefficients in a quadrant, Sibirsk. Mat. Žur., 15 (1974), 332-342.

Zemanian, A.
1. Generalized integral transforms, Interscience-Wiley, N. Y., 1968.

INDEX